Frederick L Dibble

Vagaries of Sanitary Science

Frederick L Dibble

Vagaries of Sanitary Science

ISBN/EAN: 9783337130480

Printed in Europe, USA, Canada, Australia, Japan

Cover: Foto ©berggeist007 / pixelio.de

More available books at **www.hansebooks.com**

VAGARIES

OF

SANITARY SCIENCE

BY

F. L. DIBBLE, M.D.

It is held as a fundamental principle in science that every opinion, before it is admitted as true and taught to others, should first be established by proper proofs, which must not in any way run counter to established truths, such as, for instance, that twice two are four and not five. Inferences and conclusions which are opposed to such truths are rejected by science.—LIEBIG.

PHILADELPHIA
J. B. LIPPINCOTT COMPANY
1893

THIS BOOK

IS DEDICATED WITH PROFOUND RESPECT

TO THE

WORKINGMEN

OF THE

MEDICAL PROFESSION.

CONTENTS.

	PAGE
INTRODUCTORY	7

CHAPTER I.
Sanitarians—Ancient, Mediæval, and Modern 15

CHAPTER II.
The Great Sanitary Awakening, with an Account of some of the Circumstances which attended the Birth of Sanitary Science 29

CHAPTER III.
The Air 50

CHAPTER IV.
The Air (continued) 79

CHAPTER V.
The Water 106

CHAPTER VI.
The Soil 142

CHAPTER VII.
The Sewer-Gas 153

CHAPTER VIII.
Cemeteries 181

CHAPTER IX.
Public Funerals 205

CHAPTER X.
The Meat... 215

CHAPTER XI.
The Milk... 231

CHAPTER XII.
Filth and Fecal Diseases—Typhoid Fever................ 248

CHAPTER XIII.
Yellow Fever... 280

CHAPTER XIV.
Cholera.. 292

CHAPTER XV.
Diphtheria... 329

CHAPTER XVI.
Epidemics.. 345

CHAPTER XVII.
Boards of Health..................................... 364

CHAPTER XVIII.
The Vital Statistics................................. 403

CHAPTER XIX.
Conclusion... 454

INTRODUCTORY.

About five years ago, during an hour of leisure, the author of this book took up an annual report of a local board of health and read it. It was full of general misstatements and inconsistencies, and abounded in misrepresentations of vital statistics. Yet this erratic and unreliable report was offered to scientific men, and to the medical profession, as a truthful exhibit of the value and working of Sanitary Science. On the strength of this report, the law-making power, to which it was addressed, was besought, in the name of the Public Safety, to formulate the most offensive and despotic legislation, to create new offices and to levy new taxes. The author was impelled, in a spirit of badinage, to publish anonymously in a newspaper a review of the document.

The sonorous and impressive titles of State Medicine, Preventive Medicine, Sanitary Reform, Sanitary Science, and their relation to the great question of the public health had always been deeply ingrained in his mind. The object which these titles represented surpassed in importance all considerations of therapeutics as applied to individual patients in the daily practice of medicine, as much as questions of national life exceeded those of the welfare of private persons.

The author felt that in reviewing this report in a mirthful way he had done an unholy thing: he had trifled with the public health, had mocked its ministers who were organized as a board, and who were supposed to be calmly but deeply meditating on the cause of those diseases which the medical profession had hitherto been unable to fathom, and which cause, as it was exposed to view by the light of Sanitary Science, they made known to a world sitting in darkness. The author believed that the report which he had criticised had been carelessly gotten up by its makers (they were his professional and personal friends, as friends go), and had glided from their hands without their being aware of its incongruities. He never doubted for a moment the merits of State Medicine and Sanitary Science, and in treating them lightly felt that he merited, and he expected to receive, a condign punishment.

His remorse was something like it would have been if, fifty years ago, he had laughed at his Sunday-school teacher, who in his ministrations had inadvertently donned a bad hat or worn a patched coat.

As by unexpected impunity wicked men are often emboldened to continue their transgressions, so the author's unlooked-for escape from chastisement encouraged him to go farther, and he reviewed a report of a State board of health, which was even fuller and fatter with nonsense than was that of the city board.

This kind of skirmishing went on for a year or two; shot after shot was fired in vain in the hope of drawing the ammunition from the sanitarians, when finally a reply came which in no way vindicated the absurdities nor explained the inaccuracies of their printed reports, but which was a biting taunt that the reviewer was an enemy of the public health. Besides, it was more than hinted, in a subsequent report, that the progress of Preventive Medicine was obstructed by those physicians who saw in its ultimate triumph a diminution of their own revenues.

It occurred to the author that if strictures on Sanitary Science could be replied to only with taunt and innuendo, it was fundamentally defective; so that, although his so doing implied a doubt of the infallibility of its professors, and although it might indicate an irreverent spirit and lay him open to the reiterated charge of being an enemy of the public health, he ventured to look up its history and examine its claims to be considered a science. He found that it had its origin in a kind of disorderly agitation that suddenly seized the people of Great Britain, following an inquiry into the condition and manners of living of the poorer classes in that country. Sanitary reform was not consequent to any new biological or pathological discovery; neither was it connected with any line of scientific research. It owed its rise and progress in our own land more to the fondness and habit of imitating the English than to any other cause. In both countries, although its inception was perhaps unalloyed by selfishness, speculators within and without the medical profession were quick to discern and grasp the opportunity to be cheaply lifted to fame and fortune, and, stimulated in this way by self-interest, when the excitement was well under way, its momentum was irresistible.

The theme of the whole movement was the causation by filth of infectious disease; and the phenomena of zymosis were so treated as to explain the origin of such disease. It was boldly declared by the reformers that filth—organic matter in decomposition and fermentation—was capable of exciting in the human system fermentative, zymotic, filth diseases, and these epithets were applied to all of those which had in them the element of contagion or infection. Zymotic disease was therefore preventable through the removal of filth and the hinderance of zymosis, and the science of Preventive Medicine was reared on this fantastic idea, which for the last fifty years has formed the basis of all sanitary legislation in Great Britain and America. It was never pretended that this whimsical theory had any foundation in scientific inquiry, and it never had the sanction of thoughtful and practical men in the medical profession. Dr. Farr, who was one of the first, if not the first, to apply the term zymotic to disease, was careful to say that he used it only because it was more convenient than the periphrasis of epidemic, endemic, and contagious diseases.

At the very outset of his labors the author was struck with astonishment at the almost utter barrenness, on the part of the sanitarians, of anything like scientific investigation. Not less surprising was this other fact,—namely, that if perchance—which rarely happened—an investigation which merited the name scientific was by themselves undertaken, or, what was oftener the case, if a genuinely scientific inquiry into their pretensions was made by scientific men, the result, in either case, invariably was their complete overthrow.

From time to time a sanitary orator at a sanitary convention would improvise some wild proposition about the air, water, and soil, or would indulge in some strange phantasy respecting the sewers, the cemeteries, or the markets; this would be wafted with great thoracic vehemence from the larynx of one reformer to that of another,—nobody would investigate its truth or falsity,—soon it would find an echo in some sanitary journal, and straightway would be given a place among the "settled principles of Sanitary Science."

If any resistance were offered to these vain imaginings, it was not listened to in a scientific spirit; but the opponent was censured as a foe to the public health, and if he resented this imputation, he was hushed by the reproach that he was an advocate of uncleanliness, and, as we shall see later, was said to be "content to wallow in his own filth."

The author's amazement had no bounds when, on examining, one after another, the "settled principles of Sanitary Science," he found that these had no scientific basis; that they rested on froth, noise, and panic, and that the shapeless spectres which the reformers had raised to intimidate the public disappeared when they were looked squarely in the face.

The whole sanitary movement had no resemblance to scientific investigation; it could be likened only to a political upheaval or a fanatical religious awakening. Indeed, it can be fitly compared with the imposition on mankind of those false religions whose priests have held, at different periods of the world's history, whole continents in terror by their inventions. As, in order to sustain these false religions, it was necessary that their ministers should continually reinforce them with some new dogma, so the vigor and stability of Sanitary Science depended on the ingenuity of its promoters to persistently summon up some new terror with which to frighten the people, and then proceed to caress them into tranquillity by the passage of some sanitary ordinance or by the pretended discovery of some antidote or antiseptic. These successive conjurations, combined with legislative enactments which imparted to them force, were called the "gigantic strides of Sanitary Science."

There have been times during the progress of this work when the author has doubted the evidence of his senses. More than once, in order to be convinced that his own eyes did not deceive him, he has laid before others the statements and figures of professional sanitarians which were so absurd, so self-contradictory, that it did not seem possible that they could have emanated elsewhere than from the brain of a lunatic or an imbecile. When pressed to explain their tissues of paradox and absurdity, the reasons which they gave were often so trivial that if, in a well-regulated household, similar ones had been offered to soothe the budding curiosity of a nursling, its attendant would have been visited with reproof if not with summary dismissal.

The author has often felt a sense of shame that many of those who were foisting this sanitary nonsense on the people, and on physicians, and demanding that it be embodied in statute law, were nominally of the medical profession. In the beginning of this inquiry, he had no thought of publishing its result. But, as the work proceeded, he became more and more impressed with its gravity, not only to medical men, but to the public at large. He esteems that an important point has been gained if he shall succeed in calming the fears, quieting the panics, and restoring the composure of his fellow-citizens, whose minds have been continuously excited and kept at a painful tension by sanitary reformers, concerning the dangers of air, water, soil, cemeteries, markets, public and private improvements, and if he has shown that none of these, in the conditions in which they have been set forth by the pretended guardians of the public health, are causes of disease, and especially of infectious disease.

In the main, in this work, the author has drawn no conclusions; he has submitted facts. That these facts are to his own mind demonstrative, that they have changed entirely his former belief in the etiology of infectious disease, he makes no effort to conceal. It was never anything but a "belief" transmitted to him by oracular men who had no claim to be considered scientific. If, sometimes, he has betrayed a warmth of expression, it is because of natural indignation that he had not only been the dupe of noisy men who were posing as scientists, but that through their teachings he had been the instrument of duping others.

If the objection be raised that the conclusions to which the facts here presented tend shatter the faith in Sanitary Science and leave the public health comfortless and the people nothing to lean upon, the reply is, that if the faith be false it should be discarded; that in natural science an intelligent agnosticism is better than blind credulity in error, especially when the subject is of such weighty moment to humanity as knowledge of the causes of disease. Though in theological matters it may be debatable (in many minds it is firmly settled) that mankind is happier with a false belief than with none at all, a like conviction has never obtained in questions of physical inquiry, and it will not be denied that one of the first steps towards ascertaining the truth is to expose and remove error. Neither will it be disputed that the virgin mind, untainted in its sincerity, is fitter for the reception of truth, when this is made known, than is the mind which is clogged and darkened by falsehood.

If further opposition be made, that the display of these facts will expose the pretensions of a large number of professional sanitarians, who, by playing on the fears of their fellow-men, have acquired fame and position; and that it will wound a large number of amateurs of both sexes who have been seduced into dabbling and coquetting with Sanitary Science, for the reason that it required no mental labor to become proficient in its mysteries, and who have found therein a solace for their *ennui*, the reply is, that the objection is well taken, and from a social point should be considered, but should not be sustained in view of the vast importance to the people that no errors should be confirmed regarding the public health; and, moreover, attractive fields are being continually thrown open, which offer an ample refuge to that large class of minds which seek intellectual repose in improvisation rather than in scientific research.

Those to whom truth is distasteful, lest it shall shock a life-long prejudice, are advised not to read this work,—it will only irritate them. It had better remain closed to those who fear that they will sink into depravity should they listen to evidence respecting the innocence of nature's metamorphoses. Those are cautioned not to open it who, though feeling secure in their own virtue, are so solicitous and apprehensive for the vulnerability of that of their neighbors, lest they retrograde in civilization and prefer nastiness to elegance unless their minds are steadily tortured with hideous fables of disease and death. On the other hand, those timid people who for the last thirty years have had their waking hours vexed and their sleep plagued by an unceasing procession of sanitary terrors are invited to read it. It may comfort them. Those also are invited to read it who love truth for truth's sake, and who believe themselves sufficiently steadfast to receive it, and who can survey nature's changes in decay and death, not only with the same composure but with the same poetic fervor that they view her creative and formative processes, without imagining, in the absence of all proof, that these mutations are inseparable from the explosion of epidemic disease.

VAGARIES

OF

SANITARY SCIENCE.

CHAPTER I.

Sanitarians—Ancient, Mediæval, and Modern.

THAT instinct of self-preservation, the most powerful of all instincts, which is ever active with the individual, making him timid or courageous in the presence of danger, especially the danger of epidemic and infectious disease, extends itself with no less potency to collective bodies of men for the conservation of communities, nations, and races.

Man is hardly conscious of life when he begins to be tormented with the fear of death. His hope of a continued being and a fear of its extinction have often overcome reason and judgment, and have made him the dupe of the pretender and the charlatan in all ages.

The most primitive people of whom we have any historic records, even in their transit through the wilderness, submitted to as stringent sanitary codes as any of those which have been contrived in our own time. It was doubtless a prophylactic ordinance which marked the Jew with a fleshly sign that distinguished him from the rest of mankind. By the fifteenth day of the second month of their journey the Israelites had begun to worry about their health. The cloud yet rested on the tabernacle by day and the fire blazed on it by night to guide the exiles in the desert, when their leader and prophet put forth a system of dietetics which, to a great extent, is observed by their descendants to the present day. Birds, beasts, insects, reptiles, and fish were proscribed which the experience of all other peoples has proved to be harmless; others were allowed and even commanded to be eaten from which we turn with disgust. The camel, the cony, the hare, and the swine were forbidden; they were unclean; not only were these and everything in the waters that had not fins and scales denounced as an abomination, but the decree went further, "Ye shall have their carcasses in abomination." They might eat the sheep, the ox, the goat, the deer, and the pygarg, and the locust and the bald locust, the beetle and the grasshopper, but "any creeping thing that creepeth" they should not be defiled thereby. They should not eat the blood, for the blood was the life. Neither should they eat anything that had died of itself. This they might sell to the alien or give unto the stranger. Experience and observation had probably taught their guide that the flesh of animals dead of disease was innoxious, for he had long before counselled them to love the stranger, reminding them that they, too, had been strangers in the land of Egypt. But his delicate and fastidious mind would not tolerate such food for the Jew. His people were a peculiar people, and his God was a jealous God.

Attempts have been made to show that the aim of the prophet sanitarian was to set forth in these mandates the virtues of self-restraint and temperance. This can hardly be so; for when the wanderers languished and murmured by the way, his method of infusing new courage into their hearts was by appealing to their appetites. If any message had been given to him from the flaming bush of the life eternal, he spoke not a word of it to the materialistic Jew. Many times he told his followers that he was conducting them to a land flowing with milk and honey, where they were to eat and fill themselves and wax fat; and as they approached the river where he was to lay down his leadership and find an unmarked grave, he burst forth into a victorious song and laid before them a rich but unclassified *menu*, that Jew might taste with delight and that Gentile might adore. For the Lord had made Jacob to "ride on the high places of the earth, that he might eat the increase of the fields; and he made him to suck honey out of the rock, and oil out of the flinty rock; butter of kine, and milk of sheep, with fat of lambs, and rams of the breed of Bashan, and goats with the fat of kidneys of wheat; and thou didst drink the pure blood of the grape." This sacerdotal hygiene took cognizance of the most secret relations of life. If a man or a woman had an issue, separation from the rest was enjoined for a stated period. The etiology of disease was in the fiat of Jehovah. The priest diagnosticated the malady, prescribed the treatment, made the prognosis, supervised the convalescence.

The therapeutics of the Hebrews of that day would be no more acceptable to us than their prophylactic measures and their dietetics. The blood of the trespass-offering was to be put on the tip òf the right ear of him that was to be cleansed of the leprosy, and upon the thumb of his right hand and upon the great toe of his right foot. When the wandering children were bitten by fiery serpents, they had but to look on the brazen serpent that their leader lifted up and they lived. If they were faithful to the statutes and commandments, they should not suffer the diseases that had been brought on the Egyptians. But if they did not keep the laws, "then the Lord will make thy plagues wonderful, and the plagues of thy seed, even great plagues, and of long continuance, and sore sicknesses, and of long continuance." "The Lord will smite thee with the botch of Egypt, and with the emerods, and with the scab, and with the itch, whereof thou canst not be healed."

Many centuries later appeared another sanitarian, a legislator in the person of Lycurgus. His prophylaxis was of the most radical kind. The new-born Spartan child which gave no promise of future vigor was abandoned or destroyed. If permitted to live, not long after it had been weaned from the mother breast it was seized by the state, removed from the family, and made a part of the common stock. It was subjected to influences and exercises to strengthen and indurate bone and muscle and nerve. A public table supplied by the simplest food nourished the bodies of the Spartan youth. Those mental qualities which tended to self-preservation were stimulated and encouraged. Lying was a virtue and theft a duty. When the Spartan brave made love, he sought and won his bride by stealth and violence; and the first-born of Spartan children were the product of a rape. Marriage could not be contracted by men before thirty, nor by women under twenty years of age. When the Spartan woman was likely to become a mother, the pictures of the handsomest young men were hung in her chamber, that gazing on them might produce a favorable effect on the child. That this system of legislative hygiene, which in our day would bear the pompous title of State Medicine, was effective to establish a vigorous body will not be denied. The Spartans were a healthy but a bad lot.

Five hundred years later appears a man whose deep philosophy is set forth in such sententious phrase as, "Life is short, and art is long; the occasion fleeting; experience fallacious, and judgment difficult." He displayed such practical knowledge of the art of medicine that his works have been studied and admired for more than twenty-three centuries. The modern sanitarians, in their retrospective excursions to seek in antiquity a warrant for the vagaries and chimeras of their own creation, rest fondly on Hippocrates II. or the Great. In his book on "Airs, Waters, and Places," they find, or pretend to find, a treatise on hygiene, whose title they have transformed into their own shibboleth of "pure air, pure water, pure soil," as the essentials of public health. That they have never read or that they misrepresent the Great Father is plain, for there is absolutely nothing in his book on "Airs, Waters, and Places" that can possibly be twisted to the modern sanitarians' use. Their prototypes belong to the charlatans of an earlier and a later age.

A thousand years have rolled on from the time of Hippocrates, when the mediæval sanitarian arises to teach and guide his fellow-man in the maintenance and promotion of his health. The school of Salerno, in the tenth century, emits hygienic maxims in Leonine rhymes, which afford amusement, if not instruction, to the reader of to-day. This famous body, which existed for centuries, seems to have been entirely neglected by the modern sanitarians, for we do not remember ever to have seen it alluded to in any of their writings. In some of the apothegms of this school they will find a counterpart of their watchwords, "pure air, pure water, pure soil," sometimes done into English, sometimes into French verse. In pestilential times the astute, mediæval reformer discovered the etiology of epidemics in the machinations of the Jews, and his prophylaxis consisted in first robbing and then roasting and hanging these unhappy people.

From the Middle Ages to the present time are strewn accounts of efforts to promote the public health,—all or nearly all, however, founded on false notions of the etiology of the diseases which these endeavors sought to control. It cannot be gainsaid that the Mosaic theory of the etiology of epidemics—to wit, the Divine will—can be less successfully contradicted than any invented by the sanitary reformers.

The feebleness of sanitarians and boards of health so impressed Noah Webster that he declared in his work on epidemics, a hundred years ago, that he found "no sufficient evidence that health laws ever saved a country or city from pestilence in a single instance; but abundant, positive proof of their inefficacy in a great number of cases;" and he remarks that men have perished by millions in the most salubrious regions, exposed to no local causes of disease whatever, except such as exist in the most healthy periods. In the progress of this work we expect to show that, notwithstanding the boasts of the sanitary reformer of to-day, his methods of dealing with pestilence in its relation to public health are almost exactly like those of ancient and mediæval times. But our business in these pages is to consider the rise and progress of the modern sanitarian, and particularly as he exhibits himself in our own country. Twenty-five years ago he was in embryo. The ferment about sanitary reform, which had existed in England for a quarter of a century before, had just reached our shores. There was then no occasion for alarm about the public health. There was no reason to doubt that here, as elsewhere in the civilized world, the general death-rate of human beings was diminishing; that the mean duration of life was being extended; and that, too, except through a general advancement in civilization, it was being accomplished independent of any known human agency. But the occasion was ripe for noisy, superficial men to be heard, and these commenced to bring themselves into prominence as scientists. They began to prate loudly of air, water, and soil. These elements, indispensable to life, were pestilential, polluted, contaminated. Towns and cities which had been renowned for centuries for their health were suddenly discovered to be the breeding-places of epidemic disease. Suspicion was cast on sources of water-supply which for generations had furnished a delicious and healthful beverage. The reformers told us that the very fact that these waters were clear and sparkling and grateful to the taste should arouse our distrust, for just such waters had been proved to contain the essence of contagious ailments. The soil, too, had become saturated with putridity; it was a seething volcano of disease ready to burst forth at any moment. The mysterious relation of disease to these airs, waters, and soils demanded an intercessor to negotiate the conditions of health between them and the people; this mediator should be the sanitarian; the brokerage was to be paid by a draft on the public funds.

The sanitary reformer saw, or pretended to see, the hidden principle of disease lurking in every operation of nature whereby organic substances were decomposed and their original elements set free to assume a new *rôle* in the universe. The fanciful thought that certain types of diseases, febrile, eruptive, epidemic, and contagious, which had been no less fancifully named zymotic, were, in their advent and course, analogous to the fermentative process, was for the sanitarian an attractive basis for his theory; and he went a step farther and called them "filth diseases."

He made no investigation; he relied on his nose for information: this taught him that all of those transformations of matter, those reciprocal offerings of animate and inanimate bodies, the cessation of which would bring disaster and destruction to all life on the globe, were the sources of zymotic and, as he called it, *preventable* disease.

Herein lay the septic ferment, the morbific element; and as new biological discoveries were made, which seemed to show that the principle of disease was a material object, a germ, he attempted, with a most ludicrous result, to apply it to his theory. He adapted to this discovery the parable of the sower: the filth was the soil, the germ was the seed, the harvest was zymotic or preventable disease. This jumble of fancies, words, and phrases was baptized with the name of Sanitary Science; its advocates called themselves, at first, sanitary reformers, and later, sanitarians.

We expect to show that there is no more proof of the evolution of a noxious element in the breaking up of organic matter than there is in its creation, or in the union and decomposition of inorganic substances; that in either case nature's analyses and nature's syntheses are equally innocent.

To bring themselves into notice the reformers took every opportunity to excite and magnify the fears of the people about their health; they foretold epidemics that never appeared; on the other hand, epidemics broke out of which they gave no warning.

Were a new convenience devised, whereby our houses were made more inviting and more comfortable, they created a new fright, and persuaded us that the improvement could be tolerated with safety only by the supervision of a health inspector. Were any method brought out to produce a new article of food, whereby the price of a necessity or luxury of life could be lessened to mankind, the cry of danger to the public health was raised. Did some enterprising man introduce a cheaper meat-supply, which implied the alimentation of millions, interested parties invoked the public health to suppress it, and found coadjutors in sanitarians and boards of health to so encumber its distribution that the beneficent project was often defeated. Were negro domination in a city obnoxious, "the settled principles of Sanitary Science" demanded that the State Executive should appoint its officials, and Jacksonville's autonomy must be sacrificed to maintain the public health. Were an international political crime contemplated, Sanitary Science furnished the excuse; and our self-preservation depended on the seizure of Cuba as a precaution against yellow fever.

The reformers darkly hinted that there existed among us an unprincipled body, which to advance its interests did not scruple to plot against the public health, and nothing swelled their importance so much as the system of espionage and secret accusation which they ordained and fostered, whereby every quarrelsome man and every spiteful woman, in seeking revenge against a neighbor for a real or fancied grievance, felt secure of finding a confederate in local boards of health. This was carried to such an extent that their own public reports acknowledged from thirty to forty—private admissions sometimes made it fifty—per cent. of all complaints as groundless, most of them based on ill-humor. In Boston, in one year, sixteen hundred and thirty complaints were investigated, in which no cause for action was found. The Sanitary Committee of the New York Board of Health* reports, "Many complaints upon investigation proved to have been exaggerated and in some cases to have originated in malice or a desire to secure personal aggrandizement." One city of sixteen thousand inhabitants † declared that many of the complaints made to its board of health were "the result of spite."

Bewildered and frightened men bore all this because they were led to believe that the public health was in danger. They tried to save their wives and little ones from perils that existed only in the brain of the sanitarians. Power was conferred on these to issue decrees which they called "Sanitary Codes," every one of which, if we except those pertaining to isolation and vaccination, is useless, silly, and oppressive, and has no bearing whatever on public or private health. They told the people that if they would delegate to them the authority to enforce these statutes and commandments, they would be saved from epidemic disease; but if they were unheeded, not the plagues and the botch and the emerods of the Egyptians would seize them, but worse. They prophesied the return of the mediæval pestilence,—the Black Death of the Middle Ages.

Dangers beset us at every turn; if we stayed in the

---

* 1886. † Connecticut Board of Health Report, 1886.

cities, we were poisoned by sewer-gas; it was sure death to go away from home; if we ventured a visit to the country, the water-cooler with its impure ice was in the palace-car,* or it was filled from the polluted road-side well. Milk and cream were dairied in barn-yards; the gases of decay from foul soils, cesspools, and privies were leeching into hotels and dwellings; there were in-door sanitary conveniences for taking typhoid fever from defective water-closets, " which constitute at least nine-tenths of all the hotel fittings throughout the country, not excepting even Saratoga." There were out-door sanitary conveniences over pits and vaults, increasing dangers a thousand-fold to those who were subjected to these putrid emanations.

What arrested the attention of observers was the facility with which the professors of the new-born science achieved celebrity. In every other department—mechanics, science, or art—distinction was attained only by protracted and patient industry and study. But in the very dawn of Sanitary Science, its apostles had but to foretell some great disaster, improvise a rhapsody on cleanliness, offer an essay on sewer-gas, or publish a diatribe on the grinding landlord, and they were greeted by their fellows at the next sanitary convention as the EMINENT SANITARIANS, and henceforth were to enjoy the triple dignity of prophet, legislator, and sage.

They found a powerful ally in the priesthood. Here was a body of men who were genuine lovers and promoters of the public weal. They were made to believe that the general health depended in some way on obedience to the mandates of the sanitarians; and the pulpit thundered in favor of sanitary reform.

To the tender, susceptible, and prophetic female mind, which conceived an indissoluble tie existing between phys-

---

* *Sanitarian*, vol. x.

ical purity and bodily health, the infant science offered peculiar charms. The sanitarians rehearsed the seductive tale that the sphere of woman lay in sanitary reform, and, after toying with it for a season, she embraced the new doctrine, and by her ready pen and ready utterance has probably done more to nurture and popularize the same than all the clergy and doctors put together. She was told that, while the more profound secrets of the dawning knowledge were hidden from the wise and prudent housewife, and revealed only to those who had solemnly consecrated themselves to their interpretation, still she was amply competent to grasp the minor points and be useful to herself and family. She could look after the plumbing. If there were anything here to arouse suspicion, the alert mistress had in her closet an unerring test for sewer-gas in that carminative, anti-flatulent, anti-spasmodic essence of peppermint. When this was poured into the pipes, if she smelled it, or thought she smelled it, she could rescue the household by telephoning the family plumber before it was too late. Those ladies who had a literary turn gave expression and vigor to the new science in their novels. Filth and sewer-gas as causes of disease and drainage as its preventive and cure were set forth in graphic story and threatened to displace altogether those finer particulars of obstetrical knowledge which had so often adorned their tales and entertained and instructed their readers. In Robert Ellsmere we have a most happy combination of both sanitary and obstetrical science.

Here and there the sanitarians suborned a talented member of the medical profession and subsidized him to their uses. Though the profession at large tendered no direct opposition to the current of sanitary reform, and in some cases were persuaded as public bodies to endorse boards of health, many of its members looked on incredulously, and some of the most influential denied flatly the dictates of the sanitarians respecting the origin of epidemic disease. As sensitive to the public good as to the welfare of their individual patients, and trusting that perhaps there might be a grain of truth somewhere in this whirlwind of chaff, medical men yielded to the tempest of sanitary reform, and tolerated by their silence the charlatanry of its advocates. Many of them, doubtless, believed in the tradition which had been handed down, of the danger to health from putrid emanations and the decomposition of organic matter. We purpose to show that, in every instance where this subject has had a careful and systematic investigation, medical men have acknowledged a surrender of their prejudices.

When it suited their interests, the sanitarians juggled with tables of mortality and misrepresented vital statistics. They stifled all investigation, all discussion. If any man dared, for a moment, to oppose the fury by calling for proof of the new doctrines, he was branded as an enemy of the public health. Meanwhile, by keeping the people in a continual ferment and panic, they established a veritable reign of sanitary terror.*

The kingdom of Sanitary Science suffered violence, and the violent were to take it by storm. The reformers called for "aggressive sanitation." They demanded heavy fines and imprisonment for those who should transgress their sanitary codes.

One eminent sanitarian † in a public lecture declared that the causes of infectious diseases and the means of preventing them were as well known and as readily controlled as those of railroad dangers, and he suggests that it be enacted, etc., "that every legal resident in every town in Connecticut, who shall, while residing in the town, have either of the following diseases, viz., yellow fever, cholera, small-

---

* See note at end of chapter.
† Connecticut Board of Health Report, 1888.

pox, typhus fever, scarlet fever, typhoid fever, shall be entitled to receive from the treasury of the town three dollars for each day that he is confined to his house by such sickness;" and that "every person so afflicted shall be subject to such regulations and restrictions as the board of health of the town shall determine."

Another health-officer,* in Michigan, says that the world can never be reformed by moral suasion alone; and, impatient of the slow process of the law, he advocates a prompter method, and says, "If tenants whose humble homes have been visited by the angel of death would mob the landlord and throw him into the reeking cesspool, it would do more good than the best hygienic tract on sewer-gas that was ever written. If a thousand emigrants escaping from a foul steamer would burn it up, it would do more good than an act of Congress. If the proprietor of a dairy distributing milk from premises where there is small-pox, scarlet fever, diphtheria, or typhoid fever, were set upon and hung on the nearest lamp-post, instead of having him prosecuted, this would infuse a healthy vigor into the makers of law."

The sanitarians affected a deep concern for the "humble home," and they summoned us to behold their efforts to ameliorate the condition of the poor, who were at the mercy of the grinding landlord. This self-eulogium on the one hand and denunciation on the other attracted observation and excited criticism. People asked who were these humane men and what had been their previous history. They had never before been distinguished for benevolence. A goodly number of them had had the sympathy of their neighbors for their want of success in former endeavors in life; others were second-class ward politicians. But with the help of the *dilettanti* of both sexes they organized for

---

* *Sanitarian*, vol. xi.

business, and for nearly a whole generation have held the people on the verge of a panic, reminding them that some portentous calamity was impending.

It is a suggestive and significant fact that, in order to avert this calamity, we were bidden to clothe the sanitarians with power, invest them with office, and provide them with money.

NOTE.—A factory which had been in operation for forty years, and around which had grown up a small town whose prosperity depended on it, had suffered great persecution and had spent large sums of money for defence against a band of speculators who had invested in real estate in the vicinity, and who now complained of the factory as dangerous to the public health. The author is sure that tens of thousands (maybe fifty thousand) of dollars of expense had been caused this company. He asked for information and, at first, received an exultant reply that the company had had a costly fight, but had beaten its enemies, and that when the superintendent returned, who was then absent, he would gladly furnish the details. Not so with the wary superintendent. He said that it was true that his company had suffered great trouble and expense, extending over a period of two or three years, but he asked to be excused from giving any facts; they were then on very good terms with the board of health and preferred to remain so; that to reopen the case by publishing anything which had occurred might make further trouble, and they wished to let the matter rest.

The author called in person on a water company which had endured a long and expensive contest with the health authorities on account of the pretended dangerous water it was supplying. A rival was in the field, and in order to succeed had invoked the public health. The company admitted they had suffered grievous wrong, but as they had won their case they did not wish to say anything for publication which might again stir up the matter.

The agent of a steamship company gave the author a verbal account of the annoyance and expense it had been put to by a certain board of health. When leave was asked to publish the story, the agent showed great concern lest making it known should subject his company to sanitary vengeance.

Not long ago an alarming account was given of the horrible sanitary condition of a public building in one of our large cities. Employees and visitors were in imminent danger every hour of being poisoned by sewer-gas. This condition had existed for twenty-five years. At certain seasons there are between eight hundred and one thousand people engaged in the building during the entire day and sometimes part of the night. Large numbers visit it at all seasons of the year. The author inquired carefully of many of those employed there if they knew or had ever heard of any sickness arising from the building. At every inquiry he was laughed at. One of the principal officials then told him that there was a one-hundred-thousand-dollar job in prospect, and he gave information to show that the whole sensation was got up by a ring of speculators. When leave was asked to publish these details the official wilted at once, and implored the author not to give them publicity; that he (the official) would be held up as an enemy of the public health, and it might cost him his position, but that if called before a proper committee he would state the facts.

These are only a few instances, not to mention direct charges of blackmail, which the author has met with, showing that men and corporations prefer to suffer their wrongs in silence rather than encounter the vengeance of boards of health.

CHAPTER II.

The Great Sanitary Awakening, with an Account of some of the Circumstances which attended the Birth of Sanitary Science.

About fifty years ago a report was made to the British House of Commons on the health of towns and the condition of the laboring classes in Great Britain. This document gave a sorrowful account of the labor, wages, food, clothing, and shelter of these classes. A large population lived in cellars; one room frequently accommodated two, three, and four families. Parents with children above the age of puberty occupied the same bed. The lodging-houses were yet more crowded; three and four adults were often found sleeping under the same coverlid.

Dr. Neil Arnott [*] describes a portion of Edinburgh that he visited. "We entered a low passage like a house-door, which led from the street through the first house to a square immediately behind, which court was occupied entirely as a dung-receptable (with the exception of a narrow path

---

[*] London *Lancet*, vol. ii., 1842-43.

around it) of the most disgusting kind. Beyond this court the second passage led to a second square court, occupied in the same way by its dung-hill; and from this court there was yet a third passage leading to a third court and dung-heap. There were no privies or drains there, and the dung-heaps received all the filth which the swarm of wretched inhabitants could give. The interiors and inmates corresponded to the exteriors; we saw half-dressed wretches crowding together in one bed to be warm, though in the middle of the day. Several women were imprisoned under one blanket, because as many others, who had on their backs all the articles of dress that belonged to the party, were then out of doors in the street." The common lodging-houses were resorts of the miserable of both sexes, bedded together promiscuously at night, "men, women, and children in an atmosphere odorous of gin, brimstone, and onions, and human miasms. Thirty and forty are often herded together in a couple of small rooms, four, five, and six in a bed; and should one of the helpless inmates (as is often the case) die of typhus fever, it is by no means uncommon to find the identical unchanged beds occupied on the very next night by fresh sleepers."

An inquiry into the manner of living of the laboring population in the inner ward of St. George's, Hanover Square, showed that one thousand four hundred and sixty-five families had two thousand one hundred and seventy-five rooms, and two thousand five hundred and ten beds; nine hundred and twenty-nine families had each one room, and six hundred and twenty-three each only one bed.

The Wynds of Glasgow * comprised a population of from fifteen thousand to thirty thousand persons. "This quarter consisted of a labyrinth of lanes, out of which numberless

---

* General Report on Sanitary Condition of the Labor Population of Great Britain.

entrances led into small, square courts with a dung-hill reeking in the centre. In some of these lodging-houses we found (at night) a whole layer of human beings littered along the floor, sometimes fifteen and twenty, some clothed and some naked; men, women, and children, huddled promiscuously together, their bed a layer of musty straw intermixed with rags."

Many of the houses of the poor were built around courts, with a pit in the middle to receive the filth of the occupants. "In some the whole courts up to the doors of the houses are covered with filth."* Their food was scanty; meat not at all, or rarely once a week. There were in Liverpool eight thousand cellars occupied by from thirty-five thousand to forty thousand people oût of a total of two hundred and fifty thousand. The working classes here numbered about one hundred and seventy-five thousand; a very large proportion of these had "no means of getting rid of their filth but by throwing it into the street," so that "the air is constantly contaminated by the emanations from this surface of putrefying and offensive matter." In one cellar thirty people slept every night; a hole was dug in the floor for the offal and filth of the household; in another were three cart-loads of dung mixed with the offal from slaughterhouses, and "the family in the cellar lived and slept contentedly cheek by jowl with the putrefying mass." Comparisons were made of the mortality of English towns, showing the enormous disproportion of deaths between the poorer classes and the well-to-do. These tables of the death-rates were misleading in that they took no account of the birth-rate.

The scope of those who made these inquiries and who furnished the reports seemed to be to show that the filth in some way was the cause of the disease that prevailed,

---

* Sanitary Condition of Laboring Classes in England and Wales.

and the medical officers appeared to comprehend it, and framed their answers to correspond to this aim. A very large majority took no notice whatever of the ignorance, poverty, intemperance, and imprudence of these people, but expressed their belief that the filth in which they lived was the cause of the high mortality. A few, either because they were less obsequious, or less keenly alive to what was expected of them, blurted out the fact that they found the "dung-heap and the poison, but all the inhabitants in health," qualifying the statement, it may be, with the information that this poison was waiting for "a change in the weather or temperature," and all would be sick with "headaches, constipation, small-pox, and fever; and in many cases atrophy."

In Truro, fever prevailed where there was only a small amount of decomposition. In Kent and Sussex, filth prevailed everywhere; but Dr. Tuffnal states that "throughout the greater part of these counties comparatively few diseases can be found to arise from want of sanitary precaution."

At Brighton were filth and overcrowding, but "the more seemingly unhealthy districts quote no fever." Dr. Baker says the cause of ill-health in Derby is the factory system as a whole; "because beginning with childhood, and going on to youth, it brings up puny parents of a puny race, who in their turn perpetuate and increase the evil."

In Birmingham, the river Rea is the main sewer of the town; in summer it is covered with a thick scum of offensive and decomposing matter. About fifty thousand of the people here live in narrow, ill-ventilated, filthy, badly-drained courts. Most of the houses are three stories high. In each court is an ash-pit, a privy, a wash-house, one or more pig-styes, and heaps of manure. Many of the lodging-houses are in a loathsome condition, crowded with beds occupied indiscriminately by both sexes. In the daytime these houses are thronged with dirty, half-dressed women and children; in the evening the inmates are eating, drinking, and smoking. The slaughter-houses are scattered all about the town, but "we do not find that any injury to the public health is derived from the state of the slaughter-houses." The knackers' yards, skinners' yards, and catgut-factories are extremely offensive, "but we do not find that these situations are more than others the seat of fevers or contagious disorders." Contagious fever is so rare here as to be almost unknown, and there is no part of the town where fever exists more than another. "We find it occurring in the elevated as well as in the lower situations."

Surgeon Ryland declared bluntly that locality had nothing to do with typhoid fever; that it occurred quite as much or more in the higher and better-drained parts of the town; and that undrained houses and collections of stagnant water are insufficient to cause the disease. Children here entered the factories at as early an age as seven years. Dr. Howard says the amount of fever in Manchester is not large for a town so "peculiarly fitted to promote the diffusion of contagious disease." Indeed, he says the exemption is remarkable, when the entire absence of cleanliness is considered; and he thinks that contagion is the great element, for the filth in some of the streets and courts that are exempt from disease is horrible; large, open cesspools, filthy and dilapidated privies full to overflowing, "disgusting and offensive beyond conception." "Abominably filthy places remain free from fever for long periods;" and he believes poverty and destitution are more powerful causes, and that something besides filth is necessary to generate the disease. He shows by a table that the outbreaks in Manchester for forty-five years have corresponded to periods of great distress, bad harvests, and consequent scarcity of food and work. He remarks, too, that when the number of deaths was greatest, the number of births was greatest also. The wages of the men in Manchester for ten years, ending 1841, varied from nine shillings to nine and sixpence per week. The wages of women and children averaged from five to eight shillings. Marriages were contracted early in life, and one witness testifies that "few women in his neighborhood ever marry until they have had or are about to have their first infant."

In Salop, Cheshire, and North Wales, many families have only one room; here four or five and sometimes eight or ten people sleep. The houses abound in filth. But "notwithstanding the crowded and deplorable state of these habitations contagious diseases do not appear to have generally prevailed."

In Inverness the nastiness is past endurance. "There is not a street, lane, or approach to it, that is not disgustingly defiled at all times, so much so as to render the whole place an absolute nuisance." Fever is here, but the doctor writes, "For many years it has seldom been rife in its pestiferous influence." "The people owe this more to the kindness of Almighty God than to any means taken for its prevention."* No effort was made in this report to adjust these discrepancies of the medical officers.

In the vast majority of cases, however, the filth was accused as the sole generator of disease. The piercing eye of Dr. Barham had noticed fever connected with a "near proximity to even a small amount of organic matter," and all measures for improvement he says may be neutralized "if a little nidus of morbid effluvia be allowed to remain." It was easy to show that the death-rate was higher among filthy people than elsewhere. These were generally poor, badly fed, clothed, and sheltered, often intemperate, and almost always imprudent and wasteful. They produced children in abundance at an early age; these were not and

---

* General Report on Sanitary Condition of the Labor Population of Great Britain.

could not be properly cared for, and it was well that so many of them died.

Mr. Chadwick showed that in some filthy and unhealthy districts the death-rate was sixty-eight per cent. and the birth-rate forty-eight per cent. higher than in some healthy and well-drained localities, and he ascribed the high death-rate only to the filthy condition of the people. The tables and figures which this amiable enthusiast presents must always be viewed with caution, for whenever they are analyzed they rarely sustain his conclusions. Most surprising of all is the fact that in spite of the accumulation of misery, poverty, and dirt, the human death-rate for three centuries had been steadily declining. So far as estimates can be depended on, says Mr. Chadwick, the deaths in London in 1700 were one in twenty; in 1800 they were one in thirty-nine. In 1799 the average age at death was twenty-six years; in 1830 it was twenty-nine years. All this was taking place long before the era of, and was not dependent on, sanitary reform.

During the progress of an inquiry * that was made by the metropolitan registrars into the sanitary condition of those portions of London which yielded high mortality rates, six, eight, ten, or even a dozen persons were often found sleeping in one room. The only protection from cold which they enjoyed was through "a few coals during the severe weather from the benevolent." The children were numerous; they were herded together and "walking on the cold stones or sitting at the door in all weathers." With the adults, "spirits often supply the place of lodging, food, and raiment." The houses were old and filthy, occupied by people of "the lowest description, uneducated and foul-mouthed; mendicants, costermongers, thieves, and abandoned females." Their "food consists of salt fish and other

---

* Fifth Report of the Registrar-General of England and Wales.

scraps collected by the mendicants and disposed of to the general dealers." The men spent one-third of their earnings in drink; the women mostly indulged in gin, and stupefied themselves and their babies with paregoric. The mortality from zymotic disease was 6.013 per 1,000,000 of inhabitants for cities and 3.142 for the country. The reformers ascribed this difference to the filth which surrounded the people in the cities. They paid no attention to the facts that where the mortality was highest, the people suffered in the winter from stinging cold and from stifling heat in summer; that the year round they endured a gnawing hunger, a burning thirst that alcohol itself could not quench, and a throbbing anxiety that nicotine and opium could not soothe.

A further glance would have shown the reformers that the high mortality was not alone from zymotic diseases. The same returns which gave the proportion of zymotic diseases in city and country showed that the mortality from diseases of the nervous system was 4.267 per 1,000,000 in the cities and 2.256 in the country. The deaths from respiratory diseases were 7.967 per 1,000,000 in cities and 5.327 in the country, and the mortality from diseases of the digestive organs was 1.972 per 1,000,000 in the cities and 1.042 in the country. But the reformers only saw, or only pretended to see, that the drainage was defective, and that the yards, courts, houses, and bodies of many of these people in the cities were foul. Zymosis was going on, filth was created, and the conclusion was irresistible that this was the cause of zymotic, or filth diseases; they refused to search further for the cause of the high mortality.

These disclosures of the condition of great masses of the people in Great Britain touched the conscience and aroused the compassion of the English nation. For the moment no direct opposition was offered to the filth pathology, and with a great sound of trumpets it was proclaimed that good drainage and the removal of filth would exterminate that class of diseases known as zymotic or contagious, and the cry was raised for sanitary reform.

The cemeteries received the first attention. These, both in London and in other towns, were in a disgusting condition, and common decency called for a change in the care of them and in the method of burial. The evidence against them was most revolting; but as will be shown in another chapter, not a single case of disease was proved to have originated from one of these cemeteries.

Sanitary reform, the *Lancet* said, was now the order of the day; we are "likely to be beset with blue-books, white-faced pamphlets, speechifying and figure-making, jobbery and intrigue;" but the work must go on. Doctors poured in their reports that fever occurred under their care where drainage and ventilation were defective. Dr. Stark * declared there were three hundred thousand cesspools in London, with an aggregate exhaling surface of sixty-two acres; they would make together an enormous receptacle ten miles in length, fifty feet in width, and six feet six inches in depth. In 1849 it was said that the public mind was now "on the right scent;" the cholera was making great ravages and insufficient drainage was the cause; "all cholera cases appear where the victims have been exposed to exhalations from drains, cesspools, etc."

Associations were formed to improve the sanitary state of towns throughout the kingdom. If here and there a medical man like Dr. Corrigan declared his belief that facts were in direct opposition to the new theory, he was quickly smothered in the ferment. In 1850 it was discovered that "the Thames water is polluted with every conceivable filth and abomination;" the water of all the companies "is contaminated with dead and living organic matter;" and Mr.

---

* *Lancet*, vol. ii., 1848.

Bowie, a surgeon, testified that it "is thick, muddy, discolored, putrid, and unfit for culinary or drinking purposes." In a feeble way, Dr. Alfred Taylor * tried to restore a little calmness by affirming that if the organic matter did not affect the color and taste of water it was not unhealthy; but in the state of the public mind such reassurance was in vain.

The soil, too, of London had been absorbing filth for centuries; it was nauseated with putridity; ready to vomit forth disease and death; the sewers were so constructed that they not only poisoned the air of the streets, but foul air entered through them into the houses, bringing bedrooms and nurseries in communication with the sewers; the water was polluted, the air infected, the sunlight intercepted. And then it was ascertained that enormous quantities of diseased and half putrid meat and fish were sold, which generated disease.

In 1857† the Thames was a vast sewer; the most filthy and dangerous river in Europe; "the stench for miles is intolerable; the moving mass of filth threatens the millions of inhabitants with pestilence and death;" it was a disgrace to the metropolis; a national calamity. A little later and the Thames was charged with sewer-gases; "and these gases are admitted to be poisonous." An occasional protest was made against this furor. One physician hinted that the way to ascertain if the Thames was a source of disease was not to take a steamboat ride on it and toss bits of white paper into the water and then pronounce authoritatively from that evidence, but, by a series of observations, compare the disease on its banks with that remote from it. This, he said, had not been done. Another writes that "a little calm discussion was desirable;" and "it is not proved [if it is, where are the proofs?] that it [the river] acts prejudicially

---

* London *Lancet*, vol. ii. † Ibid., vol. ii., 1857.

on the health of the metropolis;" and it was declared that there was "not one tittle of evidence springing from observation" to support the conclusions that press and public drew from the statements respecting the impurity of the river.

No heed was paid to these retorts, and the next year the London *Lancet* had a certain feeling of satisfaction in hearing that the Chancellor, Mr. Gladstone, and Mr. Cayley had been driven from committee-room "handkerchief to nose," and now "that their stomachs heave, they smell a dreadful smell; their minds expand," and there is an "epidemic diarrhœa of motions for relief." The editor says that many physicians declare that this sewage in the river is in no way bad for the health; and, what is still more strange, that some of these are men of science and known ability. The promulgation of such views at this time is a great error, for it may postpone legislation on the subject. All investigations to ascertain if the Thames had ever really been the cause of disease seems to have been carefully avoided; the reformers, however, were unceasing in stirring up panics about the river. One report* says it is truly wonderful that some plague or epidemic has not sprung out of the putrescent water. The river is one vast uncovered sewer, reeking with noxious and pestiferous abominations, and that men should be found who say that this does not injure health must "inflict the greatest injury on science, and produce in the mind of the public great mistrust of its professors." This water has been found to contain sewage, sulphuretted hydrogen, muscular fibre tinged with bile, husks of wheat, and potato cells. The danger is "immediate and imminent." Dr. Letheby declares that "the water is now in a high state of putrefaction;" it "abounds with the highest forms of infusorial life," and that which is near

---

* London *Lancet*, vol. ii., 1858.

to shore is poisonous to almost every living thing but vibrios.

In July, 1859, the filthy odor of the Thames warns us of approaching pestilence; "its noisome stench" has reached the House of Commons; much illness, and, indeed, death, are in Westminster, "owing most unquestionably to the putrid and disgraceful state of the river, aided by the intolerable heat. Lord Alfred Hervey was taken sick while attending a committee, from the effect of the horrible state of the Thames." "It is truly horrible to contemplate what may be the result."

Strange coincidence! In the same volume of the *Lancet* which furnished this information we read,* "The population of London now appears to be in a very healthy condition." "In the last two weeks deaths of persons at all ages from typhus and common fever decreased from forty-three to twenty-one; and fatal cases of zymotic disease in the aggregate from two hundred and ninety-one to two hundred and fifty-three." For one or two weeks only during this season was there any increase in mortality, and this arose not from zymotic but from local and constitutional diseases. And the summer, as a whole, was far more healthy than the average.

But all over the land it was said,† there is "one desolating germ of filth which is ever active in the fruition of disease; one accumulating poison, deadly alike in the cesspools of large cities and in the middens of country cottages." Typhus fever, cholera, scarlatina—the three great scourges of European populations—find here their nidus.

The excitement spread like wildfire. Meetings were held all over the kingdom,—in halls, school-houses, and drawing-rooms,—presided over by noblemen of distinguished lineage. A Ladies' National Sanitary Association for the Diffusion

---

* June 19, 1858. † London *Lancet.*

of Sanitary Knowledge was formed in London, with branches in different towns. This had for patrons and patronesses crown princes and princesses, dukes and duchesses, earls and countesses, bishops, right honorable and honorable, and distinguished men of the learned professions and prominent laymen, with a long list of secretaries and committees. The society's report for 1861 said there had been already spread over the country 138,500 sanitary tracts, 10,000 on "The Cheap Doctor," 8000 on "The Sick Child's Cry," and 8000 on "Never Despair." Sanitary classes were formed; lectures were delivered on catching cold, on drainage, food, air, clothing, etc.

The ladies put their sanitary reason into sanitary rhyme, and "Never Despair" found expression in these lines:

> "When times are hard and money scarce
>   And you are full of care,
> There's one thing you must never do,
>   You never must despair.
>
> "It makes the spirit faint and fail,
>   It wears the health away;
> It takes all vigor from the heart
>   And wastes life day by day."

To illustrate the danger of filth we have the following stanzas:

> "If things get worse and worse within,
>   And heaps of filth and rubbish lie
> Fermenting, steaming at your very doors,
>   How can you wonder that your children die?
>
> "Work till you've cleaned within, without,
>   And done your duty, done your best;
> Then may you claim it as a right
>   Your landlord he must do the rest."

Instructions for the baby were given as follows:

> "When baby ought to eat,
>   He'll have some teeth to bite,
> And if he must have any food,
>   Be sure it's soft and light."

Tight lacing was treated in this way:

> "Now if you press the body round,
>   The soft bones soon give place,
> And then the lungs can't freely breathe,
>   Nor the heart have full space.
>
> "Proud thoughts, high looks, and selfish ways,
>   Words that give others pain,—
> These things we all should by God's help
>   Incessantly restrain."

In a sanitary ditty on "Naughtiness and Sickness" the greedy boy is admonished that,—

> "The greedy and the gluttonous
>   Get sick and can't enjoy
> What would have been quite nice and good
>   Shared with another boy."

These hygienic idyls seemed to have charmed the people. One essayist on woman's work in sanitary reform said that these tracts in verse were "very suitable for reading aloud at Maternal Meetings;" and he called on female writers to "make imaginative literature a vehicle of popular sanitary instruction;" to tell us why preventable disease and death forever sit scattering our hopes and joys and holding a grim carnival among our loved ones." "Let us have," he cried, "a sanitary Jane Eyre, Adam Bede, and John Halifax."

One physician writes, "You ladies will do a good work, if you only bring us medical men to lecture to the people. But we cannot put ourselves forward. If your association were only to do this, you would do a good work."

Mothers' meetings were called; tea meetings were held,

to which "Bible women" were invited; missionaries were sent out; these were supplied with brooms and brushes to lend to the poor, and a special fund was set apart to supply skipping-ropes, balls, and toys to poor children. By January, 1865, the Ladies' Association had sent out more than seven hundred thousand tracts, "a moral force that implies vast extension of sanitary information." *

Out of such fury and tumult, and with the aid of such hysterical throes, the world witnessed the birth of Sanitary Science. The infant came near being suffocated in the travail by the very ladies who had assisted at the accouchement; for, in arranging one of the lecture courses, a "homœopath" had either crept in or had been smuggled in by the ladies as one of the speakers. This caused a great commotion; the *Lancet* scolded them soundly. It told them they could not have taken a surer course to throw doubt on their ability to conduct or even understand Sanitary Science; and added, "In their innocence they may imagine the sanitary conduct of a homœopath would be the same as that of a medical practitioner of any other school; but this is an error." It protested against the association lending itself to the propagation of miserable fancies. History is silent as to the result of this contest. Our own experience and observation have been that, whenever the ladies take it into their heads to boom a homœopath or any other doctor they never fail to succeed.

The prince consort died in December, 1861, of typhoid fever. Windsor Castle was, in the opinion of the best engineers, the most complete in sanitary works of any large building in the world. The pythogenic nature of this fever was now established in the minds of the reformers, and if nothing were found to account for the prince's illness the doctrine was in danger. Finally some witnesses were pro-

---

* Report of Ladies' Sanitary Protective Association, 1865.

duced who testified that they actually had smelled some smells somewhere about Windsor Castle, though why the prince consort should be the only one taken sick the sanitary reformers did not explain. They publicly exchanged congratulations, however, that the prince's illness and death had given a great impulse to Sanitary Science, and that the towns generally were sounding the alarm.

A few medical men made one more attempt to stem the torrent of nonsense. They called for calm investigation, for proof; they said the new doctrine was inconsistent with facts. They were, for the most part, overwhelmed, and were only too glad to be silent after being pilloried in the *Lancet*. They knew too well to whom the editor was pointing when he wrote that "people are yet found who literally wallow like unclean animals in their own filth and protest that it is wholesome. They foul their water sources, drink with gusto the sparkling fluid, and vow there is no water to be compared with it. By these means children are cut off by thousands; adolescents grow up with the seeds of debility and disease; adults are struck down in their prime; the sum of life is shortened; the productive and protective powers of the country are diminished."

In 1867 the Thames water had great quantities of putrescent animal matter on account of the late rains, and in the city "the pumps are spouting poison." Meantime, all sorts of patent sanitary fittings were advertised in the *Lancet*,— patent stack-pipe water-closet ventilators, filters to filter the water, smoke-stacks for sewer-gas,—all warranted sure preventives of disease. Every new fright brought out new patents.

The *Lancet* was very angry because Dr. Letheby had proved that the water-supply had nothing to do with the cholera of 1866; there were some slight symptoms of reaction, for people were beginning even to lose faith in the legend of the Broad Street pump. The analyses of the same waters by different chemists showed entirely different results, and all was bewilderment and confusion. Even the water from the chalk-wells showed contamination, and the cry was now raised that there were fissures in the chalk formation that allowed the ingress of sewage.

The *Lancet* recorded the "spread of infectious diseases, of holocausts of infant life, of death-rates shockingly excessive." "How is it that this death-page of our national history yet remains where it was years ago when our scientific knowledge was less?" Severer laws must be enacted to save the people.

Fever was reported at Barnsley; a cesspool was found here within a dozen yards of a pump, and the *Lancet* says, "There does not seem to have been any attempt to ascertain whether the inhabitants of the district were or were not systematically drinking the discharges from one another's bowels."

There certainly seems to have been no attempt made to show that this condition had prevailed, it may be for a century, without producing fever in Barnsley.

The Prince of Wales's illness with typhoid fever now gave another impetus to sanitary reform. The cause could not be ascertained any better than that of his father's sickness ten years before. It was discovered that at Londesborough Lodge, where the royal party was located, a water-closet, which had free ventilation with the outside air by an open window, was close to the prince's bedroom; but the *British Medical and Surgical Journal* said that this was the case in thousands of instances in London, yet no fever was the result. It was then found out that the prince was in the habit of riding by some carrion that was exposed to allure the pheasants; also it was proved that he had actually passed by a pile of manure which had painfully affected the nose of His Grace the Duke of Beaufort. Though the *Lancet* was reluctant to believe that the constitution of the heir to the British throne was so feeble as to collapse under so simple a matter as a bad smell from a manure-heap, yet in some one of these various ways he must have contracted the fever, because it was conceded by the most eminent sanitarians that this was the type of filth diseases.

Every new alarm that was invented was heralded as a "gigantic stride" of Sanitary Science. In 1875 the sanitary aspects of the sacraments were discussed; there was danger of contagion in the communion wine-cup. Again, blood-poisoning had been caused by licking gummed envelopes, and there was disease and death in the books of the circulating libraries. Dr. Farr now fulminated against the Thames water; it was loaded with filth. Dr. Letheby declared the alarm unnecessary, and the *Lancet* pronounced this contradiction "a scandal in the scientific world," and insisted that it was time that the richest city on the globe should know the truth about its water.

The *Lancet* said, "The demon filth which surrounds and poisons us on all sides is not to be exorcised by gentle language. No words which society will permit us to use are too coarse to hurl at the monster. It is not pleasant to drink the diluted excreta of men or even of pigs; but it is still worse to drink the eggs or germs of cholera or typhoid fever."

When the fever broke out at Wolverhampton and somebody suggested contamination of the water and that it be examined, he was withered by the reply that it was "a waste of time to analyze such water; of course it was contaminated." "A moment's thought" of the foul soakage was sufficient to indicate its character.

The epidemic in Croydon in 1876 staggered the sanitarians for a moment in their filth-theory of disease. Croydon had had for twenty-five years all the advantages of efficient sewerage, good water, and good sanitary administration, yet in twelve months there were twelve hundred cases of typhoid fever in the town, and "this seemed to render doubtful some of the fundamental principles upon which sanitary measures are founded." But the reformers soon rallied; the old theme was revived, and in 1880 the Thames was again a putrescent pond and a fermenting sewer, menacing the people with disease and death. One memorial made to the Lord Mayor on the stench of the river recited that a man fell into it a few days before and was found dead, although he had been in the water only two minutes, "having been actually poisoned by the deadly properties of the water." The Thames through London is more and more polluted;* it is no better than a common sewer. No one can go to London from Woolwich with comfort. There is a disgraceful system of house-drainage in the best parts of London. Sewer-gas is a special product of our refined system of sewerage. "We have now a perfect apparatus for treasuring it up and laying it on in our houses." Disinfectants are of no use; they are only disguisers. Poisoning by sewer-gas which has been deprived of its smell is the source of much sickness; the odor may be destroyed, but the poison remains.

In 1884 † the condition of the Thames was as bad as it could be; "a deadly, insidious odor arises in the form of sewer-gas," polluting air and water for miles. The river "can only be compared to a huge sewer-tank, putrescent and most offensive. We are living in extreme risk." "It is literally dangerous to breathe the air," and there was a continual wrangle over the conflicting analyses of the water; and so on to 1890, October 25, when we read in the *Lancet* that the progress of sanitation and increase of typhoid fever in India was an anomaly; but trust in the filth pathology was unshaken.

The florists of Liverpool were now deeply stirred about

---

* *Builder*, 1881. † London *Lancet*, vol. ii.

the public health; they implored the mayor to place restrictions on the hawkers of flowers in the streets, for many of them lived where infectious diseases prevailed. For once the *Lancet* was incredulous, for it said "the exhalations from some flowers are antiseptic," and it hinted that "the dread of competition" in the minds of the florists excited "their zeal for hygienic purity."

If the reader thinks this is a travesty on the rise and progress of Sanitary Science, let him be assured that the story is taken, not from the wild ravings of irresponsible newspapers, but for the most part from the pages of the most influential journals that were printed in Great Britain between the years 1845 and 1891. In all of this sanitary inebriety there is no trace of scientific research. If an English physician questioned the filth pathology of epidemic disease, and suggested an investigation, he was quickly silenced by the ribaldry of the London *Lancet*. Scotch and Irish medical men, like Christison, Hughes Bennett, Stokes, and Graves, maintained a point-blank denial to the new doctrine; they declared that it was antagonistic to the facts.

The stubborn truth faces us all through these forty-five years that, while the people in England were breathing this pestilential air, drinking this polluted water, living on this contaminated soil, eating this diseased and putrid meat and fish, the death-rate was not increased, but, on the whole, was steadily diminishing, as it had done for three hundred years before the dawn of Sanitary Science.

The birth of Sanitary Science in America was not preceded by the tedious, irritable, and painful gestation which heralded its advent in England. Indeed, the infant science can hardly be said to have had an embryotic existence in this country, for it was greedily accepted as it came forth from the hands of the English reformers, by a set of men here who soon found that by ingeniously exploiting its vagaries they could attain to an importance and acquire a position and emolument that had heretofore been denied them both in the community and in the medical profession. What the American sanitarian lacked in originality was amply compensated to him by the faculties of imitation and volubility of expression; and, as will be seen later, as whim and chimera one after another were hatched in and launched from the brain of the English reformer, they were seized at once by his American copyist, without examination or investigation, and were appropriated and published by him without delay as the "settled principles of Sanitary Science." The excitement began soon after the close of the Civil War, and by 1870 had reached a kind of frenzy. Whole sections were deeply moved by the elocution of the reformers. Sanitary surveys and inspections were made which showed that towns and cities, great and small, were on the brink of a dreadful precipice; their inhabitants were located on a filthy soil, were enveloped in foul air, and were drinking foul water. Sanitary conventions were held, sanitary platforms were erected, on which the sanitary orator mounted, and, with flaming eye, declaimed to his quaking audience that the conditions under which they were living invited the direst of all pestilences,—the Black Death of the Middle Ages. Strong men were aroused to action, and were resolved to save themselves and their families, if possible, from the perils which environed them, on account of the pestilential, polluted, contaminated air, water, and soil.

CHAPTER III.

The Air.

"There are indeed two things; knowledge and opinion; of which the one makes its possessor really to know, the other to be ignorant."—HIPPOCRATES.

To avoid all cavil and misconception, let us premise this chapter with a disclaimer of any desire or intent to disquiet the prejudices of enlightened mankind in respect to cleanliness. That purity of body is a necessary accompaniment of purity of mind and morals needs no argument to sustain. Indeed, this proposition loses its force in proportion as any labored reasoning is obtruded to support it. As a corollary, this other proposition may attend the previous one,—to wit, that a clean body and clean surroundings,—pure air, pure water, and pure soil,—in so far as they add to the comfort, decency, and dignity of the human family, are not only accessories but essentials as well to individual as to public health. In reality, the amiable delusion that a want of cleanliness was the cause, or one of the causes, of infectious disease might have been allowed to remain, for it was harmless, except as the violence of the sanitary reformers forced it upon us, to the exclusion of all other causes, and compelled us to examine critically the evidence with which they upheld it.

The "opinion," the belief, that decomposing organic matter, animal and vegetable, by contaminating the air, is the prime factor in the production of certain diseases, notably epidemic and contagious, has always possessed not only the popular mind, but also to a great extent that of medical men who have given the subject no investigation.

Organic matter in decomposition evolves certain gases; but no one claims, and every one who has examined the subject with care will deny, that any single known gas or any known combination of the gases of putridity is capable of causing disease. Some of them, in a concentrated form or by displacing the oxygen of the atmosphere, cause asphyxia and death. But no known element which is exhaled in the breaking up of organic compounds is any more capable of generating disease than any which is given off in the decomposition of inorganic bodies.

So it was left for the sanitarian to imagine a something which he called sometimes a "septic ferment," sometimes a "morbific element," sometimes a "subtle poison," sometimes the "mephitic gases," and later this something was designated by the Massachusetts Board of Health as the "unknown factor." Not content with these definitions, the sanitarians, as we shall see, made use of symbols which, to the popular mind, expressed the highest degree of terror. The hidden principle of disease which was given off in putrefaction was represented sometimes in words, sometimes in figures, as "a demon," "an unclean spirit," "a monster," "a snake," "an unseen vampire."

It was plain to every observer that huge masses of organic substances were continually in process of decay, giving off sometimes the most offensive gases, yet no infectious disease was present. On the other hand, infectious diseases often invaded a family and a neighborhood when no visible decomposition of organic matter was in progress.

According to the sanitarians, it was this imaginary element that was developed in filth which caused the mediæval pestilences that devastated Europe, and which they predicted would surely return to us unless we gave heed to their warnings. A large number of the writers on hygiene approve, to greater or less extent, this theory of infectious disease, and it has always been the sole stock in trade of modern boards of health.

John Howard, in his book on prisons, says Dr. Hales and

Sir John Pringle observed "that air corrupted and putrefied is of such a subtle and powerful nature as to rot and dissolve heart of oak."

Bancroft * says Lord Bacon thought that no effluvia were so infectious and so pernicious to mankind as those which issued from putrefying human bodies. Fonssagives † says the "morbific principle" that proceeds from decomposing animal matters can produce the gravest maladies. Thus, he says, bursts out a pestilence after a battle, and the decomposition of matters in privies causes dreadful septicæmia.

Tardieu ‡ says that the centres from which putrid emanations are disengaged are the types of unhealthfulness. This can hardly be otherwise, for it is not only by their composition (*nature intime*) that they act, they strike the senses, and produce on the most delicate organs "*une impression pénible*," which seems the announcement of a real danger. Galen, he says, assigns the cause of pestilential fevers to putrid air from bodies left on battle-fields, and St. Augustine reports a cruel pestilence caused by masses of decayed locusts. Tardieu relates a number of similar antiquated stories to prove the noxious character of putrid emanations; but when he discusses certain trades, like tanners, curriers, catgut-makers, etc., he says that though they are foci of disagreeable odors, they are not unhealthy; and when he treats of *voiries*, although these contain everything conceivable that is foul, he declares they not only do no harm, but he has known them to re-establish health that has been impaired.

All of the noisy and influential English sanitarians have persuaded themselves that filth is the great cause of contagious disease. In the reports of the medical officer of the Privy Council, London, 1874, is a recital of forty-two in-

---

* Yellow Fever. † Hygiène Navale.
‡ Dictionnaire d'Hygiène, article " Emanations Putrides."

spections of towns where occurred in 1873 outbreaks of small-pox, scarlatina, typhus and typhoid fevers, and diarrhœa, which are ascribed to polluted water and accumulations of filth. The reports are not complete, inasmuch as they do not state how long these towns had been exposed to these conditions and had been free from such diseases. In 1874 Dr. John Simon published "Filth Diseases and their Prevention." He ascribes the disease-producing power of filth, not to any known gases which may arise from it, nor to the fact that it is disagreeable to the senses, but to a "septic ferment" which, he says, is generated in filth. The diseases caused by this "septic ferment" are diarrhœa, enteric fever, cholera, erysipelas, pyæmia, and diphtheria. He gives a short account of one hundred and forty-three outbreaks between 1869 and 1873, where privies, cesspools, and defective drains were discovered, though he does not state in what particular the sanitary condition of the afflicted towns differed from all others in Great Britain where no epidemics occurred.

Parkes, in his elaborate work on hygiene, though he gives numerous examples where disease has been imputed to foul air, water, or soil, is careful not to commit himself to the filth-theory of disease, and makes no allusion to the "septic ferment," which Mr. Simon considers the generator of infectious disorders.

Miss Florence Nightingale, in her work on nursing, says, "I was brought up, both by scientific men and ignorant women, distinctly to believe that small-pox, for instance, was a thing of which there was once a first specimen in the world which went on propagating itself. Since then I have seen with my eyes and smelt with my nose small-pox growing up in first specimens, either in close rooms or in crowded wards, when it could not have been by any possibility caught, but must have begun. Nay more, I have seen diseases begin, grow up, and pass into one another. I have seen, for instance, with a little overcrowding, continued fever grow up; and with a little more, typhoid fever; and with a little more, typhus; and all in the same ward or tent. Dry dirt is comparatively safe dirt; wet dirt becomes dangerous. Uncleansed towns have been made pestilential by having a water-supply."

In our own country the filth theory of disease has taken deep root with the sanitarians. Hardly an epidemic has occurred anywhere in this country which their official reports have not loosely ascribed to filth. If an ancient cesspool, a profligate privy, a suspicious well, an untrapped sink, or the contents of a dish-pan on the surface of the ground could be found anywhere in the vicinity of an epidemic it sufficed to explain its origin, although the same condition might prevail in millions of cases where no sickness occurred.

In the *Sanitarian*, April, 1878, a lady writes that dirty dish-cloths are a cause of typhoid fever; she had smelled a whole houseful of typhoid in one dish-rag; she is sure it caused four cases in one family where she "ran in" to assist.

In "Home and Health: A Cyclopædia of Facts," by C. H. Fowler, D.D., LL.D., and W. H. DePuy, A.M., D.D., article "Fever Infections," we are told that all fevers, like typhus, small-pox, etc., arise from a subtle poison. "This poison has been actually condensed out of impure air poisoned by filth and decay, and appears in the form of a dirty-looking, half-solid, half-fluid, half-gelatinous stuff, a few drops of which inserted into the veins of a dog will inoculate that dog with typhus fever."

Mr. Waring* says that as to the exact causes of disease we know comparatively little, but there are certain well-established truths. "One of these is that man cannot live in an atmosphere that is tainted by exhalations from putre-

---

* On Sanitary Drainage, etc.

fying organic matter without danger of being made sick,—sick unto death." Professor S. W. Johnson says the filth of vaults and cesspools "may long remain simply disagreeable without being dangerous, and may again of a sudden, *in a way whose details have as yet escaped our observation* [italics ours], become the seed-bed or the nursery of the infection that breaks out in fevers and dysentery." "Cleanliness," says Professor Lindsley,* "public cleanliness, is the highest aspiration of the public hygienist. Filth in any form is the fatal foe of human life,—a foe unsparing, insidious, unceasing, malignant, and deadly." "Already," he cries, "the city of New Haven has had to erect *a special police prison in its most unsanitary ward.* Cleanliness is the grand aim of the sanitarian's efforts."

Says a distinguished sanitarian,† "We thus see that all the great epidemics of mankind originate in filth and are propagated by filth. Of all the forms of filth, none are more active than the direct and indirect products of animal life. Man is a poisonous animal; his touch brings rottenness. Human filth is the hot-bed of epidemic disease." As a breath fans the fire, "so the fever-germ may be carried a long distance and, falling into some magazine of unsanitation, may explode into a frightful epidemic." "These fever-nests seem to attract the wandering epidemic germ, just as the depot of nitro-glycerine seems to fascinate the foolhardy hand which shall wake up this bottled earthquake."

"Does death end all?" he asks. "Certainly not. There remains the funeral. What is the funeral of the yellow-fever dead? As soon as death is certain, the body, in all its filth and with the garments it last wore, still warm with the remains of life, is seized by some hired grave-digger, thrust into a coffin, hurried into some cart, and toted off to the

---

* Connecticut Board of Health, 1883.
† Eighth Report Michigan Board of Health.

graveyard; and, when its turn comes, it is dumped into its grave and covered from sight forever. 'Mud to mud' would close the fitting burial service. Death is terrible; such a death and burial are horrible."

Another distinguished sanitarian,* when discoursing on decomposing organic matter, says, "It is perfectly safe to say that a foul-smelling air is a dangerous air. Every now and then Death makes a visit to the household, carrying off its brightest members, ruthlessly slaying father, brother, sister; the strong man, the feeble infant. Why this sacrifice, —this ruthless slaughter? Who are the invisible monsters invading this happy circle?" "Let us look around," he says. Smell informs him that there are decaying vegetables in the cellar, "pouring forth into the air deadly emanations;" this "stagnant, poison-laden air" finds its way into the lungs of the occupants of the house. "We ascend to the kitchen;" here are what every one recognizes as "kitchen-smells." In one corner is the wood-box. Turn the contents on the floor, and, "shade of Hygeia, what a smell!" Rotten bark, apple-cores, odds and ends," making "a putrescent conglomeration teeming with filth, redolent with putrefaction, and crawling with vermin." In the pantry are fragments of mouldy bread, "a magnificent place for germs of every description to hold high carnival." The beautiful carpet in the sitting-room "conceals beneath its delicate shades a conglomerate accumulation of contributions from every source of impurity within the dwelling and without." The parlor is still worse. Above are closets, chambers, and garret, "charged with the most virulent enemies to health." In the yard myriads of insects are crawling where the dishpan has been emptied. "A few feet distant is an edifice which we are at a loss to know how to describe;" but the air from it "is freighted with the agencies of death." Then there

---

* Tenth Report Michigan Board of Health.

is the well. "Only think of the condition of a family with death enthroned in the well and daily dealing out his poisonous draughts to its members!" "Some one may say the picture is highly colored; but the experienced sanitarian will certainly say we have not told half the truth." If we only had microscopic eyes we should see in many of these houses " not an army of brave soldiers coming to our rescue from disease and death, but the emissaries of death in countless numbers, intent upon our destruction, ready to pounce down upon us at the first favorable opportunity, rack us with pain, and finally devour us." This " experienced sanitarian" does not say whether he is describing his own residence or the typical Michigan habitation.

Nothing short of these liberal quotations could possibly illustrate the native grandiloquence of our sanitarians. Lest the reader shall exclaim that this is a parody on Sanitary Science, and that we have here introduced the insane ravings of some backwoods revivalist on the terrors of the judgment day to an audience temporarily bereft of reason and powers of comparison, we here affirm that these were the declarations of eminent sanitarians who have the prefix of professor and the affix of M.D. to their names; that they were addressed to bodies of sanitarians in convention assembled, and that the subject of the discourses was Sanitary Science.

In all of this improvisation—and sanitary literature everywhere is overflowing with similar harangues—there is not a particle of evidence to sustain the filth-theory of disease, except such as is offered by the field-of-battle story of Galen, the locust story of St. Augustine, mentioned by Tardieu, the positive testimony of Miss Nightingale's eyes and nose, and the no less positive evidence of the lady respecting the dish-rag origin of typhoid fever.

It would be unbecoming in us to deny the direct testimony of the two Doctors of Divinity in the "Cyclopædia of Facts," regarding the "subtle poison" of fever which they so minutely describe as "dirty-looking, half solid, half fluid, half gelatinous," and which must have been seen by them alone; for it is mentioned nowhere else than in their book, and, so far as we have inquired, it is not to be found in any of the chemical or biological laboratories.

The remaining proof which sanitarians have to offer concerning the noxious character of putrid emanations lies only in the occasional coexistence of these and disease. There is no more scientific proof that putrid emanations cause infectious disease than that they are the sources of acute or even chronic inflammatory complaints; and it needs only a little higher flight of sanitary imagination to pronounce the latter filth-diseases, because they happen in a neighborhood where a cesspool, or privy, or drain may be found.

The "opinion" of many of the most thoughtful medical men has been often expressed, that this coexistence of putridity—filth—with disease has not the relation of cause and effect. The author has been unable to find a single instance, where this subject was scientifically investigated, in which the proof was not overwhelming that organic matter in any of its chemical changes was incapable of producing epidemic disease.

Dr. Ferguson says, "It is, in truth, unnecessary to multiply facts and illustrations to prove that putrefaction and the matter of disease are altogether distinct and independent elements; and that, however frequently they may be found in company, they have no necessary connection."

Dr. Chisholm,* in an article on contagion, after enumerating whole tribes and nations who maintain their health in the presence of the worst putridity, cites the towns of Bristol, Bermondsey, Conham, and Bitton. In these cities, bone-boiling, glue-making, fat-rendering, and tanning are going

---

* *Edinburgh Medical and Surgical Journal,* vol. vi.

on, which fill the atmosphere for a mile around with a disgusting fetor. In Bristol the streets are narrow, the houses crowded and ill-ventilated; yet the harmless nature of these exhalations is daily verified. Neither the workmen nor the inhabitants are made sick; the reverse is the fact; they are the most healthy of any of the laboring poor. After a careful survey, Dr. Chisholm, who almost apologizes for saying so, concludes that the theories of ingenious chemists regarding the power of animal effluvia to produce disease receive no support from practical knowledge or the economy of nature.

Dr. Nathaniel Bancroft * says, "I have no desire to weaken any of the prejudices which tend to promote cleanliness in civilized countries, any further than is absolutely necessary for the manifestation of truth on a question of great importance to mankind;" but he declares that there is no connection between offensive smell and nastiness and contagious fever. That putrefaction, which is but a natural separation of organized matter, is the servant of chemical attraction, and the products are as certain and constant as the combination of soda with muriatic acid.

Dr. Graves † says, "The causes of epidemic disease escape the scrutiny of both nostrils and vision. Filth is the outward and visible sign of poverty, and, like poverty, is itself an evil; it oftener accompanies than causes disease."

Dr. Stokes quotes the report of a sanitary inspector of an Irish town of four thousand inhabitants. Every part "was teeming with effluvia from such decayed substances as are admitted to be of the most noxious kind, but this town has always been a remarkably healthy place." Dr. Pratt declares that if fever were caused by decomposition of animal and vegetable matters, Ireland would have been depopulated long ere this from sea to sea. He gives this

---

* Yellow Fever.   † *Dublin Quarterly,* vol. vi.

opinion on the experience of twenty-five years' practice of an Irish dispensary officer among the agricultural classes, where the house-yards are often heaped up to the very door with manure, and are the receptacles of slops and refuse of all sorts, which ferment and "produce a green and stagnant pool," where gases are generated which burst on the surface. Dr. Pratt says, "In such places a case of fever of any type rarely occurs, the average length of life is as high as elsewhere, and illness except common colds is almost unknown." Dr. Stokes says, in his lecture on public health, "There is no proof that dirt ever in itself caused a case of specific disease. The real antagonistics to any successful preventive medicine are poverty, destitution, ignorance, apathy, insufficient and improper food, filthy habits, overcrowding, bad ventilation, insufficient clothing, the living in ruined and neglected tenements, the destruction of proper pride and the blessed influence of home."

Dr. J. C. Warren * made a thorough study of the influence of putrefaction in the production of disease. The workmen in those trades most exposed to putridity were most exempt from infectious disorders. He says the whaling-vessels are saturated with putrid animal matter, in hot as well as cold climates; the odor is sometimes intolerable to those not habituated to it; but the seamen on these ships are more healthy than those engaged in any other service.

Dr. Davis † records his studies of the influence of occupation on health. He found that the health of brewers, distillers, tanners, curriers, glue-makers, tallow-chandlers, soap-makers, of all of those, in fact, who were exposed to putridity in its worst forms, was never impaired by these occupations, and though the stench from glue-making is a nuisance to the entire neighborhood, "many assert that on

---

* *Boston Medical and Surgical Journal*, 1829.   † Ibid., 1833.

entering this employment they experience a marked increase of appetite and health." Notwithstanding that those who are engaged in tanning are exposed to masses of putridity, dampness, and dust, experience has shown that this trade is a benefit to consumptives.

One of the most searching investigations of the influence of putridity on health was made by Parent Duchatelet in 1832.* This distinguished hygienist was guided in all of his inquiries by the love of truth. Unlike the sanitarian of a later day, he sought neither to exalt himself nor to increase his revenue by exciting a panic in the public mind. If an imaginary danger appeared, he strove to allay the fears of his fellow-citizens. The village of Montfaucon was just outside of Paris. Here was the depot for all of the refuse and fecal matter of the great city, and here the fæces were converted into poudrette. Here were also bone-boiling, glue-factories, and the rendering of dead and diseased animals. Very often the authorities were appealed to for their suppression on the ground of danger to the public health. When the cholera was approaching Paris in 1832 these petitions increased, signed largely by medical men, who represented that if Montfaucon did no harm in ordinary times, it was only waiting for the spark of an epidemic to explode in pestilence; that localities like this invited the disease; that here it would centre and diffuse itself over the country. Parent was appointed to investigate and report. He says the odor here was " insufferable, indescribable, insupportable." But if we interrogate the workmen, they, without exception, answer that, far from being noxious, these odors are beneficial to their health ("*contribuent à leur bonne santé*"). Parent says that in every direction where he inquired of the inhabitants and of workmen in other employments in the vicinity, though they complained

---

* Hygiène Publique.

loudly of the nuisance, they admitted that their health never suffered.

The children here were in the best of health; so were the women. One of these particularly attracted the notice of Parent on account of her fecundity; he says she was "*habituellement enceinte;*" her children had great force and vigor. He had seen this mother nurse her infant and toss it afterwards into the carcass of a horse for a cradle. When the cholera arrived in Paris it appeared to shun Montfaucon; the mortality from that disease here was very small in comparison with the rest of Paris.

The village of Noisy-le-Sec received the refuse of the Montfaucon establishments. M. Dumoissin, the mayor, told him that those nearest the animal remains suffered the least during the epidemic; that his observations had destroyed all his previous opinions respecting these putrid matters, and that the peasants who handled them, far from thinking them harmful, believed that their fermentation purified the air.

Fleury,[*] after reviewing the evidence respecting emanations from putrid animal substances, says, in view of no increased mortality nor increase of infectious diseases in the presence of immense masses of putrefaction, we not only recognize their innocuousness, but perhaps must admit that these emanations exercise a favorable and prophylactic influence.

Dr. John Snow [†] says the mortality of persons engaged in any occupation is the best criterion of its salubrity. The death-rate of males of twenty years and upward in London for eighteen months ending July, 1856, was 241 per 10,000. The death-rate for those males of twenty years and upward who were employed in trades where organic decomposition was continually going on was 201 per 10,000 in the same

---

[*] Cours d'Hygiène, 1852.
[†] London *Lancet*, vol. ii., 1856.

time. Dr. Snow estimates that a man working with his face one yard from offensive substances would breathe ten thousand times as much of the gases given off as a person living a hundred yards from the spot. He says the health of persons employed in any occupation is necessarily the measure of the effects of such occupation on the public health. Bone-boiling, skin-dressing, and other offensive trades are carried on at Lambeth; this part of London contains many of the other causes that are supposed to promote cholera; the ground is low, covered by a poor, crowded population; yet the deaths here from cholera in 1854 were 29 per 10,000, whilst in London at large they were 45 per 10,000. Dr. Snow concludes that "the science of public health, like other branches of knowledge, may be as much benefited by the removal of errors which stand in the way of its progress as by direct discovery."

Dr. R. F. Foote writes to the *Sanitary Review* * that outside the doors of the houses in Constantinople is a heap of animal and vegetable rubbish; dogs are the only scavengers; apartments are overcrowded, several members of a family living and sleeping in one room; there is no drainage; large cesspools are near the houses, and the privy is on the ground-floor beneath the living-room. "With all these deficiencies, we are bound to consider Constantinople a healthy city." The result, he thinks, is due to "climate," and whatever may be the causes which tend to affect the public health, it is not the less true that it is more flourishing in Constantinople than in any other of the large towns in Europe.

Dr. William Budd † says nothing proves the falsity of the pythogenic theory of fever more than the summers of 1858-59 in London. Here "an extreme case, a gigantic scale in the phenomena, and perfect accuracy in the registration of

---

* Vol. iv., 1857.   † *British Medical Journal,* 1861.

the results—three of the best of all guarantees against fallacy—were all combined to make the induction sure." "For the first time in the history of man the sewage of two million people had been brought to seethe and ferment under a burning sun in one vast open cloaca in their midst." Stench so foul had never ascended to pollute the air. "Never before had a stink risen to the height of an historic event." The law courts were broken up by the noxious vapor; the steamers lost their usual traffic; day after day were passionate appeals from "those condemned to live on the Stygian banks." "Members of Parliament and noble lords, dabblers in sanitary science, vied with professional sanitarians in predicting a pestilence." "Meanwhile, the hot weather passed away; the returns of sickness and mortality were made up, and, strange to relate, the result showed not only a death-rate below the average, but as the leading peculiarity of the season, a remarkable diminution in the prevalence of fever, diarrhœa, and the other forms of disease commonly ascribed to putrid emanations."

Dr. Letheby, in his report, said, "With all this condition of the Thames, however, the health of the metropolis has been markedly good; in the corresponding period of 1857 the cases of fever, diarrhœa, and dysentery attended in the city by the medical officers of the unions amounted to two hundred and ninety-three of the former and one hundred and eighty-one of the latter; but during the past quarter, the quarter of the stench, there were only two hundred and two of the former and ninety-three of the latter." "So that while pythogenic compounds were poisoning the air with a forty-thousand fever-power, pythogenic fever, so far from rising in proportion, fell vastly below its average."

Dr. McWilliams, the medical supervisor, said that not only was the general sickness less in 1858–59, but he adds, "As regards the type of those forms of disease (including diarrhœa, choleraic diarrhœa, dysentery, etc.) which in this country noxious exhalations are supposed to originate, we find the additions for the four hot months from this class of complaints 26.3 below the average of the three previous years, and seventy-three per cent. less than that of 1857." Dr. Budd says, " Before these inexorable figures the illusions of a half-century vanish in a moment."

The irrigation of lands with sewage on a large scale had been in operation in Milan and Edinburgh for one hundred and fifty years before the dawn of Sanitary Science. It was really nothing more nor less than manuring fields, a process which had been going on by nature and art since the foundation of the world. The proposal to so dispose of the sewage of cities was a grand occasion for the exercise of sanitary fancy to awaken anxiety about the public health. As usual the sanitarians made no investigations, no inquiries. They simply let their imaginations have play, and cried aloud that the emanations from these farms would produce pythogenic diseases; sewage would percolate into the wells; the sewage was charged with the ova of entozoa, and was certain to cause tape-worm and other entozoic diseases in man and animals; and "the effects of sewer-gases were never so bad as when sewage was spread out on the land;" that the propagation through sewage "of certain epidemic diseases, especially cholera, enteric fever, and diarrhœa, among communities is one of the best-established facts in Sanitary Science." Horses and cattle that would eat the grass grown on these lands would be diseased; milk and butter would be poisoned, vegetables grown here would not only absorb the sewage but the germs of disease.

At Gennevilliers, where a farm took about one-half of the sewage of Paris, they said that the grass and the vegetables took up the filth.* "You can break them and squeeze water out that has a decided smell of sewage." One sani-

---

* Sewage Disposal, Robinson.

tarian said he was now "in a position to explain the occurrence of typhoid fever from the use of milk."

These predictions of disaster retarded and hampered the profitable undertakings. The hue and cry about the public health, sometimes prompted by fear and sometimes by self-interest, was so great that governments were forced to investigate these farms; always with the same result,—that this broadcast flowing of sewage was proved to be innoxious. It was shown in some cases that there had been a decided and constant fall in the death-rate among the inhabitants on and about these fields after the flow of sewage had begun. In Norwood district, where there is a sewage farm of sixty acres, the death-rate had fallen * from 18.6, the average of three years ending 1865, to 13.4, the average for the seven years ending 1872. At other farms the death-rate was equally low. At Breton one hundred and twenty-one acres take the sewage of five thousand people; the workmen here are uniformly healthy; the water is drained five or six feet from the surface; it is clear and free from odor, and is proved to be of good quality. The Aldershot field of one hundred and twelve acres takes two hundred thousand gallons daily of sewage. At Banbury, Rugby, Carlisle, and other places, although some of the grounds are reported in a condition of negligence, the workmen and inhabitants are all in good health.

Dr. Arnott had explained exactly how the Edinburgh farm would cause epidemic disease. Professor Christison said his prejudices for years had been against the irrigated meadows there, but he had been forced to surrender them. The nuisance was not denied; it was sometimes intolerable; but he was satisfied that neither enteric fever nor diarrhœa nor dysentery nor diphtheria, either in epidemic or non-epidemic seasons, was found in and around them more than

---

* London *Lancet*, 1873.

elsewhere. Dr. Arnott had said these fields must cause pestilence. Professor Christison declared that the fact remained that they had been there for two hundred years and no disease could be ascribed to them; indeed, if any part of Edinburgh was freer than another from zymotic disease it was these sewage-irrigated meadows.

M. Durand Claye * reports on the farm at Gennevilliers; he says the plain here is a proper filter to absorb and purify impure water; that under the thin layer of earth, where the sand and gravel begin, there is no trace of organic matter; a proof that the superficial layer of earth is a complete filter (*un filtre énergique*) for the sewer-water. From a well in the midst of the fields the water was pure, limpid, and tasteless (*parfaitement pur de matières fermentiscibles*). The water from the drains is equally pure. The death-rate here had been steadily declining since the farm was established.

Professor Corfield says that when a sewage farm is a large filter and the effluent water is collected in subsoil drains, this water is perfectly fit to drink. He knows where it is usually drunk by the workmen. The farms have not caused by noxious emanations any injury to the health of neighborhoods where they have been placed; there is no evidence that the sewage affects the wells; and he does not think there is a single case where sewage farms, "badly conducted as many of them are," have caused the least injury to health.

At Sherburn, Massachusetts,† the water that leaches from drains five feet deep beneath the surface of the sewage farm is as clear as spring-water, and was selected as such by an expert.

In 1888,‡ M. Ogier reported to the French government on the condition of the farm at Gennevilliers. His conclusion was that no better or safer way of disposing of the Paris sewage could be devised.

---

* Annales d'Hygiène, 1875.   † *Sanitarian*, vol. xiii.
‡ Annales d'Hygiène.

In the summer of 1890, Dr. Henry J. Barnes * visited the sewage farms of Edinburgh, Birmingham, Leamington, and Croydon in England, and Gennevilliers, near Paris. The health officers at Edinburgh and Leith declared their hostility to the sewage irrigation; but admitted that the district was "most healthful," and that there was no evidence of any injury resulting from the sewage. During the cholera epidemic at Leith and Edinburgh in 1865-66, although these farms received the cholera evacuations, not a case occurred on them or in their vicinity. At none of the others that Dr. Barnes visited in England was there any disease reported as springing from the sewage. He inspected Gennevilliers on a hot day in August, and drank of the effluent water of the sewage, being assured of its purity. The inhabitants here were engaged in raising funds to erect a monument to M. Durand Claye, who, regardless of threats of personal violence, of persecution by lawsuits, and opposed by nearly all the *savans* in Europe, had caused the transformation, through this flow of sewage, of sixteen hundred acres of barren sand into a beautiful garden. The town had increased in population thirty-five per cent. since 1868. The death-rate for five years preceding the irrigation was 32 per 1000; for the last five years it has been less than 25 per 1000. In 1882, when typhoid fever raged in Paris, there was no increase in Gennevilliers; and not a case of cholera was here in 1884, when this disease was epidemic in the city.

In 1881 a pecuniary loss was reported † from the sewage farm at Reading; but "looked at from a sanitary point of view, the farming operations had been a great success." Reading had never before been in such an excellent sanitary condition.

In *Chemical News*, ‡ Mr. Hope, an enthusiastic advocate

---

* *Boston Medical and Surgical Journal*, 1891.
† London *Lancet*, vol. ii.   ‡ 1880.

of sewage farms, suggests a convalescent hospital for diseases of the chest in the midst of his sewage fields, " guaranteeing to the patients a certain number of irrigations in the month with genuine unadulterated London sewage ;" " where London beauties might come to recruit their wasted energies at the close of the season, attired in a *costume de circonstance*, with coquettish jack-boots; and would perhaps at times listen to a lecture on agriculture from the farmer himself, while luxuriating in the health-restoring breeze."

Through this system of sewage-disposal much waste land had been brought under cultivation, which had afforded employment and subsistence for a large number of people. The sanitary terrorists had done all they could to prevent it.

Dr. H. Gibbons, Sr.,[*] says that the origin of enteric and kindred fevers has been associated with filth ; to deny this is rank heresy. Nevertheless, facts will sustain the assertion that the specific causes of infectious disease have no odor; that in the majority of cases the localities where the vilest filth exists are free from infectious diseases. " Objection may be made to this view, that it is a defence of carelessness or filthy modes of life. But it is simply a defence of truth. There is a seeming policy in holding before the eyes of common people the spectre of death lurking in filthiness and foul odors. But cleanliness is virtue in itself enough to be cherished for its own sake, and not enforced by false terrors. Still more should the error be avoided by the medical inquirer."

Dr. Gibbons says that in San Francisco the health authorities require the Chinese to have five hundred cubic feet of air to each occupant. The public health is supposed to demand this law. If they are found with less, they are fined and imprisoned. " I have seen in the county jail as many as thirteen Chinamen confined in a single cell about twelve feet

---

[*] *Pacific Medical and Surgical Journal*, 1882.

square, containing by computation less than ninety cubic feet of air to each; and this for violating the five hundred cubic feet of air ordinance."

The city of Oakland, built on a sand flat, has until within a few years obtained its water from shallow wells dug fifteen or twenty feet deep, whilst all the sewage and filth of the city was deposited in the soil about them. "But Oakland is and always has been a healthy locality, having much less of typhoid fever and diphtheria than scores of cleanly, well-drained, and hilly places on the Pacific coast." After citing other cities similarly situated to Oakland, Dr. Gibbons concludes:

"*First*. The specific poison of infectious diseases has no odor.

"*Second*. It may be associated with offensive odors, and it may not. In by far the larger proportion of cases it is not.

"*Third*. Offensive odors cause no permanent disease.

"*Fourth*. Certain emanations offensive to smell, notably if sulphur be the basis, tend to destroy organic germs.

"*Fifth*. Water which has been polluted by excrementitious or other organic matters is mostly drunk with impunity; the gastric juice digesting or destroying the organic germs. The production of disease by such water is the exception and not the rule."

Professor M. P. Cazenave,[*] of the faculty of Lyons, gives an account of his visit to Tunis. "There is," he says, "a net-work of sewers here without any regular slope, open to the air, filled with excrement and animal and vegetable remains, the accumulation of years. The ground-work of these sewers is not protected by masonry, so that the soil is impregnated with filth and the atmosphere is loaded with fetor. The cemeteries are in a dreadful condition, the coffins being placed only a little below the surface. What

---

[*] Annales d'Hygiène, vol. xvii., 1887.

becomes," asks Professor Cazenave, "of that aphorism of public hygiene of Fonssagrives, that a city's health is in proportion as its sewerage is perfect and its cemeteries well arranged? In spite of its condition, Tunis enjoys as good health as the greater part of European cities. Here is a striking example (*confirmation éclatante*) of the falsity of the opinion that ascribes a dangerous quality to a repugnant and detestable atmosphere." Unwilling to yield all his prejudices, the professor says, when these sewers are remade and this old infiltrated soil is stirred, where are sleeping legions of infectious microbes, an explosion of typhoid and cholera may result. "But for the moment let us register this fact:" the "vast sink that represents the subsoil of Tunis engenders odors, but does not engender epidemics." Typhoid fever is rare here. It is our troops who are lodged often on the heights under apparently satisfactory conditions that furnish the greatest amount of typhoid fever (*qui payent le tribut le plus large*).

The professor wished to believe that the people of Tunis had some hereditary resistance; that there had been some natural selection whereby the fittest had survived and the feeble eliminated; but the constant stream of European immigration had not been the occasion of any epidemic, and he says we must conclude with Bouchardat, that these odors are innocent; and he asks, May not sulphuretted hydrogen and its compounds, which are constantly disengaged from these places, be antiseptic and destructive to certain classes of microbes? *En résumé*, he says science demonstrates that our senses are not competent as guides in hygiene.

The testimony of travellers and residents in China—missionaries and physicians—is abundant and uniform as to the disregard in that vast empire of everything which is considered with us as sanitary precaution. Dr. Wilson *

---

* Medical Notes on China.

says the Chinese import no manure and derive no benefit from agricultural discoveries in Europe. Close to houses are placed large earthen vessels which receive all decomposable refuse, to which is added urine; in the corrupting mass at a heat from 80° to 86° multitudes of maggots form, giving to it life and motion. These pans are arranged along the public walks; when putrefaction is at its highest point, the contents are carried in open buckets on the shoulders of men to the fields. Human ordure is carefully collected and highly valued. The water in the canals is filthy in the extreme; the banks are lined with ooze and abound in animal and vegetable products; the offal dams up the canals and the stench is unbearable.

Dr. Wilson reports that the English crews had fever and ague, but little dysentery, and typhoid or continued fever is not alluded to.

Williams * says the streets in Canton are twelve, many of them only eight, feet wide. Public necessaries and offal are borne through them; they pollute the air; the sewers often get choked and exude their contents on the walks. The ammoniacal gases generated by this filth aggravate ophthalmic disease, and it is matter of surprise that cholera, plague, or yellow fever does not visit this people. The Chinese transport cakes of human manure through the streets, thus creating an intolerable nuisance. Tanks are dug by the wayside; pails are placed in the streets for the convenience of the people. Urine, fæces, soot, bones, fish, mud from canals, offal of all kinds are thrown into them. Fermentation takes place, maggots swarm, and the disgusting mass is conveyed through the streets.

Mr. Williams says that, in general, the Chinese enjoy good health. They are as long-lived as other people. Ophthalmia, dyspepsia, cutaneous and digestive diseases, and where

---

* Middle Kingdom.

there is rice-culture, intermittent fevers prevail. Cholera has raged in some parts of the empire, but has never been epidemic at Canton. When it has appeared in China, it has invaded some towns and passed by others just as obnoxious.

Surgeon John Rose[*] says that cholera at Shanghai is sometimes seen sporadically, but is seldom or never epidemic. He says it is remarkable, considering the filthy condition of the canals, that violent epidemics do not more frequently come, for the canals which supply the people with water for domestic use are the receptacles of all sorts of refuse.

Dr. F. Wong[†] says that during a residence of more than ten years in Pekin, he has seen only two cases of typhoid fever among foreigners. This is more curious, he says, as the conditions usually supposed to be productive of that fever are here in full operation, and large numbers of people use water and inhale air charged with the impurities of human excreta.

In 1873,[‡] Dr. Dudgeon writes of the remarkable exemption of the inhabitants of Pekin from fever: "If foul smells create fever," he says, "there ought to be no immunity here." The condition and mortality of Pekin explode the belief that offensive odors are harmful. Hundreds of people live in and around and above cesspools, yet look well and healthy. He adds, "Many diseases prevail here, as in the West, without the agency of this reputed cause,—noxious odors; and the causes exist at all times here without producing such diseases."

The most convincing proofs of the sanitary condition and freedom from zymotic disease in China are found in the reports for twenty years of the medical officers of the Impe-

---

[*] London *Lancet*, vol. i., 1862.
[†] Ibid., vol. ii., 1872.
[‡] Ibid., vol. ii.

rial Maritime Customs. There is a remarkable uniformity of statement from these physicians, scattered over a thousand miles in extent of this country. These reports are made half-yearly, and contain an account of the health and prevailing diseases in China among natives and foreigners. Except small-pox and occasional limited epidemics of cholera and one of diphtheria, which occurred in Pekin in 1866, no outbreak of any importance is recorded for twenty-five years.

From Ichang, the doctor writes, "Why disease is not always with us, not why epidemics should sometimes come, is the problem," and the immunity from disease "must be ascribed to meteorological conditions." "In Corea small-pox, ague, and cholera are reported; typhoid ought to be common, as all the conditions are favorable to its existence; as yet (1885), however, no cases have presented themselves." In 1886, Dr. McFarland reports the sanitary condition of Ichang as most deplorable, but there "is an absence of epidemic disease."

At Han-Kow, in spite of the filthy condition, the health of the city compares favorably with that of any port in China.

Dr. Lowry writes that at Parkoi animal and vegetable substances lie decomposing on all sides; privies are open in the most exposed places; the houses are little better than the streets; in some of them an open gutter runs through into which all filth is thrown; the floors are saturated with excrement and the stench is vile. The diseases that he mentions as occurring here are scrofula, skin-diseases, small-pox, and syphilis. He makes no allusion to typhoid fever or diphtheria.

Dr. White [*] reports no serious illness at Ching-Kiang; on account of the filth "the wonder is that a man, woman, or child survives."

---

[*] 1886.

Dr. Deane wonders that there is so little sickness at Kuing-Chow. For twenty years these reports have been made, all uniform in tone, with a remarkable similarity in language, giving disgusting details of the vile sanitary condition of the country and its immunity from contagious disease.

But Hebrew prophet, Christian martyr, Mohammedan fanatic, or Mormon saint never clung to his faith with greater tenacity than do these English medical men to theirs. Their reports are full of expressions of surprise that epidemics do not come; they seem lost in wonderment; but never express a doubt of the filth-theory of these epidemics.

In 1886, however, Dr. Daly, the only dissenter in the group, in reporting on the health of Ning-Po, says we live in houses near large *kongs* filled with old and putrefying fæces, the accumulations of months, especially during the summer, when manure is not wanted for the fields. In spring and autumn they are removed in boats, which travel the canals where the natives wash their vegetables. "I have often seen women doing this within a few feet of a night-soil boat. On the banks of the canal on which the largest traffic of these boats occurs is situated the dairy that supplies most of the foreigners with milk, and in this canal the dairy-folk wash their utensils. No precaution is taken with the excreta from cholera or from fever patients." Dr. Daly seems in a little doubt about the pythogenic theory of the origin of typhoid fever, for he says, "If" it be a right one, it is not universally applicable; "if" it were, the disease would rage here, for everything to favor fecal decomposition exists,—warmth, stagnation, accumulation, and partial seclusion,—" yet no case of typhoid has occurred for many years among the foreigners, and it is an extremely rare disease among natives."

The editor of the *China Medical Missionary*, September, 1888, asks, "Do sanitary measures limit disease in populous cities?" He says a lesson can be learned from the condition of Canton and hundreds of other cities in China, where generation after generation has passed without any benefit from sanitary measures. In other countries millions of dollars are spent under the direction of the ablest scientific men to ward off disease. But in China no attention is paid to the subject. "Wherein do the results differ?" he asks. Canton has 1,500,000 inhabitants, and there are 333,333 persons to the square mile; the streets vary from five to eight feet in width, a few being twelve or fifteen; the actual space for each person in the city is fifty-five and a half square feet. There are no drains in Canton; there are ditches loosely walled up, but there is no fall to carry off the water. The ditches are generally choked with filth, and offensive gases are poured forth. The water is derived from wells, the river, and springs. The impure river water is used by a small part of the population. The largest part of the water-supply is from wells four to ten or fifteen feet deep; a great part of this is the refuse water which has been used by the 1,500,000 people. "It percolates through the filth of the sinks and ditches, and then through soil which has been saturated for centuries with animal, vegetable, and saline deposits." The custom of burning incense at the house-doors morning and evening is supposed to exert some counteracting influence on noxious gases; but the writer truly says it does not differ from ordinary smoke, and only adds so much more carbon to the air in minute particles. Here is Canton, he says, on the border of the torrid zone, destitute of all the sanitary appliances which modern science pronounces essential for the public health; with a population three times as dense as that of any western city; with impure water and abounding in filth and offensive smells, so that it is a by-word with travellers. A practice of thirty years convinces the editor that Canton is not more unhealthy nor is it any more subject to epidemics than Western cities.

W. K. Burton,* professor of sanitary engineering at Tokio, gives an account of the sanitary condition of Japan. All fæces and urine are carefully stored in receptacles until fermentation is accomplished. The mass is then diluted and poured on the fields. The farmers and peasants who live near these putrid masses and handle them "enjoy remarkably good health." The professor says, "It is not too much to say that the soil on which most of the large towns in Japan stand must be sewage-sodden, with the result, of course, of contaminating the air both inside and outside the houses." But the worst of all is, "the water for domestic purposes is drawn from shallow wells dug in this soil. The water, of course, is simply dilute sewage." The death-rate for the whole empire is 19.33 per 1000, while in the least healthy districts it is only 24.22 per 1000 of the population.

The death-rate of Tokio, with its filthy air, water, and soil, and bad drainage, is 21.08 per 1000.

The sanitarians assure us that the two most sensitive and unerring tests of the hygienic condition of a city or country are the proportions of infant mortality and mortality from zymotic diseases to the total number of deaths. These relations are so intimate that, given the hygienic condition, they can predict the mortality, and *vice versa*.

Professor Burton says the infant mortality is very low in Japan, his explanation being that "there is great attention paid to the children by their mothers." The mortality from zymotic disease is also very low. He gives a table, which, he says, is furnished by Professor Bealz, M.D., who has practised in Japan for thirteen years. The table shows that in 1888 there was a total of 756,367 deaths. Of these, 45,715 are set down to infectious or zymotic diseases.

In 1887 there was a total of 753,855 deaths, and 113,696

---

* *Sanitary Record*, 1890.

were of five years and under,—only about one-third of the amount of infant mortality that occurs in our large cities. Professor Burton attributes this comparative immunity from infectious diseases to "climate." It seems never to have occurred to him as a possibility that the sewage-sodden soil of Japan, contaminating the air inside and outside of the houses, and the diluted sewage forming the drinking-water of the people may have nothing to do in the production of infectious disease.

He says that Japan has acquired the title of "the Sanatorium of the East."

His statements are corroborated by the statistical tables of births and deaths shown at the second national industrial exhibition at Tokio, in 1881. The percentage of zymotic to total mortality for the year ending 1879 was 13.66, about one-half what it is in our cities which have the most careful sanitary supervision. The following year the percentage of zymotic to total mortality rose to 19.13. The general death-rate for the empire was 17.01 per 1000 of the population. The percentage of mortality under ten years of age in 1879 was 28.70; in 1880 it was 25.76. For the year ending 1880 there were 2863 deaths from typhoid fever in the empire, and 534 from diphtheria.

Professor Corfield * says that the large undrained cesspool and the latrines at Belgaum, in Bombay, have existed from time immemorial, with the wells in close proximity. "The whole ground in and about the lines is pregnant with abomination," with a mean temperature of 74° F. Yet, "as usual, at Belgaum there was comparative immunity from fever," and "Belgaum has long possessed a reputation for salubrity." If Professor Corfield had not told us, we might have thought the salubrity of Belgaum depended on "climate," as did that of China, or that "great attention was

---

* Treatment and Utilization of Sewage.

paid to the children by their mothers," as in Japan. It was neither. Professor Corfield does not seem to be indulging in raillery when he says that the reason for Belgaum's salubrity, surrounded by such putridity, is, "The town is laid out with some regularity, and the principal streets are kept in good order."

CHAPTER IV.

The Air—(Continued).

IF we turn to the reports of boards of health in our own country, the proof is no less abundant and conclusive that putrid emanations corrupting the air are not competent to cause disease. Those towns and cities which in these reports suffer the most obloquy on account of their filthiness appear in the mortality tables to enjoy the largest measure of health. In fact, our sanitarians offer the interesting phenomena of laying down certain inviolable principles of public hygiene, supporting them by the most flaming eloquence, and then industriously and with apparent unconsciousness collecting an overwhelming mass of statistical testimony to show there is not a word of truth in the original propositions.

The first report of the Massachusetts Board of Health contains an account of an inspection of the slaughter-houses at Brighton. "The stench about all of these places so kept is horrible; and, although the day of inspection was a fine, dry one, with a free northwest wind blowing, the odor of some of them could be observed for more than half a mile very strongly." "Here is a putrid mass consisting of blood, which decomposes almost as soon as it falls upon such material, the excrement of the animals killed and of the hogs, the half-digested food contained in the entrails, and the offal itself, covered with decomposing matter; in this filth the hogs wallow. The surrounding ground is filtered with decomposing matter. The slaughter-house hog not only eats flesh, but flesh in a state of putridity; and is, therefore, entitled to be regarded as the carrion beast. If he is good to eat, so is the crow and buzzard. Few persons would be willing to eat him if they saw him in his putrid sty, with wreaths of entrails hanging about his neck and his body smeared with blood. Human instinct (which is sometimes better than reason) *recoils from such food.* The slaughter-house pig-pens are filled with animal matter, with rotting blood, mingled with excrement, and are, *therefore,* a source of *danger to public health."*

Close to this description the board admits that Brighton is not an unhealthy town; that in 1865–66 its death-rate was less than that for the State; and there is no record, nor any charge, that anybody was ever made sick by this putridity.

Amazing coincidence! In the next report is a table of deaths of persons above five years of age from typhoid fever in Massachusetts for the ten years ending 1868. There are twenty-eight towns reported, which have a population of between three and four thousand each.

Towns.	Population.	Number of deaths from Typhoid Fever.
Dennis	3592	43
Harwich	3540	57
Provincetown	3470	18
G. Barrington	3920	36
Salisbury	3609	29
Watertown	3779	19
Greenfield	3211	56
Palmer	3080	32
Dartmouth	3435	29
Easton	3076	30
Groton	3176	28

THE AIR.   81

Towns.	Population.	Number of deaths from Typhoid Fever.
Medway	3219	30
Wrentham	3072	22
Milbury	3780	29
Grafton	3961	58
Ware	3374	46
Amherst	3415	33
Rockport	3367	26
Ipswich	3311	23
Deerfield	3038	40
Hollister	3125	26
Stoneham	3298	29
Wakefield	3244	19
Webster	3608	62
Braintree	3775	23
Canton	3318	13
Leominster	3313	21
Spencer	3024	30

The average population of the twenty-eight towns is three thousand three hundred and ninety-seven; the average number of deaths from typhoid fever for ten years is thirty-two. Brighton's population is three thousand eight hundred and fifty-four; the number of deaths from typhoid fever for ten years ending 1868 is nineteen. In proportion to population, Brighton, with all its filth, so vividly described by the board of health, had only about one-half of the amount of typhoid fever that the other twenty-eight towns had, which were not reproached for uncleanness. Yet, in the same book, Dr. Derbe says that typhoid fever is "born of impurity." As Dr. Budd would say, "Before these inexorable figures, the illusions of a half-century vanish in a moment." The board ascribes the salubrity of Brighton to its favorable location and to the good condition of its people; it says nothing about the location of the other twenty-eight towns or the condition of their people.

The Massachusetts Board (1879), in discussing the cause of typhoid fever, says, "Since the investigations of the board

f

have been directed to this point, it has repeatedly been shown that immense amounts of urine and excrement,—oxydized, incompletely oxydized,—and as they come from the bladder and intestines, are consumed in drinking-water and inhaled at the rate of nine thousand litres of contaminated air a day, and this for years by young, old, and middle-aged persons, without any disease resulting that may be attributed to filth."

The same board's report for 1880 contains an account of an inspection of schools in that state by Mr. E. W. Bowditch. Some of the schools in New Bedford have standing water in the cellars; a great many bad smells are reported; disgusting urinals and privies scarcely fit for animals, much less for human beings. In one school peppermint showed ten leaks in the plumbing; the drains were not tight in another; in another a "sickening smell" pervaded the building, and untrapped wastes connect with sewers; at another, sewers and private drains ventilate into the building, and some of the schools are supplied with water from a "suspected well." These buildings have a seating-capacity of four thousand six hundred and twenty-six; this year there had been altogether one hundred and fifteen cases of scarlet fever and diphtheria, some measles, and in one school "some sickness."

The Lowell schools were in equally bad condition, and in equally good health. The privies in one school were in the worst order he had ever seen.

This eminent sanitarian is always so rich in details that we cannot help giving them in his own words. "At one school the rear of the building on the boys' side seems to be used as the common urinal; on the side and front walls of the building were counted eight different places urinated upon and wet at the time of inspection." Mr. Bowditch is kind enough, however, to add, "It is stated that this was doubtless owing to boys from outside the school." At

Bartlett Street school "the boys have urinated on the walls of the building within five feet of the entrance-door!" Mr. Bowditch seems to have lost all patience with these audacious and depraved youths who are violating the fundamental principles of Sanitary Science; for he withholds his absolution from the boys of the Bartlett school, and very likely would not believe the young rascals if they laid their tricks to "the other boys."

The Cedar Street school has six rooms, with seats for about three hundred pupils. The ventilation is by windows only. There is standing water in the cellar within two feet of the cellar floor. The water-closets are in the basement, and the ceiling is neither lathed nor plastered. "There is a sickening smell up to the top of the basement stairs." "Peppermint showed that the drains were not tight in the basement." "There is a pig-pen on the opposite side of the street, and a neighbor's privy and poultry-yard seventeen feet of one side of the building, and fourteen school-room windows look out on them." "There have not been many absences lately; not over twenty from sickness; but two deaths, and these happened during vacation." In the High School, " cellar air" is used for the steam radiators; a number of these are so situated that "water-closet air instead of cellar air" is introduced into certain rooms. No sickness is reported in the High School. Mill Street school is in very bad sanitary condition, but Mr. Bowditch says "there is very little sickness of any kind." At Maxwell Street school is an untrapped sink. "Smells here come up the sink, but it is not known whether from a cesspool or privy." "One case of sickness only" is reported at this school.

Mr. Bowditch * inspected the summer resorts of Massachusetts. As far back as 1869, Dr. Derbe, he says, had predicted that Martha's Vineyard would "sooner or later be

---

* Massachusetts Board of Health Report, 1879.

visited by pestilence." Here he found privies, "shallows in the ground or on the surface," and the drinking-water is reached at from eight to twenty-five feet. The population in summer is about fifteen thousand, and at the height of the season it reaches double that amount. There are two ponds here that are veritable cesspools for the filth created by the thousands of visitors; rank marsh grass, masses of refuse sewage, "everything that is nasty," can be found in these ponds.

All over the town, sinks deliver their water from six to fifty feet from wells; the privies are not much farther off; and there are no traps to the kitchen wastes. At one large hotel the swill-barrel almost touches the well; across the street is a mass of filth several inches deep. The "proprietor states it is hoed out every autumn." One public-house has "several privies within a radius of fifty feet." The old well has six privies within twenty-five feet. Another house has its well within a foot of a sink-drain. The proprietor admits that the drain is a little too near the house, for the stench keeps him awake nights. The well at another house is seventeen feet from a cesspool; it is such "beautiful water" that the neighbors come for it.

Another house has a well within twenty-five feet of one cesspool and four privies. This well is used by five families. Another house has twenty-four privies and thirteen cesspools within a radius of one hundred and forty feet; the well here is used by a number of families. At another house fourteen privies could be counted, "all within range." Of forty-two wells that were examined, the water of twenty-two was found impure.

Mr. Bowditch mentions only two cases of sickness as having occurred at Martha's Vineyard.

Into the large boarding-houses sewer-gas was flowing; in one, peppermint poured into a basin permeated all the others; and it is the habit to empty chamber-slops into the basins.

In one of these houses there had been "one case of sore throat during the past summer." In a hotel for one hundred and fifty guests, none of the joints in the plumbing were tight; here, Mr. Bowditch states, "there has never been a serious case of illness in the house for sixteen years." Another house takes one hundred and seventy-five guests; the soil-traps here are all unventilated. A field one hundred and eleven by forty feet received all the soil-pipe matter. Outside the house is a two-story privy, with twelve sections; the pipes between the second and third floors leak sewer-gas. "There has been no sickness in the house." Visits to other resorts showed an equal amount of filthiness, and a corresponding amount of good health.

We cannot sufficiently admire the school-girl *naïveté* of the eminent sanitarian who records these inspections. The humor of the situation lies in this, that he seems entirely oblivious of the fact that he is accumulating the best of evidence to show that the filthy condition of these places had nothing to do with the production of disease.

The Rhode Island Board of Health Report for 1883 narrates an inspection of thirty-nine summer hotels in that State. These were reported in good sanitary condition; and it is comforting to note that just as large a measure of health prevailed in them as in those which were reported in bad sanitary condition by Mr. Bowditch.

But nothing which the reformers themselves have collected can illustrate the harmlessness of filth like the sanitary history of Newport. It is recorded in the *Sanitarian*, vols. ix., x., and xi. As early as 1872, Dr. Sims was taken ill at Newport, and ascribed the attack to impure water; but it was 1878 before the public was aroused, or even suspected the dangers which menaced the city with destruction. It was now found out that blasting epidemics were ready to explode, on account of the filth which had been "deposited here for two centuries." The wells were fed "by springs of dead water that has passed through the bodies of a whole neighborhood." In summer, when there was the greatest accumulation of refuse, there was the most obstruction to the outflow of sewage, and there is an "accumulation of festering filth." "Newporters are sleeping over a smouldering fire which will ere long break forth like a volcano!"

In winter the houses are ventilating-chimneys, "sucking in from all directions foul air from the soil;" and there is no barrier from the organisms which swarm in the ground-air around leaching cesspools, leaky drains, or the filthy-made ground." Newport was destitute of sanitary authorities. Of all the well-water analyzed in 1881, only one specimen was good. By December of the same year it was found that Newport's ice was contaminated with sewage, and the public water-supply was polluted. Professor Raphael Pumpelly has proved that water filtered through one hundred feet of sand will retain its impurities.

It was affirmed that Newport's Common Council "gloated in municipal dirt; its sensibilities were so blunted by municipal filth that it was incapable of appreciating the advantages of cleanliness."

The pest-wagons of night-scavengers and leaky swill-carts traverse the streets, dribbling their filthy contents all over the city. Back-yard excrement-storage, cesspools, piggeries, manure vaults, and various factories add to the soil-poisoning and welcome "the patrons of fashion and fever." "Filth reigns supreme." The waters of the wells differ only in degrees of impurity. Newport's water is doubly poisonous; it has too much solid matter, and is also poisoned by leakage from cesspools. Its air is laden with cesspool effluvia. The city is blind to dangers depending on foul water, air, and soil. It is now discovered that barn-yards and privies drain into the pond where ice is cut. The City Council is obstructing sanitary reform; it has permitted certain portions of the city to become a perfect nuisance.

In spite of the winter's rain and snow (1881), "the stench from the polluted soil is very evident."

Consulting sanitary experts were hurriedly sent for. Dr. Bell, of New York, said Newport's greatest evil was a water-logged soil; so that in winter, the cellars being warm, they soak in all moisture which cannot be absorbed; this mingles with the vapors and is carried with them up through the houses; these cellars are malarious nides; and he told the dreadful truth that the dangerous constituents of cesspools were absorbed into the soil, while the inert, harmless residue was carried away. Dr. Bell said that if, in addition to soil-dampness, further dampness were added, the danger increased, making a most offensive condition, "dark, mouldy, and to be likened only to the clammy sweat of death."

The experts found things so bad as to allow no possibility of doubt as to the relation of cause to effect in producing disease,—soil-pipes broken and cracked, joints with no cement, some pipes plugged up entirely, and a backward flow of sewage into cellars and under basement-floors, "poisoning the water, though to the taste it seemed excellent." The experts found only one house in Newport that had not sanitary defects, and in this the furnace took the air direct from the cellars. Several of the houses were "mere death-traps."

"Go into any of the hotels, from the greatest to the smallest, and you will find the air laden with cesspool effluvia," wrote Dr. Sims. Dr. Peters said he had never seen a street so fair and yet so foul as Bellevue Avenue; and he described other streets of approximate filthiness, where were smells from drains, sewers, and out-buildings; out-houses were clustered so closely as not to allow circulation of air; and he did not believe that delicate persons could recover their health here.

In 1882, Dr. Storer writes,* "The excitement in Newport,

---

* Proceedings American Public Health Association.

which was already blood-warm, has risen far above fever-heat." A meeting was called of the summer residents, representing a capital of twenty millions of dollars. The clergy took an active part; the ladies joined the Sanitary Protective Association, and by vote and voice cheered the movement on. Dr. Storer writes that in 1876 nearly one-fourth of all deaths in Newport were from zymotic diseases; in 1877 these were one-sixth of the whole, in 1878 more than one-sixth, and in 1879 nearly one-fourth. "What more startling evidence than this," he cries, "of all that the sanitarians have asserted concerning Newport can be asked for?" Besides, there are eleven undertakers here; and now, to add to Newport's horrors, opium-taking, whether as a cause or effect of the filth, is going on to such an extent that "there are thirteen habitual opium-drunkards," who use over thirty-four thousand grains of opium each month.

In all this sanitary delirium only twice is there any record of any one having tried to face the danger with anything like composure. At one of the meetings of the Sanitary Protective Association, Mrs. Lieber asked what sanitary dangers were in Newport that did not exist elsewhere. And Mr. Davis timidly inquired if, theory aside, there had ever occurred any serious sickness from these causes among summer visitors. Both were quickly squelched; for the clergy and the doctors reasserted the dangers, and we hear no more from the remonstrants. They were very likely glad to keep still, lest, like the London obstructionists, they might be pointed out as "wallowing in their own filth."

The sanitarians who reported on China, Japan, Belgaum, and Brighton told us why these places were so healthy; none of them, however, condescend to explain the cause of Newport's salubrity.

Mr. E. W. Bowditch, in his report of inspection at Newport, says, "Bearing in mind that the city water is undesirable to use, that perhaps a majority of the wells are already tainted, and that the sewer system is of little or no value save to concentrate nuisances at certain points, that no precaution is taken against zymotic disease except to placard front doors, the apparent freedom from preventable sickness is *remarkable*."

Newport has always enjoyed a world-wide reputation for healthfulness. Registration showed always a smaller death-rate than that of the State of Rhode Island at large; and it was admitted that this low rate was augmented by the death of many invalids who were conveyed there in the last stages of illness. For ten years ending 1880 the death-rate for the State of Rhode Island was 15.9; that for Newport was 14.5 per 1000 of population in the same time. It may be thought that this low death-rate depended on a low birth-rate. This is not so; for Newport's birth-rate for ten years ending 1879 was 24.2 against 22.3 per 1000 for the State. In 1880 the per cent. of zymotic disease to total mortality was 28.09 for the State and 17.35 for Newport; and nearly every year when this subject is mentioned in the Rhode Island reports, Newport is recorded as possessing advantages over the State.

The period of sanitary activity began in Newport in 1880 with the formation of the Sanitary Association. The period of sanitary law was established in 1885, when a board of health was organized. The board reports that a large amount of work had been done in removing unsanitary conditions. The air, water, and soil have been made purer by the improvements. Sanitary codes have been passed; sanitary inspections without number have been made. We naturally turn to the vital statistics to ascertain the effect of these measures on the health of the people. For the ten years ending 1890 the average general death-rate was 15.7, against that of 14.5 for the ten years ending 1880. For the six years ending 1890, the period of sanitary law, the average general death-rate was precisely the same as the previous six years,—namely, 15.73. If we look a little closer for the effect of sanitary law and sanitary measures on zymotic disease in Newport, we find that for the five years ending 1884 there were 91 deaths from cholera infantum, and 114 deaths from the same disease for the five years ending 1889. There were 22 deaths from croup for the five years ending 1884, and 30 from the same disease for the five years ending 1889. For the five years ending 1884 there were 26 deaths from diphtheria, and for the next five years there were more than double that number,—namely, 56 deaths. There were 46 deaths from typhoid fever for the five years ending 1884, and for the next five years there were 51 deaths from the same disease. In the mean time there was not much change in the population of the city. The State census made it about 19,500 in 1885, and the Federal count was about the same number in 1890. So we have the interesting phenomena at Newport of the removal of unsanitary conditions, and not only a marked rise in the general death-rate, but a decided increase of those diseases which the sanitarians tell us are caused by such unsanitary conditions.

Inexplicable circumstance!

A few months after the culmination of the febrile sanitary excitement at Newport, Dr. Storer read a paper * before the Sanitary Protective Association of that city to show that it is one of the best places in the United States for consumptives. He does not tell us how it can profit a man to gain immunity from phthisis only to be suddenly cut off by diphtheria, typhoid fever, and dysentery. The water-logged soil which the consulting sanitarian said was Newport's greatest evil, and from which was being sucked into the houses not only poisonous air but "organisms" of disease, has suddenly undergone a change. Dr. Storer says in his paper that one of the reasons of Newport's comparative ex-

---

* *Sanitarian*, January, 1883.

emption from consumption is the absence of "soil-moisture." Without a single word of warning against zymotic disease, or about "the clammy sweat of death," he entices not only the consumptive to Newport, but others, and that, too, in the winter, when the danger, according to the sanitary experts, is at its height. He recommends its "soft, balmy, soothing, sleep-inducing climate in the amelioration and cure of nervous diseases." For nine months in the year "it is a veritable haven of rest."

The citizens of New Haven,[*] says the health officer of that city, who have often been told of their danger, "go on year after year, rivalling a miser in storing up their own excrement and every other species of repulsive and loathsome nastiness in receptacles as near as possible to their own houses." "Where, then, is their boasted intelligence and their prudent regard for their families?" And when, in the interest of these "suffering and stricken families," attempts are made by the wise and good sanitarian to deprive them of their "odorous and odious subterranean accumulations, they dispute their power." "They cry out, 'Oppression! Sacred rights of citizenship invaded!' and resistance to the utmost is threatened." "Look among the houses of the working-classes, and see how often the industrious mechanic and laboring-man is wronged and made the victim of his landlord's power." There is no property, he says, which makes such large returns as that which is rented to the laboring-classes. "They need the protection of constituted authority." "This suffering and afflicted portion of our fellow-citizens have rights." In the masses of filth, which are stored so near dwellings, germs of disease find the most favorable conditions for existence. "Cesspools, filthy drains, and filth in any form afford a fertile soil for the reproduction of the typhoid-fever germ." In 1884 he reminds the people

---

[*] Connecticut Board of Health Report, 1883.

of their danger from cholera. He says a single evacuation from a cholera patient, thrown into a privy or cesspool, will infect a whole neighborhood.

The people of New Haven had long neglected the warning voice, and had often been rebuked by the sanitary officials for their stinginess in not providing money for these authorities. More liberal appropriations were now called for in view of the approaching epidemic. Their health-officer told them that never before had such timely notice been given to a people; the fulness of time was come; there was sure to be much sickness during the next summer; "whatever is done must be done quickly;" "the occasion will admit of no delay." And now, as if ravished by the spirit of prophecy, the sanitary seer exclaims, " The cholera will come to the country!" "The cholera will come!"* The impenitent citizens of New Haven huddled together and waited for the result, but seem to have held on to their money. That year, which was to bring such disaster to them, the general death-rate was 17.9 against 17.55 per 1000 for the year before, and the proportion of zymotic to total mortality was six per cent. less than that of the previous year.

New Haven went on storing up its filth, so that in 1885, when a census of the privies and cesspools was taken, it contained more than 12,000 of these structures, and there were not fewer than 5000 untrapped basins and kitchen-sinks "and otherwise defective plumbing." In addition there were nearly a thousand wells in daily use, most of them "less than twenty-five feet from some leaching filth-pit."

In 1887 † the health-officer of that city reports to the State board that the city board of health has been restricted in its

---

* Connecticut Board of Health Report, 1884.
† Report of New Haven Board of Health.

action by having no sanitary ordinances. It has to learn of the "existence of infectious diseases as best it can." It is "still further hampered by parsimonious appropriations" from the City Council. In short, the board has neither power nor money. The city's air is pestilential, its water polluted, its soil contaminated, and there is no remedy.

In the State Report for 1890, Dr. Osborn compares the death-rate of New Haven with what it was twenty-one years ago. In 1869 it was 23.37; in 1889 it had fallen to 17.50 per 1000 of the population. He compares the proportion of zymotic to total mortality for the five years ending 1873 with the five years ending 1888, and finds that the percentage of zymotic to total mortality in the first period was 29.44, and that of the second period was 21.68.

As the death-rate of Newport, both general and zymotic, went up with the removal of causes of disease, so did that of New Haven steadily decline as these causes were multiplied. It is fortunate that we are not left to conjecture the reason of this decline. We could never have guessed it. Like China, New Haven had a "climate;" the mothers there "take good care of their children," as in Japan; like Belgaum, it is regularly laid out. If we had not been told the reason, we might have been contented with the fact that it was "remarkable," as at Newport. The health-officer of New Haven is not the man to leave us in the dark on this momentous subject. In a communication to the State Board of Health he says New Haven's sanitary condition is due to "moral suasion."

The Connecticut Board of Health reports an inspection of the borough of Stamford in 1884. A stream here pollutes the air with a disagreeable stench, and there is no method to dispose of sewage. In a thickly-populated street there is a large blind-ditch, an elongated cesspool half a mile in length; the ground on each side is infiltrated with sewage; the ground under the houses is saturated with it, and the gases pass up through the buildings. The citizens of Stamford are living over a "vast underground retort, generating its deleterious gases to corrupt the soil about its dwellings, poison the water of its wells, and defile the air of its houses." Stamford had been noted for its health for two hundred years. That very year its death-rate is recorded as 14.6, against 15.6 for the county, and 16.6 per 1000 for the State. The next year frightful epidemics of diphtheria * broke out in Greenwich and New Canaan, two neatly-kept adjoining towns, in hourly communication with Stamford. Stamford escaped the disease. Was it the "vast underground retort" that protected it from diphtheria that year?

The same year an official inspection of summer resorts was made in Connecticut. Some of the places were in a very unsanitary condition; others were reported irreproachable; all were equally healthy.

A number of the jails were also inspected; all were very filthy, but the inmates were in excellent health. At the State prison is a large open filth-pit, just outside the walls, which is infrequently cleaned and never disinfected; the excreta of the prisoners is received in pails, and there are no flues whereby the cells are ventilated. There is very limited airspace, and the building is damp; the walls and floors are wet from the condensation of moisture. "The health of the prisoners does not seem to suffer from these influences." "It appears to be true that the prisoners often have better health than before they were committed."

The New York State Board of Health Reports give no less positive evidence of the harmlessness of filth. Newtown Creek and Hunter's Point, on Long Island,† are so foul that Mr. F. Law Olmsted smelled the odor from them as high as Forty-sixth Street, in New York City. Dr. Agnew smelled

---

* Connecticut Board of Health Report, 1886.
† Third Report New York State Board of Health.

it at Madison Avenue; in fact, the testimony proves that the stench spread its way from Fifteenth to Seventy-eighth Streets in the city of New York. Odors of putrescent animal matter, putrid fish, offal, fish-scrap, bone-boiling, common manures mixed with tarry smoke, kerosene smells, forming "a series of stenches that have acquired a magnitude that can hardly have been witnessed elsewhere in the world." The Fourth New York Board of Health Report says "the greatest of all stench nuisances is the creek itself;" its borders are crowded with nuisances. The governor is appealed to because "the security of life and health, as well as the value of property in said town," is at stake. It is not claimed that any case of disease or any epidemic has ever arisen from these foul nuisances.

The author has visited Hunter's Point many times; has conversed with physicians who have practised there many years; with druggists and citizens generally, to ascertain if disease, especially zymotic disease, prevails there more than elsewhere. The replies are singularly uniform, to the effect that none has ever been suspected to have arisen from this foulness. The citizens are aggrieved that such vile nuisances exist, but they do not believe their health is impaired thereby.

In the Fourth Report of the New York State Board of Health is an account of an investigation of the Glen Cove starch-factory at Sea Cliff. Two hundred and twenty-five people, mostly summer residents, signed a petition to the governor for its suppression. The petition said the foul stench and gases so permeated the atmosphere as at times to render respiration oppressive and produce nausea; that they interfered with the enjoyment of life and property, and caused great physical hurt; that they were prejudicial to the sick and destructive to the comfort of those in health. Dr. L. deposed that he and his family had suffered from nausea and sense of suffocation, and that the "aforesaid exhalations and the vast quantities of bacteria and other forms of germ life engendered by said refuse are exceedingly deleterious to health." Dr. F. deposed that a patient of his complained that the odors took away his appetite, and "in my estimation said exhalations are and must be deleterious to health." Mrs. D. affirmed that her hotel accommodates two hundred and fifty guests, besides transient visitors; that they have frequently been compelled to leave the table on account of the stench, and that this renders the enjoyment of life and property uncomfortable. Dr. N. testified that the refuse kills the fish, and these decaying cause "a new and increased danger to the life and health of those inhabiting the vicinity;" and, besides, the property is decreased in value, and the holders are harmed in health and in peace of mind.

The permanent population at Sea Cliff is seven hundred; in summer it is about four thousand. None of these affiants, doctors or others, specified any epidemic, or even a single case of disease, that had been caused by the starch-works. A weaker firm than the Glen Cove Company would probably have gone under; but this concern did not propose to wind up its business without a struggle. Five hundred and thirteen, mostly permanent, residents and property-holders in and about Glen Cove testified that they had lived there and near the factory "for the period set opposite to our respective names," and that the charge that it is prejudicial to health was, in their belief, "untrue and without any foundation of fact to support the same." Then followed special affidavits of people who had lived in Glen Cove from five to fifty years, who declared that they never knew or heard of any disease being caused by these odors; and a large number of fishermen who had fished here from three to twenty years deposed that the refuse did not in any way affect clams, oysters, or fish; that these were as plentiful and in as good condition as the same found anywhere. Five physicians of the place testified to its healthfulness, and to the harmlessness of the company's operations.

At Lawrence and New Brighton similar nuisances exist, but no disease is reported from them. Cortland, Rhinebeck, Harrison, and Canajoharie are in an unsanitary condition, but no sickness reported. The stream that flows through Mt. Vernon has been foul-smelling and disgusting for several years; the health-officer says it "cannot but be a source of disease;" but no disease is reported.

The Eighth Report of the New York State Board of Health narrates the sanitary inspection of Tivoli. Here the majority of the people use water from shallow wells; the drains are so arranged that the sewage of one family is turned on to the premises of another, "any way to be rid of it for a time;" two slaughter-houses are near the centre of the upper village; "one of the leading residents has his well below and within less than fifty feet of two pig-pens in a filthy condition, two privies ditto, and one barn;" one well that supplies nine families is ten feet deep, and the water is so foul that the tenants "don't think it healthy to use." No sickness is reported here. But with an unshaken trust in the filth pathology, the saintly reformer who reports on Tivoli sees in its health the interposition of the Divine hand, for he says, "There is, no doubt, a special Providence watching over these people, or they would not now be alive."

The Ninth Report of the New York State Board of Health details the sanitary inspection of Tonawanda, Chatham, Fishkill Landing, Matteawan, and Rye; in these towns the soil and water were badly polluted, but the people were in good health. Rye was afflicted with great nuisances to sight and smell. From a very offensive pond here ice is cut. The death-rate of Rye that year is recorded as 10.75 per thousand of population. The Pennsylvania Board of Health reports (1886) an inspection of the soldiers' orphans' home at

Mercer; it is overcrowded in every department; the buildings are badly located, privies in bad condition, clothing deficient, and much of it filthy. The "general health has not been bad;" only two deaths the past year, one from sunstroke and one from croup; the only one sick at time of inspection was "a boy with toothache."

The foundation of the city of New Orleans is made by the *débris* of the Mississippi River; its soil is like a sponge; it has and can have no drainage. Dr. Holt, formerly president of the Louisiana Board of Health,* says that there is hardly a privy whose contents have not free access to the soil, to saturate the ground. Water ninety-five feet from the surface has yielded a large percentage of urea and organic matter. The soil is saturated with human excrement; the people of New Orleans live on a dungheap, and it may be said that they have a privy in common.

Dr. Joseph Jones says that the main drains and canals of New Orleans are blocked up with offal, presenting a green, seething, putrefying mass of filth, belching forth noxious vapors. Large numbers of the people sleep on the ground-floor of houses badly constructed, badly drained, situated on land which is saturated with water, which is the *seepage* from privies and foul drains. Dr. Jones writes to Mayor Shakespeare † that examination and measurement show that there is a mud deposit in all of the drainage-canals in New Orleans, varying in depth from four to eight feet. Fermentation and the evolution of foul gases are constantly going on in this immense mass of filth. "Every known and unknown combination and product of the putrefaction of vegetable and animal matter can be found in these foul reservoirs." On page 212 of the same book, Dr. Jones says one-third of those dying in New Orleans die in poverty

---

* *Sanitarian*, vol. vii.
† Louisiana Board of Health Report, 1880–83.

and are buried at the public expense; "one-sixth of those who die in New Orleans perish in silence and misery;" their deaths are certified to by the coroner. The general death-rate, white and colored, for the last four years has been 26.43, 25.02, 23.41, 23.92, an average of 25.19. The death-rate of the whites in those years was 23.59, 22.36, 22.90, 21.27, an average of 22.53. The general death-rate of New Orleans is greatly augmented by strangers, sailors, and laborers on the river. Dr. Jones says * the death-rate of the whites—exclusive of foreigners and strangers and laborers on the various lines of railroads, who crowd the hospitals and prisons—would not exceed fifteen per thousand of the inhabitants per annum.

If we apply the sanitary touchstones to this city,—the proportions of zymotic and infant mortality to total number of deaths,—we find that for the four years ending 1889 the percentage of zymotic deaths to total mortality † in New Orleans was 15.04, 16.7, 16, 18.7, an average of 16.61. The percentage of infantile deaths, five years and under, to total mortality for the same years was thirty, thirty-two, twenty-nine, thirty-three, an average of thirty-one. The average number of those who died in public institutions yearly was twelve hundred and fifty-five for the four years ending 1889. In the same time four thousand one hundred and thirty-nine deaths were certified to by the coroner, a yearly average of one thousand and thirty-five.

The city of Washington contains twelve thousand less people, white and colored, than New Orleans. It has an abundant supply of good water, it is well sewered, its streets are broad and kept scrupulously clean, its plumbing is carefully supervised, it has sanitary regulations without number; its sanitary inspectors are in emblazoned uniform; to the

---

* Louisiana Board of Health Report, 1882.
† Louisiana Board of Health Report, 1886-89.

observer it is the ideal of the sanitarian, in direct contrast to New Orleans. For the last twelve years the infant mortality of Washington has been forty per cent. of the total mortality. Its mean death-rate * for thirteen years has been 23.88. The percentage of zymotic disease to total mortality for the four years ending 1889 was 19.9, 21.50, 22.19, 24, an average of 21.89, against that of 16.61 for New Orleans during the same time.

That the general population of Washington is better supplied with the material comforts of life than is that of New Orleans is plain to the most superficial observer. This is proved by the deaths in the public institutions and by those under the head of violence and neglect, which are probably coroner's cases. For the four years ending 1889 there died in the public institutions in Washington, including the government hospital for the insane, where occur over a hundred deaths yearly, six hundred and ninety-seven, seven hundred and twenty-three, eight hundred and twenty-two, eight hundred and nineteen, an average of seven hundred and sixty-five. Under the head of violence and neglect there died in the same years two hundred and three, one hundred and eighty-nine, one hundred and sixty, one hundred and fifty-four, an average of one hundred and seventy-six.

Dr. Hatch † reports on the sanitary condition of Sacramento: The drainage here is defective; waste-water from kitchens is thrown on the surface; in the large majority of cases the privy is a mere hole dug in the ground; when full, it is covered over and another is dug by its side; percolation from cesspools still further pollutes the soil. This, he says, has been going on in a low alluvial soil for twenty-eight years. The hotels and houses are so imperfectly

---

* Board of Health Report, District of Columbia, 1889.
† Board of Health Report, 1879–80.

plumbed that sewer-gas enters them. There is a slough but a few steps from the principal business street, which is daily and hourly befouled by filth. Dr. Hatch says, "Yet with these very evident defects, these violations of hygienic rules, the sanitary condition of the city is good and the death-rate by no means discouraging." That year, 1880, it was unusually high,—19.7 per thousand,—but then the percentage of zymotic disease to total mortality was 14.6, and the percentage of infantile to total mortality was 24.7. The next year the death-rate in Sacramento fell to 18.2 per thousand.

The New Hampshire Board of Health Report for 1891 says that Carroll County jail, which was reported in 1889 as the worst in the State as regards sanitary condition, is in no sense a decent place for the detention of criminals. There are about three hundred and fifty inmates in the Hillsborough County almshouse; in this institution there is no system of sewerage worthy the name, and the old vaults are always in an unsanitary condition. No sickness is reported at either of these establishments. The sanitary condition of Grafton County jail is the most abominable to be found in the State. "It would be difficult to devise a more filthy and disgusting arrangement than is here to be found." In one story "the cells are directly connected with the soil-pipes." "The sanitary condition of the entire institution is such as to jeopardize the health of all those living within its walls." "It is not to be wondered at that the jailer lost a son from typhoid fever." The board writes to the county commissioners that "a fatal case of diphtheria has recently occurred at the jail." The commissioners reply, "There has been no case of diphtheria at the jail." "There is no sickness amongst those confined at the jail." "The jailer's son contracted typhoid fever elsewhere, and came home and died there." Here was an imported case of typhoid fever into an institution which was in the vilest "sanitary condition."

The reputed causes, not only for the spread but for the generation of that disease, were here in the greatest abundance. Yet we find no case occurring there but the imported one, and "no sickness amongst those confined at the jail."

In *Science*, vol. x., 1887, Mr. William Glenn writes that the Back Basin at Baltimore is a nearly stagnant pond two hundred by five hundred yards. It receives the drainage of about eighty thousand people. In Back Basin are the sediments of this drainage. "Here they undergo fermentation and decay, at times giving off odors offensive indeed." Mr. Glenn says he has been in quite constant communication with the workmen who dredge this sewage, and who pass their days stirring about it. "They live in an atmosphere loaded with offensive gases." "And what of their health? With singular unanimity they declare that the occupation is a healthy one." Among a hundred men engaged in that work he had not heard, for nearly three years, of any case of zymotic disease. Men, he says, engaged in this business ought to sicken and die. "Curiously enough, they do not, more than men in other occupations." Mr. Glenn says he has no knowledge of what filth-diseases are or are not, and he has no suggestions to offer; he simply states the facts.

In the report[*] of the Board of Supervisors of San Francisco on Chinatown, headed, "Startling report of the hideous and disgusting features of Chinatown," the board says, "In a sanitary point of view, Chinatown presents a singular anomaly. With the habits, manners, customs, and whole economy of life violating every accepted rule of hygiene; with open cesspools, exhalations from water-closets, sinks, urinals, and sewers tainting the atmosphere with noxious vapors and stifling odors; with people herded and packed in damp cellars, it is not to be denied that, as a whole, the

---

[*] San Francisco Daily Report, July, 1885.

general health of this locality compares more than favorably with other sections of the town, which are surrounded by far more favorable conditions."

That portion of New York City which is bounded by Broadway, Fourteenth Street, and the East River contains about four hundred thousand people. There are streets in this district which are more densely inhabited than any other part of our globe, except portions of Naples and of the large cities of China. Here are the most filthy wharves, slips, streets, lanes, yards, houses, clothes, and persons to be found in the metropolis. When the inhabitants who live here bathe, they bathe in filthy water. Mr. Riis has told us about this part of New York in "How the other Half live." While the mass of the people in this quarter may be just as kind in their dealings with their fellow-beings, and as industrious as the masses are elsewhere, it is none the less true that this locality is the resting-place and abode of the most dissolute tramps of both sexes, and the lair of the most brutal loafers on the face of the earth.

In 1889 the New York City Board of Health put forth a remarkable document. For convenience it divided the city into six districts. The first, south of Fourteenth Street and east of Broadway; the second, south of Fourteenth Street and west of Broadway; the third and fourth extended north from Fourteenth Street on either side of Broadway to Fifty-ninth Street and Harlem River; the fifth and sixth, north of Fifth-ninth Street to the end of the island and the Twenty-third and Twenty-fourth Wards. The death-rate for the city at large was 26.33 per thousand. The death-rate of the filthy and crowded district south of Fourteenth Street and east of Broadway was 22.55 per thousand. The third district, the next most crowded and filthy, had a death-rate of 22.10, against the death-rate at large of 26.33; and the highest death-rate in any of the filthy and crowded localities was but 26.60, against 26.33, which was the death-rate at large for the city.

There is probably—there *must* be—a gross error in this document. The author of this book does not pretend to be able to detect it. It is not the low birth-rate; it may be in the *large* number of people in the first and third districts that are between the ages of fifteen and forty years. It is strange if the New York Health Board does not know better than any one unconnected with it where this error lies. If there be no error (which God forbid!) in the figures of the pamphlet we are considering, then the unblushing fact will not be conjured down, say what we will, that filth and overcrowding and intemperance and carelessness are more conducive to good health and a long life than are cleanliness, an abundance of fresh air, temperance, and prudence.

The sanitarians were confounded at the results of their own reports, which pulverized their doctrine of the filth-generation of disease. Resolute not to abandon it, for to do so involved their annihilation and put out of commission their boards of health, they proceeded to shift their ground, and the discovery of the microbe of disease momentarily afforded them a refuge. This, they said, unlocked the mysteries of Sanitary Science; after all, it was not the filth that caused the disease; it was the microbe that found in the filth the pabulum for its growth and the stimulus to its self-fecundation. If the germ was absent, the filth was inert; if the filth was destroyed, the germ withered and disappeared.

Herein lay the essence of Sanitary Science; and to prevent the disastrous conjunction of the germ and the filth the supreme efforts of the sanitarians should be directed. They made no investigations; but the improvisations of amateurs, amatrices, and professors of Sanitary Science teemed with accounts of the antics of the newly-found germ in its beloved filth. Vain delusion!

By and by some scientific men investigated the behavior of the disease-germ in the presence of putridity and decay. M. Miquel * says he will prove, contrary to the opinion of many authors, that vapor from masses in putrefaction is micrographically pure; that the gases proceeding from it are always free from bacteria; that the air itself from putrefying meat, even in its intensest putrefaction, distended by gas and giving off an insupportable odor, far from being charged with microbes, is entirely pure, if it is in a certain condition of humidity.

The cholera microbe, when tested by its discoverer, was found to speedily disappear in the presence of putrid bacteria. Mixed with well-water, the cholera bacilli retained their vitality † thirty days; in the Berlin canal, six to seven days; mixed with fæces, twenty-seven hours; and in the contents of cesspools they could not be demonstrated after twenty-four hours. Flügge ‡ says that the refuse of cattle-stalls, kitchen-water, and general filth are excellent conditions for the putrefactive bacteria, but "they are totally unsuitable for the growth of infective agents." "We see in all waste-waters, in putrid fluids, etc., that the facultative (disease) parasites, even when they are sown in enormous quantities, die in a few hours, or, at most, in a few days."

In the *Centralblatt für Bakteriologie*, vol. vi., 1889, Dr. Justyn Karlinski records his experiments with the typhoid bacillus in sewage. He put two hundred cubic centimetres of fresh typhoid stools in a quart of filth from a privy which was rich with bacteria; the bacilli in the stools were small in proportion to those in the specimen from the privy. Forty-eight hours later not a typhoid bacillus could be found. Four times this experiment was tried, with the same result. He sterilized two hundred cubic centimetres

---

* Les Organismes Vivants, 1883.
† *British Medical Journal*, vol. i., 1886.
‡ Micro-organisms.

of sewer filth and mixed with it ten cubic centimetres of typhoid fæces; for a whole month he was able to detect typhoid bacilli, though in small numbers compared with those in the typhoid stools. In sixty different observations he found the contents of the sewers acid; the lower strata of some privies were almost feebly alkaline. He put fifty cubic centimetres of typhoid stools, which had more than two thousand colonies of typhoid bacilli to each cubic centimetre, with fifty cubic centimetres of privy fæces, with alkaline reaction. After the whole was carefully mixed he could find only a single typhoid to four hundred and sixty strange bacteria; after five days, not one in nine hundred; after ten days, not one in three thousand. He repeated and varied these experiments, always with the result that the more water and the more filth, the sooner did the typhoid bacilli disappear.

"*Je mehr Kanaljauche und wasser, je grösser die anzahl von Faulnissorganismen, desto schneller gehen die sonst widerstandfähigen typhus bacillen, die mit den Dejekten in die Senkgruben gelangen, zu Grunde.*"

CHAPTER V.

The Water.

THE tripod on which Sanitary Science rests—to wit, "pure air, pure water, and pure soil"—breaks down completely when we consider the second element of which it is composed.

Regarding this, about the only point on which the scientists are agreed is that pure water is an ideal substance, that it does not and cannot exist in nature, and that waters differ only in degrees of impurity. So they have arbitrarily laid down certain formulæ, and have classified waters as Pure, Usable, Suspicious, Impure, according as these were found to contain a less or greater amount of foreign matter; and, based solely on chemical analysis, there is a wide divergence of opinion about what constitutes a safe or a dangerous water for domestic uses.

The satisfaction which the sanitarians derived from the anxiety and distress they had caused about the air we breathed was transposed into a riotous joy as they beheld the pangs they awakened after they had thoroughly aroused our suspicions about the water we were drinking. If they succeeded in having one condemned and rejected which was abundant, clear and sparkling to the eye, and grateful to the taste, and which the experience of a century had pronounced wholesome, Sanitary Science had made a "gigantic stride." They delighted to tell us that the brilliancy and stimulating taste of a water were perhaps the tokens of its impurity, and that such water might contain unspeakable filth. Our only safety lay in making it insipid by boiling. After it was boiled we added ice to relieve its tastelessness, and for a while went along in perfect security, when Sanitary Science made another "gigantic stride" and discovered that the ice was contaminated. Boiling the water, to be sure, would destroy the morbific principle; but freezing it, instead of removing impurities, actually concentrated them and made them more dangerous, and the disease microbe that lurked therein was none the less active and virulent. The very prismatic glow of the ice was the sign of danger. We had safely passed Scylla, but had gone to pieces on Charybdis. We were ready to sink in despair, when the countenance of the gracious sanitarian assumed a more benignant aspect as with outstretched arms he told us of the inexhaustible resources of Sanitary Science, which he was ready to invoke for the safety of ourselves and our families. He would inspect the ice.

The sanitarians teach us * that pure water, or water that is pure enough for potable use, should contain less than six grains of solids to a gallon; there should be no indications of nitrites; the amount of nitrates and free ammonia should be slight; and of albuminoid ammonia there should be less than 0.0035 grain per gallon. It should be colorless, odorless, and tasteless.

Without tiring the reader with technical descriptions of the gradations of "Usable" and "Suspicious" water, it is said to be "Impure" if there is decided smell or taste; if it has over fifty grains per gallon of solids, and over four grains of destructible organic matter, with nitrates, nitrites, and ammonia to any great extent. It is conceded, however, that ten times this amount of these identical solids, with nitrates, nitrites, and ammonia mixed with our food, would be harmless, nor is it claimed that any artificial fluid containing this amount of proscribed solids, nitrates, and ammonia would have the least effect to produce disease.

Sanitary literature is overrunning with "beliefs" and "opinions" about the different diseases which are caused by drinking-water. All forms of malarial fevers have been ascribed to it; also yellow, typhoid, and scarlet fevers, erysipelas, diphtheria, cholera, dyspepsia, boils, goitre, ulcers, elephantiasis, diseases of the bones, diarrhœa, dysentery, calculi, bronchial catarrh, and the entozoic diseases. The statements, however, respecting its agency in the production of disease are so conflicting that Parkes † says, "The exact connection between impure water and disease does not stand on so precise an experimental basis as might be wished;" but he adds, "Apart from actual evidence, we are entitled to conclude that abundant and good water is a prime sanitary necessity." When we consider the devout "belief" of the English sanitarians regarding the infective

---

* Parkes, Hygiene. † Hygiene.

power of what they say is impure water, these admissions of Parkes, who seems to have carefully weighed the evidence, are all the more striking. The most acute German observers have never found any relation of water to disease. To use the words of the Massachusetts Board of Health, "The idea is essentially English."

It is extremely doubtful whether there is any scientific proof that the water of any spring, well, pond, lake, or stream, which has been in use as potable water, ever caused in the human system a specific disease, or anything more than slight or temporary disorder, unless such water was contaminated, deliberately or accidentally, by some irritant poison. Marsh-water is accused of causing malaria; but Professor Colin* cites portions of Italy and Algiers where marsh-water is drunk and no malaria follows. Marsh-water is largely drunk in Holland and Hungary, but it does not produce malaria. Surgeon-General Lawson says that in Florida, where it was used by the United States troops, malaria was less severe than in those military departments where the water-supply was from other sources.

In 1881 the work of draining Lake Okeechobee, in Florida, was commenced. Colonel J. M. Kreamer, the engineer, testifies that during its progress, extending over a period of four years, more than five hundred men were employed,—whites,—most of them unacclimated. They labored in the swamps summer and winter, a good deal of the time immersed to their waists; they drank no other than marsh-water. Colonel Kreamer declares that during these years not a single case of malarial disease occurred among them. He says he conferred with medical men in advance, who counselled him regarding prophylactics; that he supplied himself with them, but had no occasion for their use. It was prophesied that after these swamps were drained

---

* Parkes.

malarial disease would occur, but in a letter, dated May 2, 1891, Colonel Kreamer says, "Residents living on the reclaimed lands are exempt from chills and fever. Settlers on these lands, who suffered from chills and fever annually in Kansas, have had no recurrence from this malady since locating in Florida;" yet during all of these years they have continued to drink marsh-water. Rohé* says that, in his opinion, "the instances in which malarial fevers are due to drinking-water are very rare."

The most puzzling, and at the same time the most ridiculous, disagreement exists among analysts of water. Dr. F. L. Phipson † says we possess no positive data whereby to condemn a sample of drinking-water, and cannot possess such data until physiological experiments have been made to prove when and why a given kind of water is bad or good. Our text-books for the last thirty years—English and foreign—have stated when the total impurities of a water should condemn it, but this is a mere assertion that has been copied from one book to another. Mr. Charles Ekin,‡ F.C.S., on Water Analysis, says, "A fair trial of the different processes leads to the conclusion that all are absolutely worthless, so far as distinguishing between organic matter that is innoxious is concerned." "As giving any indication of the wholesomeness of a water they are useless." Fox § relates that the water of the same well being analyzed by five different chemists of repute, the first opines that the water is of good quality; the second, that it is surface-water and is bad; the third, that it contains so much organic matter as to be unfit for drinking; the fourth, that it is a perfectly pure water; the fifth, that it is unusually pure. Fox tells also (p. 179) of two waters from neighboring

---

* Hygiene, 1891.
† *Chemical News*, 1879.
‡ *Journal Franklin Institute*, 1880.
§ Sanitary Examination of Water.

pumps which were examined by a health officer; one was pronounced pure and the other quite unfit for use. He says the confidence of the people was somewhat shaken in their health officer when it was discovered that "both pumps derived their water from one and the same well." Mr. Reuben Haines * says, "The most diverse opinions have occasionally been expressed in regard to the wholesomeness of the water-supply of a city by chemists of established reputation."

Dr. J. A. Tanner,† in a lecture on water-analysis, gives a table of the analyses of twenty specimens of artificially-prepared waters containing sewage from different sources, dejecta from typhoid fever, and black-vomit from yellow-fever patients. Three of these samples were pronounced good by the three different processes of examination,—the combustion, the ammonia, and the permanganate. Eight of the twenty specimens were condemned, while in nine there was no agreement. Lake Drummond, which has always been in high repute among sea-captains for long voyages, and thereby proved beyond a doubt to be good, was pronounced by the three methods "impure," "foul," "impure." Dr. Tanner says, "Viewing the subject impartially, it seems we must conclude that such examinations are as apt to condemn a good water as they are to commend it, and to commend an impure water when they should condemn it; and that we know of no chemical method by which *the ethereal-like substance* causing disease when in water can be recognized at present. We are at sea, with an unreliable compass to guide us." Professor Mallet ‡ says it is not possible to decide on the wholesomeness of a water by the use of any of the processes of examination for organic matter.

---

* *Journal Franklin Institute*, 1882.
† *Sanitarian*, 1888.
‡ National Board of Health Report, 1882.

Dr. C. G. Currier * says chemistry "affords us no sufficient test of the freedom of a water from the infectious principles which cause serious disease or which lessen the sum-total of the vital forces." In a report on the sanitary examination of potable waters, Mr. Elwyn Waller † concludes that, in reviewing the methods of water-analysis, our knowledge of the whole matter falls very short of what is desirable. "The means of determining, even approximately, the safety of a water are at present extremely crude and unsatisfactory."

Dr. Macé ‡ says that chemical analysis furnishes little information respecting the character of a water, and often none at all. The inferior beings that breed and feed on azotized matters in water do no harm.

The bacteriological examinations of water give no more satisfactory results. Dr. Link § showed that "the attempt to put forward bacteriological examination as a decisive criterion for the character of a water is devoid of a satisfactory basis." Dr. E. K. Dunham ‖ concludes "that the bacteriological examination of water cannot, save under exceptional circumstances, pronounce a direct verdict as to the sanitary value of that water." Dr. T. M. Prudden ¶ says, "It is already known that in some cases the results of chemical and biological analyses do not coincide." Louis Parkes ** says, "Chemical analysis is powerless to deal with those cases of infinitesimal pollution of a pure water." "Cultivation tests are equally powerless to cope with such cases." Though chemical analysis can in the majority of cases determine the amount of organic pollution, "there is no possibility of ascertaining whether the water thus polluted is potent for evil, or whether it may not be entirely

---

* *American Journal Medical Sciences.* † Ibid., 1883.
‡ Annales d'Hygiène, 1888. § *Chemical News*, 1886.
‖ *Medical Record*, 1889. ¶ Ibid., 1887. ** *Practitioner*, 1887.

harmless." "The only way of ascertaining the probable effects on the human system of drinking such water is for the operator to perform the experiment on his own person." C. E. Cassal* says that " in the present state of knowledge no chemical analysis would justify the assertion that a water was likely to cause a particular disease," and " no process of examination whatever" will prove the noxious character of a water. " The counting of micro-organisms in some hands, even those of eminent persons, yields results which are wholly ludicrous." C. E. Pellew † says, with our present knowledge a satisfactory microscopic examination of water is " hardly possible, even for one thoroughly skilled in such investigation," and "the question of the purity or impurity of a water cannot be satisfactorily settled by bacteriological tests alone." Yet with these humiliating confessions, publicly made, in the one hand, the sanitarians in official capacity— and with an impudence which passes all comprehension— continue to offer with the other their farcical analyses of water, which are often made at the general expense, and solemnly parade them before the people as if they possessed a scientific value. On the strength of these analyses they have not only destroyed many valuable public water-supplies, but they have arbitrarily invaded the premises of private citizens, and ordered the closure of wells which have been proved by scores of years of experience to be healthful water-supplies.

We are forced to the conclusion that there are no better tests for drinking-water than our first parents possessed; and that the instincts, taste, and experience of a committee of farmers, mechanics, or intelligent housewives are more to be depended on in the selection of a public water-supply than are the so-called scientific tests of the sanitarians.

---

* *British Medical Journal*, November 5, 1891.

† Manual of Practical and Physical Chemistry, 1892.

There is no better way to ascertain the effect of so-called polluted water on the human system than by studying its influence on individuals and on large bodies of people. Emerich * drank for fourteen days from one to two pints of sewer-water which was disgusting to the eye, and which chemical analysis showed was highly polluted; it contained refuse of blisters from human skin, pieces of salad, animal hairs, particles of excrement, and much other non-appetizing material. (*Nicht gerade appetitliche Dinge mit sich.*) He drank this water while he was suffering from intestinal catarrh, and it had no unfavorable influence on the disease. Again, when he had a severe gastro-enteritis, though the disease had its course, it subsided while he continued to drink it. He gave the same to two of his patients (with their consent) who were suffering from diseases of the digestive organs; the diseases were not aggravated.

A large part—seven-eighths—of the people of London are supplied with water from the Thames and Lea. Only since 1852 has the law compelled the companies to filter it. It is declared that filtering is not competent to relieve it of its dangerous organic impurities.

The sanitary literature of England is afire with accounts of the pollution of the drinking-water of London. In 1867 † Professor Frankland reports that it has received great quantities of putrescent matter; some of it is totally unfit for domestic use. In 1869 there is unquestioned evidence that the water is infected with animal matter; in 1870 it is excessively nasty as to taste and smell, and the health of seamen and others on shipboard is in danger. In 1872 it is in a very filthy condition. In 1874 the water of one of the companies is so nasty that it resembles "pea-soup." In 1875 seven out of eight companies in London are furnishing water from pestilent sources. It is still bad in 1882, and a

---

* Handbuch der Hygeine. † London *Lancet.*

new idea is brought out; Sanitary Science has made another "gigantic stride;" the chief danger from a water is now not from its usual, but its accidental, pollution. In 1885 * "it is difficult to keep from despair" on account of the disgusting pollution of the Thames. "It is a cesspool throughout its tidal regions," and in 1887 the Thames is in a "horrible and dangerous condition," and in the winter "it is fouler than ever." This year "the owners of house-boats, steam-launches, and other craft drain their sewage into the chief drinking-water supply of London."

Mr. G. Phillips Bevan † at a meeting of the "Balloon Society," in a lecture on the Thames and public health, said that instead of getting rid of our sewage, "it is churned backwards and forwards on us every day." When the essay was ended, a resolution was unanimously passed, "that in the opinion of this meeting the system adopted by the people of Kingston, Richmond, and adjoining districts, of drinking the water into which they put their sewage, is filthy, foul, and abominable, and barbarous in the extreme, and calls loudly for an immediate remedy." Dr. G. Vivian Poore,‡ President of the Section on Sanitary Science, at the twelfth meeting of the Sanitary Institute, stated that the public was becoming alive to the fact that causes which poisoned the surface-wells of London were equally poisoning the Thames and the other sources of London water. "No thinking being could feel easy about the London water-supply." For it is impossible for the Thames to be pure. The whole of the sewage of all towns between Gloucester and London was emptied into it, and the greatest portion was drunk by the inhabitants of the metropolis. The *Medical Press* § says, "For years the condition of the Thames and Lea in every way proved the unfitness of these

---

* London *Lancet.*
† *Builder*, August 2, 1884.
‡ *Sanitary Record*, 1890.
§ April, 1891.

rivers for drinking by the people of London," and Dr. Blaxall reports in the same journal (May, 1891), that the Thames drinking-water is fouled by excrementitious matter.

A most remarkable coincidence, that during these forty years that the millions of people in London have been drinking this polluted water, the death-rate has never permanently increased, and since 1880 has steadily declined, and especially from the class of diseases that polluted water is said to generate.

London must have lost its sense of humor entirely, for there is no record that the town burst out laughing when it read in the *Medical Times and Gazette* of July 27, 1867, "It is notorious that just during the months when the organic constituents of Thames water are at their highest, diarrhœa is at its lowest; and *vice versa.*" In the *Chemical News*,* Professor Tidy, in his report to the society of medical officers of health, says, "I have diligently compared and considered the death-rates, and also, as far as possible, the causes of death, in different parts of the metropolis supplied by the Thames water, the Lea, and the water from the chalk-wells of the Kent Company, respectively. I have failed to discover any difference worth noting in the death-rate, or any evidence whatever that any special class of diseases has been prevalent from drinking the water of the Thames and Lea, or absent from the use of chalk-water; indeed, what difference there exists is in favor of the Thames and Lea waters over that of the chalk-wells." Professor Tidy † says that in 1879 the water in the Thames showed a larger amount of organic matter on account of floods; "yet, notwithstanding this, the death-rate in London for 1879 is the lowest on record." Professor Tidy says that while he admitted that disease might have been produced by impure or polluted water, it is seen from an examination of statis-

---

* 1878.   † *Chemical News*, 1880.

tics that the death-rate of towns in which water is obtained from wells is practically identical with that in towns supplied by rivers; and that in London, as regards mortality, there is very little to choose between districts with the water from the chalk-wells and those supplied by river-water. In the *Chemical News*, 1883, Messrs. Crookes, Odling, and Tidy say, "Taking a series of years, and relying solely upon the water analyses supplied to the Registrar-General, it does with singular perversity happen that the years in which the metropolitan rate of mortality is exceptionally high are the years in which the proportion of organic impurities in the water is relatively low; while the years in which the metropolitan rate of mortality is exceptionally low are the years in which the proportion of organic impurities in the water is relatively high." In the years 1869–70–71, in each of which there was the exceptionally high rate of mortality of over twenty-four per thousand, the mean proportion of organic impurity in the Thames water was represented by nine hundred and twelve; while in 1872–80–81, in each of which there was the exceptionally low rate of mortality of considerably less than twenty-two per thousand, the mean proportion of organic impurity in the Thames was represented by the number eleven hundred and seventy-one; the proportion in 1868, with its mortality of 23.5, being represented by the number one thousand.

M. Hueppe, in a report* to the International Hygienic Congress at Vienna, said that after comparing the figures of *morbidité* and mortality by typhoid fever and cholera in cities provided with *eau potable* and sufficient sewerage, with those from cities not so provided, it was impossible to decide *oui ou non* whether drinking-water had any influence on the propagation of contagious diseases.

In 1839 the city of Boston petitioned the General Court

---

* Annales d'Hygiene.

for power * to introduce water. Though no positive cases of disease were adduced as having been caused by the use of native water, notwithstanding its proverbially bad reputation, the physicians of the town seemed to concur that the introduction of a plentiful supply of pure soft water would be a good thing for the public health. In 1844,† Dr. Walter Channing made a plea for pure water in Boston. He says it is not pure now and is unfit for use; it is insufficient in quantity, and great inconvenience and danger are the results. "An abundant supply," he adds, "promotes health and longevity and as surely tends to diminish or prevent pauperism." In Boston, previous to 1848,‡ wells and privies were about equally numerous and in close proximity; but the suppression of wells had little to do in lessening typhoid fever. "Our Lowell correspondent speaks of a well used by at least a hundred families, containing fifty-two grains of inorganic matter and twenty-five grains of organic residue to the gallon; and yet the people using it seem to be even less liable to typhoid fever than those using water of a better quality." "Old Boston previous to 1848 was riddled with wells and privies side by side, all over its limited and very crowded territory. The water must have been continually charged with the products of decomposition." The report says here is a test on a grand scale. "There is a diminution of typhoid fever, but in no striking degree."

In the eighth report of the Massachusetts Board of Health is an account of an investigation into the causes of diphtheria in Gloucester in 1876–77. The wells were suspected and the waters of thirty-four of them were examined. They were so filthy that Professor Nichols, who analyzed them, exclaimed, with surprise, "Do people actually drink

---

* *Boston Medical and Surgical Journal*, 1839.　　† Ibid.
‡ Second Report Massachusetts Board of Health.

these waters?" The ammonia in them varied from 0.0037 to 2.35; the albuminoid ammonia from 0.0040 to 0.0723; inorganic residue from 3.20 to 198.9; organic residue from 0.6 to 10.7; total solid residue from 3.48 to 205.4; chlorine from 0.70 to 55.04 in 100,000 parts.

The professor says that no one of these waters is fit for drinking, and some show evidences of very great pollution. To eighteen of them, it is distinctly said, no illness could be referred. In a house supplied by one there has been much illness, "a large part of which, however, can be traced to very wet cellars and inability to provide the comforts of life." In seven households there had been illness where the water had been used from seven of these wells; of the remainder, no mention is made of their having caused sickness. "Of the worst specimens (Nos. 22 and 6), both as valuable for manure as the sewage taken from the Pittsfield sewers, No. 22 is from a well which has been dug for a century and used by three families, who are always well." "A family of robust children, types of health, have been brought up on No. 6, without any illness that could be fairly traced to the use of the water." "As regards diphtheria in connection with these thirty-four specimens of water, the worst cases occurred where the best water was used." The conclusion arrived at was that the water did not cause the epidemic of diphtheria in Gloucester in 1876–77.

"The thirty-four samples of well-water probably represent fairly the wells used for drinking purposes in Gloucester." "No. 7 is worth twice as much for manure as ordinary sewage," but "the use of the water has not been shown to have caused any illness."

The condition of fifteen wells that were near cemeteries in different parts of Massachusetts is detailed in the sixth report of the Massachusetts Board of Health. Five of these had drains and privies within fifty feet; they were from thirty to forty, and one was a hundred years old. Of none was there any mistrust that the water had caused disease. Only one of the fifteen was suspected. These waters when analyzed showed great pollution,—chlorine as high as 13.20 in 100,000 parts, and total solids as high as 138.60. It will be remembered that both the well- and public-supply of Newport, also its ice, were grossly polluted; yet the death-rate was not augmented thereby.

In 1881 Boston's water was examined by Professor Edes, who said, " It contains pollution seventy per cent. above the limit of health;" he found it "abounding in decomposing organic matter." Some of it, Dr. Barnes says, was the color of good coffee. In the tank in his own house, Dr. Barnes found a mud deposit three-fourths of an inch deep, that was very offensive when removed. Dr. Minot considered "the water-supply of Boston a disgrace to the city, and a source of danger to the health of the community." Professor Leeds, who examined the water of towns on the Hudson and elsewhere, said that Boston's supply was the worst of all, and was "absolutely dangerous." Dr. Bowditch had to give up drinking it, or bathing in it, without filtration. Yet the Massachusetts Board of Health * say, " For months a very large portion of the water-supply of the city was wholly unfit for any household use." It adds, " Nor does the use of this water appear to have had a noticeable effect upon the death-rate of the city or upon the health of the inhabitants." If we seek for something more than this general avowal in the vital statistics of Boston, we find that the average annual proportion of zymotic to total mortality in that city for the preceding ten years was 28.40 per cent.; and the proportion of infant to total mortality for the same time was 40.75 per cent.; while during the year that the three hundred and fifty thousand people of Boston were

---

* 1881.

drinking water polluted seventy per cent. above the healthy standard, "abounding in decomposing organic matter," "absolutely dangerous," and "wholly unfit for any household use," the proportion of zymotic mortality declined to 26.87, and the infant mortality (a still more sensitive test) declined to 36.75 per cent.

It must not be forgotten that if any citizen of Boston had possessed a well supplying such water, the Board of Health would have forcibly entered his premises and closed it. If he had resisted the invasion he would have been arrested, fined, and imprisoned as an enemy of the public health.

The National Board of Health [*] reports on the sanitary condition of Baltimore. Though pumps have been mostly discarded in that city, yet water from them is used in the first district on two hundred premises. The report adds, "The comparative immunity of the people from disease under the most trying sanitary conditions is extremely remarkable, but should not lead us to the conclusion that filth and bad water are conducive to health." Seven families used water from an old pump which must have been polluted by a privy; but, in spite of this, "no sickness or death" had occurred for a year. At another point, eight, at another, four, and at another, three families were all using "bad" or "very bad" water from wells close to privies in bad condition; yet the inspectors could not find that a single case of·disease, except one of consumption, had occurred for a twelvemonth in any of these families.

In the Connecticut Board of Health Report for 1885 is a paper by Dr. Wolfe on "Sanitary Examination of Drinking-Water." The Bridgeport water is polluted by animal and vegetable matter; New Haven's water is polluted by vegetable decomposition. The supply of Hartford coming from Brandy Brook is so contaminated that it is unfit for public

---

[*] 1880.

use. The people of Hartford are drinking the drainage from one of the filthiest swamps in New England. Such a water is a "*vera causa* of diarrhœa, dysentery, and malarious diseases." New Britain's supply is similar to that of Hartford. Yet that year, in the same book, Hartford reports, in "Health of Towns," malaria as decreasing, and makes no mention of diarrhœa and dysentery; New Britain makes a similar report; in New Haven the malaria is less than the average, and dysentery here in an epidemic form had not been known for thirty years.

Surgeon McKee, U.S.A., reports [*] that the water in the tank at the garrison in San Francisco, which contains twenty-seven thousand gallons, and which was supposed to be secure from pollution, had for some weeks a bad odor and a sickening, nauseous taste. An investigation disclosed that a polecat had crawled into the tank through the overflow-pipe and had been drowned. The carcass was in an advanced stage of decomposition, and must have been there some months, for fragments of the rotting body were floating in all directions. The water had been drunk by eighty people,—men, women, and children. Dr. McKee looked for the development of some "filth-disease," but no serious results followed. After the tank was cleaned some persons complained of "a sense of goneness" and a loss of appetite! The doctor says this occurrence is opposed to the theory that diseases originate in filth; and he asks, "Might not this have been almost too acute and overpowering to have attained a lodgement, when a milder poison would have been more insinuating and permanent?"

In the seventh report of the New York State Board of Health is the story of the Jamestown water. It is supplied partly by driven-wells and partly from the outlet of Chautauqua Lake, which is contaminated by sewage. The board

---

[*] *New York Medical Journal*, November 3, 1883.

reports that it is "a menace, and one affecting the life and health of the inhabitants." Of seven samples taken in December, 1886, all were grossly polluted. Nine affidavits declared that Jamestown's water was unfit for use; five of these were made by doctors. One physician testified that he had seen it "loaded with decaying and putrid animal and vegetable matters," and that it was offensive to sight and smell. Many other citizens say it is not acceptable to the community. Only one physician attributes any disease to its use.

The water company had a good deal of property at stake, and they made a gallant fight to protect it. Seven physicians swear that since the introduction of this water there has been a less amount of zymotic disease, and that it has not and does not endanger the security of life and health where it is used. Mr. Hall swears that he never heard or knew of any disease, illness, or death caused by it; others testify to the same effect. Statistics of deaths which occurred at Jamestown before and after its use are shown, which prove the diminution of zymotic ailments. The water-works were established in 1882.

In 1883 there were 11 deaths from typhoid fever.
" 1884 "   "   3   "   "   "   "
" 1885 "   "   2   "   "   "   "
" 1886 "   "   0   "   to August 15.

In 1883 there were 4 deaths from scarlet fever.
" 1884 "   "   2   "   "   "   "
" 1885 "   "   1   "   "   "   "
" 1886 "   "   0   "   "   "   "

The records of vital statistics and other evidence showed that "while the city has been rapidly increasing in population, the number of deaths from this class of (zymotic) diseases has been no less rapidly diminishing."

A sanitary inspection of the city of Allegheny in 1885 * found the water-supply grossly polluted by slaughter-houses, bone-boiling and soap-factories, besides "six sewers were found discharging their noisome contents, under the burning September sun, directly into the river above the mouth of the influent-pipe." There are also piles of garbage "festering on the river-bank." "The committee returned from this tour filled with amazement at the fact that with all these multiplied and horrible sources of pollution, the water supplied the city of Allegheny should, except when thick during freshets, be a not unpalatable and, so far as indicated by the average low rate of mortality in the city, a not unwholesome beverage."

In an official report,† Elwyn Waller, chemist, of New York City, says of the Croton water that it has a bad taste and odor, described as filthy; a green-grass color; and, on standing, a green-grass scum rises to the surface. The cause was attributed to decomposing fish, sewage, and other poisonous material. Although it continued in this filthy state for six weeks, no cases of illness were assigned to it. The same condition was present in 1859, in 1873, and in 1874. When these impurities were most abundant and most unpleasant to eye and taste, the water was not unwholesome.

The waters of sixty other places in different parts of the country were examined; all of them stank and were disagreeable to the taste and disgusting to the eye, and the odor was described as "fishy," "woody," "like cucumbers," "like dead fish," "musty." One had the odor of a horse-pond, three were distinctly classed of a pig-sty odor, and some had the smell of putrid fish. The bottoms of some of the reservoirs were filled with decaying stumps and vegetables. This state was present generally in summer, and

---

* Pennsylvania Board of Health Report, 1885. † 1881.

lasted from six weeks to five months. In some of these places the water was so filthy that it had to be abandoned. Here were sixty cities, with populations varying from five thousand to one million two hundred thousand, drinking this filthy water, which is classed as "impure" and "dangerous" by the sanitarians; yet this report concludes that the filthy conditions of these waters "never as yet proved deleterious to the health of any community where they have occurred, as far as any records go."

Dr. Griffin [*] says, "Several years ago a former incumbent of the Department of Health caused to be removed all pumps from the city, in the hope that the number of cases of typhoid annually occurring would be diminished. It has been *supposed* that the saturated soil had polluted the water, and that the drinking from the wells and pumps during the warm summer weather was in a great measure responsible for the increasing number of typhoid cases which made their appearance in the fall. Their removal, however, has not exercised any apparent effect in diminishing the number of those attacked, as the average has remained about the same during a series of years after, as before, *this sanitary improvement.*" (Italics ours.)

Bouchardat [†] says the well-waters of old cities like Paris contain the products of organic decomposition, and, besides, organic matter incompletely decomposed. They are so disagreeable that they cannot be used as potable waters; but the bakers and brewers use them, and pretend that they favor *panification* and the fabrication of beer. In spite of the repugnance one feels at their use, he says, "No fact has come to *his* knowledge to prove that they are noxious." Bouchardat, in his "Conclusions," p. 197, designates as potable waters all of those natural waters which are agreeable to drink; and declares that the only way to pronounce on their

---

[*] Report of Brooklyn Board of Health, 1889.   [†] Traité d'Hygiène.

salubrity is to observe their effect on the health of people who have long used them.

The history of the Croton water-supply and its relation to the public health of New York is without a parallel in the pages of sanitary literature. In 1885 it received * the drainage of 20,000 persons, 1879 dwellings, 602 barns, 9453 cows, 1224 horses, 1501 pigs, and, besides, grist-, saw-, and cider-mills, condensed-milk factories, cemeteries, and slaughter-houses drained into it. During freshets, which often overflow the river-banks, "all privy-vaults within that range are overflowed and washed out" into the stream. In 1888 it was shown that 25,000 people and 33,000 domestic animals drain into this supply, besides the drainage from the roads. The cemeteries add to the danger. The editor of the *Sanitarian* says, "The subsoil currents which penetrate the coffins take up everything that is soluble during the process of putrefaction and convey it into the water-courses at all seasons." "What," he asks, "are the sixty-two deaths of the Park Place disaster, in comparison with the daily stream of disease and death among the million and a half of people of the city consequent upon neglect to protect the water-supply?" "*The recent chemical analyses have startled the City Board of Health by the discovery of nitrites,—the indubitable evidence of animal putrefaction.*" Then follow more italics. "The putrefaction of human dead bodies on the Croton Valley water-shed sufficiently accounts for the presence of nitrites in the water, independent of the other six thousand one hundred and forty-six special nuisances and the surface filth of thirty-three thousand domestic animals."

"Will the people of New York be longer content to drink such an infusion?"

A special inspection made by the Board of Health in

---

* *Sanitarian*, September, 1891.

August, 1891, showed that the Croton water was grossly polluted by stables, privies, kitchen-drains, saloons, cesspools, factories, and pig-pens. Chemical analysis showed that "all of the water, therefore, below Harlem is contaminated."

The danger was so imminent that the pulpit took up the cry. "Whence this poisoned water?" exclaimed the Rev. Dr. Dixon in a sermon. The city was threatened with an epidemic of diphtheria; it had already broken out in his own congregation. "The water-supply of the city has been declared by experts to be unfit for human use. To let this fountain of life become a pool of filth marks the outermost limits of the pendulum of social degradation." The official report in the newspapers had the heading, "No wonder zymotic diseases prevail!" We seek in the vital statistics of New York City for the effect of this poisoned water. More than one million six hundred thousand people were drinking it. The culmination of filthiness and danger was reached in the summer of 1891. Throughout the performance of this comedy, the solemnity of the New York Board of Health never relaxed for a moment. Without a blush, or a word of explanation, it issued reports for the thirteen weeks ending September 12, 1891, showing that the number of deaths

From diphtheria was 281
" typhoid fever was. 107
" diarrhœal diseases was. 2527
" malarial diseases was 66

The average number of deaths for the corresponding weeks for the previous ten years

From diphtheria was 392.1
" typhoid fever was. 120.7
" diarrhœal diseases was 3373.9
" malarial fevers was 124.0

Sanitary Science had condemned this water; it had recevied the censures of the press and the maledictions of the clergy; yet, in New York City, zymotic disease had steadily declined since 1885, and during the last summer was the lowest on record, and then the decline was notable in those particular diseases of that class which the sanitarians say are especially caused by such water as the people were drinking. After their terror had been raised to the highest pitch respecting the drinking-water, the Board of Health began to relax its rigors and to soothe their fears. As the acme of the paroxysm from which the patient was suffering had been attained by successive steps, so the antidote must be applied in graduated doses. From day to day the condition of the water was reported in the papers; it was now a little better; again there was a slight relapse; but, on the whole, a steady improvement was announced, and soon notice was served on the people that it was safe to drink the water,— the nitrites had disappeared. It was not claimed that a single one of the dreadful sources of pollution had been removed; the dead bodies were continually accumulating in the graveyards; "the subsoil currents which penetrate the coffins and take up everything that is soluble during the process of putrefaction" were still flowing into the water-supply of New York City; it was still taking the overflow of privies, drains, stables, kitchens, and saloons. But the people were told that the Board of Health were hard at work examining this water; they were evaporating it to dryness; they were redissolving the remaining solids; they were distilling it in alembics; they were torturing it with reagents in test-tubes; they were straining their eyes peering at it through microscopes. And now, as if the mockery was not quite complete, about one month after the proclamation that the water had become purified, typhoid fever broke out in New York City to such an extent that it was little short of an epidemic. "Dr. Edson * said there had not been so

---

* New York *Sun*, September 22, 1891.

many cases of typhoid fever in the city since 1889, when there was a small epidemic."

We must search for the cause of this fever. It is not in the water, for that has been undefiled for a whole month. We must probe deeper. It happened, fortunately, that the New York Board of Health contained a sanitarian of prodigious discernment. He sounded the depths of Sanitary Science to divine the cause of this fever. We read in the New York *Sun* of September 22, 1891, "One reason why there is so much of the disease about just now is, according to Dr. Edson, that boys go in bathing off the docks near where the sewers empty into the river." The people of New York, lay and professional, seem to have received this solution of the mystery with infantine docility, and we will not presume to call it in question. Presuming that the essays on contaminated water will be as copious in the future as they have been in the past, let us hope that some enlightened sanitarian will condescend to explain how boys going in bathing off the dock can cause an epidemic of typhoid fever among adult men and women.

The Connecticut Board of Health reports (Second Report) that in that State ice is, in many instances, cut from polluted ponds; "but no cases of disease were traced to the ice."

The New York State Board of Health for 1886 says that Onondaga Lake receives daily twenty-five million gallons of sewage from Syracuse. This lake is a great ice-field. Biologically, all of the specimens of water and ice taken from this lake were ranked as "unsafe waters for potable purposes." But no evidence is offered to show that it had ever caused any disease. The Pittsfield ice-field * is contaminated by refuse from dwellings and two factories. The water-closets of one of these, which are used by one hun-

---

* Seventh Report Massachusetts Board of Health.

dred and fifty persons, empty into the lake, from which ice is cut. "The water is sometimes foul in appearance." One specimen contained ammonia 0.0140 and albuminoid ammonia 0.0348 to each gallon. No disease is reported from the use of the ice.

In *Popular Science Monthly*, vol. xxxii., Dr. Prudden says a considerable quantity of Hudson River ice is cut just below Albany, where the stream is so contaminated as to be absolutely filthy; typhoid here occurs frequently while ice-cutting is going on; and "the complacency with which we swallow the frozen filth affords a spectacle of self-abasement as melancholy as it is disgusting."

In *Medical Record*, 1887, Dr. Prudden says that most of the ice for New York and Brooklyn is cut on the Hudson River between Troy and Poughkeepsie. Troy empties daily eight million gallons of sewage into the Hudson, already charged with contributions from Cohoes and Lansingburg, to say nothing of the impurities which the Mohawk brings down. Albany furnishes millions of gallons daily of sewage; and the river takes the refuse of thirty thousand inhabitants of Hudson, Catskill, and Kingston besides. The Hudson River ice, "thoroughly contaminated with sewage at the upper part of the ice-fields," is further defiled by smaller towns on its borders. The two bacterial diseases—blood-poisoning and typhoid fever—are almost constantly present in Albany. Dr. Prudden is informed that "there is no systematic disinfection of the typhoid discharges, which, therefore, enter the sewers with their myriads of bacteria in a living condition."

Here, he says, they proliferate for an indefinite period, so long as sewage is present. They will live in ice for at least one hundred and three days, perhaps longer; and when thawed out, they go on proliferating just as they did before being frozen. On this river are cut three million tons of ice, supplying not far from three million people. The ordinary number of bacteria is so great that a glass of melted ice would contain five hundred thousand, which is forty times as many as are allowed for wholesome water. Dr. Prudden estimates that the excreta of at least fifty typhoid patients pass directly into the river from Albany between December and March. At one point, where twenty-five thousand tons of ice are stored, the water shows from twenty thousand to fifty thousand bacteria per cubic centimetre.

The masterly presentment of this indictment against the Hudson River ice, and the adroit manner with which it has been surrounded with a plexus of circumstantial evidence, seem to render its conviction sure. At the critical moment, however, Dr. Prudden surprises us by allowing it to slip through his fingers through an *alibi* which the elder Mr. Weller would have rejoiced to obtain as an adornment for his extempore discourses on criminal jurisprudence.

He examined one hundred and fifty-three specimens of Hudson River ice. "It will naturally be asked," he says, "if typhoid bacilli have ever been detected in the ice? They have not," he replies. Further, although it contains *staphylococcus pyogenes*, its application to wounds does not produce blood-poisoning. He anticipates that "those calling themselves somewhat ostentatiously common-sense people" may apply the "ancestor or experience argument," to show that there is no danger.

It is to the bar of common sense and experience that test-tube and microscope must come at last for final judgment. As soon as common-sense people have recovered from their consternation about "swallowing the frozen filth," they will ask why the typhoid fever bacilli were not found in any of the one hundred and fifty-three specimens, if they are poured out of Albany and Troy in the manner described, and go on proliferating so long as there is sewage, and are not destroyed by the freezing process? They will ask why, if this ice is continually receiving an increased amount of typhoid poison, and the consumption of the ice is increasing enormously every year, should the cases of fever be steadily declining in number? The free consumption of ice begins about the middle of April, attains its height by the middle of June, and drops off in September. As soon as they are over their fright, common-sense people will ask why it is that when the most of this "frozen filth" holding the typhoid poison is consumed, fever is comparatively rare; and why it is that it is nearly always rife—when it occurs at all—some six weeks after the great consumption of ice has ceased. Common-sense people want to know if the law in Sanitary Science is invariable that two and two make five, and if the less contains the greater? They will ask why the thousands of people who navigate this river in canal-boats and other vessels just in the months when typhoid fever is usually at its height, and who take their drinking-water from it, are much less subject to the disease than those not so exposed? They may go a step further and inquire if the cause of true science would not be better served and the general welfare promoted if scientific men, instead of exciting panics in the public mind, were to exert themselves to allay unnecessary fears of the people.

In the next volume (p. 32) of the *Medical Record* is a Paris letter which relates that Dr. Thoinet " had made laborious researches to discover the typhoid bacillus in the foul water of the Seine, but found that the pure water of the Vanne contains the bacilli in much larger numbers;" and that M. Pouchet, in a paper read before the Academy of Medicine, says, " The growth of the typhoid bacillus is arrested in media rich with organic matter, whatever be its nature." " It is preserved and developed much better in clean than in foul water."

The Massachusetts Board of Health * (1889) reports an

---

* Report, 1889.

investigation into the ice-supplies of that State. Thirty places reported ice taken from polluted sources. The third question which was put by the board to its correspondents was, " Could any cases of illness be assigned to the polluted ice?" The answers to this question were, " No."

Dr. A. W. Nichols reports an outbreak of sickness at Rye Beach, which was ascribed to ice. This was partaken of by five hundred individuals: twenty-six manifested "grave, continued, and characteristic symptoms" of intestinal irritation, diarrhœa, nausea, giddiness, etc. The sickness was confined to one hotel; another, an eighth of a mile away, and the neighboring cottages, escaped. The milk, water, food, and sewerage in the invaded hotel were excellent; and suspicion fell on the ice, which had been cut from a pond that received sawdust and other refuse. The report does not say whether the other hotel and the cottages were supplied with the same ice, or whether the hotel where the sickness occurred had obtained its ice from the like source years before; neither does it say how long those who were attacked had been using it previous to their seizure.

The report makes this statement,—a *remarkable* one, truly, when we consider how susceptible children are to intestinal irritants,—" It is worthy of remark that no person under the age of ten years was known to be affected by the impure ice." The analysis of the water in the pond showed that it was not as impure as many waters that people habitually drink. The symptoms and course of the disease proved that it was non-specific. It resembled more those mysterious and, as yet, unexplained seizures which in various countries have followed the ingestion of certain animal foods. But so much noise had been made about ice, with no results whatever, that the reformers grasped this occurrence and hailed it as a triumph of Sanitary Science.

Overwhelmed by evidence of their own collecting that the waters which they had classified as " filthy," " impure," and "dangerous" had no power to impair in any way the health of the millions who had consumed them for generations, the sanitarians sought to extricate themselves from the dilemma by the aid of the germ. As usual, they made no researches, but they displayed their powers of improvisation to a degree that had before this had no parallel. After all, they said, it was not the filthy water that caused the disease; it was the germ that found in filthy water, as in putrid solids, its *habitat* and its subsistence.

In the *Chemical News* * is recorded a series of brilliant experiments by Messrs. Crookes, Odling, and Tidy, to ascertain the necessary conditions for the existence and propagation of morbific microbes purposely introduced into water. The results were in accordance with the conclusions of other investigators,—namely, that in the struggle for existence, " the microbe forms proper to running water outgrow and starve out the introduced morbific forms." These experiments were made with the anthrax bacilli. It was found that their existence in the unfiltered water of the Thames was so brief as to be practically without danger, while in sterilized water they retained their vitality longer than in any other. In the intermediate conditions between unfiltered and sterilized, the duration of their existence was in proportion to the purity of the water. As has been seen on a previous page, the experiments of Thoinet and Pouchet with the typhoid bacillus corroborate those of Crooks, Odling, and Tidy.

Jules Arnould,† after reviewing the labors of the bacteriologists to ascertain the power of the different microbes to penetrate the soil and defile water, concludes: 1st. That bacteria of any kind pass with difficulty through the earth, even if it be permeable, either from above downwards or

---

* Vol. liv., 1886.
† " L'Eau et les Bacteries," *Revue d'Hygiène*.

horizontally. 2d. That water, as it presents itself in nature, even if rich in organic matters, is *antipathique* to pathogenic bacteria.

A few months later, Kraus* recorded his experiments with the typhoid, cholera, and anthrax bacilli. He criticises the methods of Meade-Bolton and others, who, in their experiments with pathogenic bacteria, sowed these in sterilized water at a temperature of from 68° to 71°, and even 96° F., a condition in which drinking-water is never found. In his experiments Kraus took three samples of water at a temperature of 51° F. The typhoid bacilli disappeared more rapidly in the more filthy specimens; and as soon as the water bacteria appeared in perceptible numbers the former were destroyed, so that after five days they were no longer to be found. The second day the cholera bacilli could not be seen, and the anthrax had disappeared by the third. Although the pathogenic bacteria at the beginning of the experiments were from one to two thousand times more numerous than the water bacteria, the latter soon got the advantage and overcame the former. The experiments proved that there could be no development of any of these three forms of pathogenic bacteria in water of the temperature at which it is generally used for drinking, but rather a speedy destruction; yet, in spite of the low temperature, the ordinary water bacteria proliferated rapidly. Indeed, Bolton's experiments had shown that the tendency of the pathogenic bacteria, even in water at from 68° to 72° F., was to diminish and not to increase.

In the *Archiv für Hygiene*, 1890, Dr. Justin Karlinski records his observations with the typhoid bacillus in water as the result of a series of experiments no less brilliant and convincing than those of Crookes, Odling, and Tidy, and Kraus with the anthrax and other bacilli.

---

* *Archiv. für Hygiène*, 1887.

Dr. Karlinski found that just in proportion to the filthiness of the water and the number of putrefactive bacteria was the existence of the typhoid bacillus curtailed; that in extremely filthy water, holding large amounts of chlorine and nitrates and rich in common bacteria, the typhoid bacilli began to disappear rapidly in one hour after they were added, and at the end of forty-eight hours were no longer to be seen. In water that was comparatively free from organic pollution and putrefactive germs, they were to be found for twelve days; but here, from the moment that the water-bacteria got the upper hand, the typhoid bacilli disappeared completely from the water.

Flügge * says, "We see in all waste-waters, in putrid fluids, etc., that the facultative (disease) parasites, even when they are sown in enormous quantities, die in a few hours, or, at most, in a few days." To insure their growth all other germs must be carefully put out of the way, and the pathogenic bacteria must be maintained at a temperature of 72°. He adds, "Typhoid bacilli have never yet been demonstrated outside the human body except Pfeffer's demonstration of their presence in the dejecta of typhoid patients." "Nor," says Flügge, "can any of the pathogenic bacteria multiply in water even where the temperature is favorable, because they absolutely require a certain though small quantity of the best nutritive material." On the other hand, he says they retain their vitality in sterilized water a comparatively long time; but in unsterilized it has been proved that they rapidly disappear, and he declares that "pathogenic bacteria have never yet been demonstrated with absolute certainty in any water, unless they have been purposely sown therein."

The thorough and exhaustive experiments of Dr. Vallet †

---

* Micro-organisms.

† Le Bacille Coli-Communis.

with the typhoid bacillus in privy contents show the impossibility of contaminating wells with it through the contiguity of privies. Every one of these experiments with the typhoid bacillus showed that it not only could not multiply in privy contents, but that it speedily died when mixed with them.

There have been two outbreaks of typhoid fever in this country which have confirmed the belief in the minds of many medical men that this disease may be communicated by drinking-water. The first is related by Dr. Flint, and occurred at North Boston, New York, nearly fifty years ago. At the time Dr. Flint did not regard the water as the cause of the epidemic, though later he seems to have accepted this view. It is supposed that every one who was seized in this epidemic drank the suspected water.

In 1843 a passenger in a stage-coach arrived at North Boston, and, on account of illness, could proceed no further. He died at the tavern a few days later of typhoid fever. Before this outbreak the disease was unknown in that neighborhood. Twenty-three days after the arrival of the stranger, some members of the tavern-keeper's family were taken sick with typhoid fever; other cases soon followed; in a month more than one-half of the population, twenty-eight persons, had been afflicted, and ten died. Only one family escaped; and that one, owing to a quarrel, did not use the suspected water. Dr. Flint, at the time, attributed the exemption of this family to non-intercourse with the others. To make the chain of water-contamination complete, he says,* " Within a few years I have learned from a physician residing at the time of the epidemic in the neighborhood, that a privy used in common was in close proximity to the well." A remarkable feat of mnemonics; for most physicians would be likely to forget the particular location of a privy in the course of two-score years!

---

* 1880.

The rehearsal of the North Boston legend has so often quickened the awe of the neophyte, and strengthened the faith of the devotee, of Sanitary Science, that we desire to approach the subject with a proper degree of reverence. But there were some circumstances about this epidemic which awake scepticism. It did not reach its height until the end of a month. One after another was seized, between October 14 and December 7, so that, though all took the water simultaneously, there was an important difference of time in the seizure of the first and the last case. We must suppose a wide range for the incubative process, or else that the water was not uniformly affected. We must also believe that the typhoid germ—which is so sensitive that in the laboratory, in order to keep it alive and to induce it to multiply, it must have the cleanest quarters and the daintiest food, must be coddled at a temperature not below 75° F., and which has always shown such an aversion for bad company that it would refuse to perform its usual functions in its presence—on this occasion preserved its integrity immersed in the contents of a privy, and travelled with them to a somewhat distant well, so that in ten days it was competent to infect people who drank the water.

The disease began to decline without any abandonment of the use of the water. It may be justly said that all of the people in North Boston who were susceptible to the disease had been attacked, and that the fire went out for want of fuel. But here was a roadside well—an appurtenance of a tavern—whose water, very likely, was every day partaken of by as many people as were seized by the fever at the hamlet. The tavern was the stopping-place of the stage-coach; it was natural that travellers in wagons (there was no railroad there then), on horseback, and on foot should refresh themselves with this water, not only during the time it was poisoning the people in North Boston, but for months afterwards. There is no record that any of these travellers were seized, and that they in their turn infected other unregenerate privies, which later infected other blameless wells, which again infected other people.

Is it not quite as consonant with reason to consider that the cause of the epidemic at North Boston, like hundreds of other circumscribed outbreaks of typhoid fever, is unknown?

The other epidemic occurred in Plymouth, Pennsylvania, in 1885. The story bids fair to become a sanitary classic. Four cities and one society sent committees of medical men to investigate the cause. All arrived at the conclusion * that the cause was to be found in the contamination of the mountain-stream by the typhoid excreta of a patient. This had been poured on the ground, had been frozen, and suddenly thawed out about the last of March, and washed into this stream. It was admitted that no proliferation of the typhoid germs went on before the thaw, on account of temperature; therefore none penetrated the ground. This mountain-stream, which came leaping down over rocks, was proved by a number of analyses to be as perfectly pure as water could be; absolutely free from sediment or organic matter. So it was necessary to abandon temporarily the theory which had been proclaimed by the sanitarians, that the typhoid bacillus must have organic matter in decomposition, sewage, to subsist on or it would die. From April 12 to May 16, between one thousand and twelve hundred persons were seized, out of a population of eight thousand.

Here, also, we are obliged to admit a difference of from twelve to forty-nine days in the incubative period of different persons for the first one thousand or twelve hundred cases, or to assume that there was no uniformity in the water. A very large majority of the cases used the mountain-stream water, though, unlike the North Boston epidemic, quite a respectable number were seized who drank other

---

* Dr. Davis, *Medical Record*, vol. i., 1885.

water than the mountain-supply. If the cause of these cases was ever investigated, it was never reported. One house on Franklin Street * contained two patients, one of whom was a school-girl whose school was supplied with the mountain-water; but it is not reported whether this girl drank of it between the last of March and the date of her seizure. Dr. Taylor says that many cases exist where well-water is used; but it is found that the patients first attacked are those accustomed to drink hydrant-water away from home. It is not stated, however, whether any of these had drank it away from home, subsequently to the last of March and before their seizure. Again, different streets supplied by the same water were differently affected. Ackley Street, Dr. Taylor said, had but few sick, while Franklin Street had a great many. This is accounted for by the fact that Ackley Street is supplied by a smaller pipe, and Franklin Street by one of the mains!

The most remarkable fact of all is that the typhoid bacillus was not even looked for in the water. The works were thoroughly overhauled by the water-company, but the epidemic continued through the summer and into the autumn. This is explained by the germs getting into the privies, and through them defiling the wells from which the people drank. This throws us back upon the original theory,—that the typhoid bacillus must have sewage to subsist on; and now we must believe that during April and May it was so delicate as to require perfectly pure water at a low temperature in which to proliferate; while in the summer it parted with its chastity, and was ready to carry on the nefarious work in company with the vulgar bacteria of the privies! The mean temperature at Wilkesbarre † for the twelve days between March 30 and April 12 was 34°,

---

* *Medical News*, May 16, 1885.
† Dr. F. B. Hodge.

48°, 44°, 40°, 50°, 32°, 42°, 38°, 43°, 41°, 29°, 33°, omitting fractions. The temperature on the mountain must have been lower, and it is not unfair to assume that of the water coming down over the snow to have been not much above 32°. We leave it to the bacteriologists if it was possible for the typhoid bacillus to proliferate, or even to live, in such water.

Dr. Torrey dared to call in question the "mountain-stream" theory of the epidemic, and was sharply reproved by the press. This goaded him to investigate the subject more thoroughly, and he says that every step confirmed the conviction in his mind that the mountain-stream theory is both "inadequate and misleading." He hints that Dr. Taylor's report was "instigated" by the Plymouth Water-Company, which, having supplemented its mountain-supply with the dirty Susquehanna, had been threatened with legal proceedings to invalidate its charter, and was strongly interested in having the blame for the epidemic laid on the careless disposal of the typhoid-fever patients' excreta. Dr. Torrey does not absolve altogether the mountain-supply, but recognizes three other factors,—the filthy river-water, impure air, and contaminated milk.

If either water was at fault, the experiments of Thoinet, Pouchet, Kraus, Karlinski, Flügge, and others with the typhoid bacillus would favor the mountain-stream theory; these having shown that the purer the water, the longer is the life of the typhoid-fever germ. The acceptance of their experiments as truths, however, overturns all previous fables,—and they are numbered by thousands,—of typhoid fever being caused by dirty water, or by wells contaminated by privies.

CHAPTER VI.

The Soil.

If we continue to decompose the trinity on which the established faith of the sanitarians is grounded, and to analyze its third element, we find that what they call an impure soil is no more competent to produce disease than what they call an impure air or an impure water.

From the beginning of the sanitary excitement great stress has been laid on the foulness of the soil. The reformers said this filth had been accumulating for centuries, wherever people had congregated in cities and towns. Privies and cesspools had corrupted the earth to an indefinite and unlimited extent; it was saturated with the elements of disease, an overcharged volcano ready to burst forth at any moment and sweep us into eternity.

Beneath, as above the ground, was a continual circulation of air; this air, poisoned by the filth, was changed into mephitic gases, which were drawn into our cellars, and, sucked through floor and ceiling, they permeated our bedrooms and infected us while we slumbered and slept. For long years people went tremblingly to bed, oppressed with these terrors. In winter the danger was tenfold, and our wives and children were in the greatest peril. It was of no avail for some who were not paralyzed by fright to try to assuage these fears by calling attention to the difficulty of securing atmospheric currents above ground, and to the improbability that air confined between particles of earth could acquire power and velocity, which the most acute inventors could not obtain on the surface, in order to ventilate our dwellings.

The sanitarians said the warm-air ascending currents made

our houses so many chimneys, sucking out the ground-air * contaminated with decaying organic matter which was in the soil and with the germs of specific disease.

If the ground-air in cellars is impure,† it gives rise to typhoid fever, diarrhœa, dysentery, cholera infantum, diphtheria, sore throats, and "numberless conditions of ill health which cannot be described under any particular name." Cesspolls, filthy vaults,‡ and filth in any form afford a fertile soil for the reproduction of the typhoid-fever germs; and the " gases generated in these vaults are also loaded with the same germs, and rise into the air to be inhaled, or to pass by underground currents into the neighboring cellars, where they are sucked as in a chimney, to poison the air of the whole house. Whatever theory we hold, we come back necessarily to filth, decomposing organic matter, as the agency in the production of disease."

The Connecticut Board in 1883 reported an epidemic of typhoid fever at Waterbury, that was due " to the saturation of the soil" with filth; and it had before informed us that the typhoid-fever germ retained in the earth its vitality and infective power "for several years at least." Professor Lindsley, on " Sanitary and Unsanitary Conditions of Soil," says, " The stagnant air, contaminated with the gases of decay from the filth with which the ground is overcharged, is sucked through the cellars of houses, and house-poisoning is the result. Where else can go the noxious gases?"

In the town of Malone,§ " the filth from privies and cesspools is now contaminating the ground-air about the dwellings;" this circulates freely, passing into the cellars and rising into the houses. "The warm-air ascending currents

---

* Fifth New York State Board of Health Report.
† Massachusetts Board of Health Report, 1881.
‡ Connecticut Board of Health Report, 1883.
§ Fifth New York State Board of Health Report.

make the houses so many chimneys sucking out the ground-air." No sickness, however, is reported at Malone.

Throughout the reports of the New York State Board, dismal accounts are given of the filthy condition of the ground in that State; this suffices as a reason for any uncommon sickness in any town, for when no other cause is apparent, the statement is usually made, "it (the sickness) is probably due to saturation of the soil with filth."

Dr. Ezra Hunt * says, "Foul air, foul water, and foul decompositions come from the ground-air;" and if we were to take four feet of the city soil, with its overladen decomposition, and compare it with that of the open country, we would at once detect manifold differences in sickness and mortality, and we would not wonder that sanitarians call it "typhoid ripe." And before the same association,† Dr. A. B. Segur says, "The amount of ground-soakage in old towns has hopelessly contaminated our soil;" and he wants the association to declare "that typhoid fever, cholera, and yellow fever are fecal diseases;" as if a proclamation of this kind would be binding on the medical profession.

It should be kept in mind that none of these sanitarians, who declaimed so loudly about the dangerous condition of the soil, pretended that they had made any investigations to sustain their oratory. Yet this improvisation not only imposed on the mass of the people, it actually influenced the medical profession. Neither the public nor the profession could conceive that these men, who were posing as sanitary reformers, scientists, educating the people, as they said, could have the effrontery to intrude on them such trumpery as scientific facts without ever having made the slightest research.

If Lemuel Gulliver, "first a surgeon, then a captain of

---

* Proceedings American Public Health Association, 1874-75.
† 1875-76.

several ships," after his return from Liliput and Brobdingnag, had set forth his claims as a learned traveller and anthropologist, on account of his labors in those countries, he would have been as much entitled to recognition as a scientist as were the *improvvisatori* who displayed their *capriccios* and *fantasias* as Sanitary Science.

They never offered a jot of evidence to prove their assertions about the soil. There was none to offer except such as practical men could afford, who had labored in and observed the earth, in their mechanical and building operations. These called us to witness that the contents of a privy or cesspool never penetrated the earth beyond a few inches; that the organic matter soon formed a stratum on the inner wall of either, which was impervious to air or water even in a loose sandy soil; that less than one foot outside of this stratum the ground was as sweet to the senses as that under the snows of Mont Blanc.

But in the excitement and frenzy of sanitary reform people were in no condition to take the evidence of their senses. If any man was so intrepid or so uncautious as to call for proof of the declarations of the sanitarians, or for the exercise of common sense in investigating them, he was stigmatized as an enemy of the public health.

To arrest the destruction of life arising from soil-pollution, sanitary codes were made more stringent, inspectors appointed, prosecutions commenced, and the sanitarians mounted the witness-stand and told judge and jury that they were Sanitary Experts; and swore, "So help them God," that the polluted soil contained "mephitic gases;" that these generated disease and the germs of disease, and that the people were in jeopardy thereby. On this evidence, the judge *ex cathedra* anathematized the enemies of the public health; the jury never failed to convict; sometimes it gave its decisions without leaving the box.

No! says the reader; out of zeal for the new-born science,

and maybe inspired by vanity on account of the importance its apostleship conferred, these professors might, on the platforms and in their journals, exaggerate and magnify its principles; but they would surely stop short of the witness-stand, and would not hurt their innocent fellow-citizens by swearing on them not only damages but contempt; before they did this, they certainly must have investigated the soil by careful experiment, and have found the "mephitic gases" and the germ.

Let the incredulous reader be assured that they did nothing of the kind; neither in the fugitive pieces nor in the *chefs-d'œuvre* of these Eminent Sanitarians can any evidence be found that they ever undertook any investigations, nor any trace of knowledge that any one else had ever made any inquiries respecting the nature or condition of the soil.

By and by some scientific men directed their attention to the subject and published the result. Fodor* tested the air from four places,—from a spot in the University court at Klausenberg, sixty feet from a privy; from the University cellar twelve feet below the surface, and from the court of the hospital, where the well was thirty feet from a privy. The fourth specimen was taken from the top of a mountain. The only difference in the air from the soil in these various places and elsewhere was in the large amount of carbonic acid and the deficiency of oxygen. The average amount of carbonic acid six feet below the surface, on the mountain-top, was greater than that at the same depth in the cellar of the University or in the hospital-yard. He found that the earth which is richest in carbonic acid is freest from organic matter, and that the quantity of this acid is no criterion of the cleanliness of earth. Sulphuretted hydrogen was not to be found in the ground-air in Klausenberg, and the "mephitic" gases are not mentioned. Dr. Beutzen's experiments led to similar

---

* 1875.

conclusions. He says it is doubtful whether the carbonic acid in the soil comes from the humus or from other pollution; that if the soil be defiled the carbonic acid does not fix the pollution; that no one can determine the sanitary condition of a place, or its likelihood to have disease, by the quantity of carbonic acid or the analysis of the organic matter. Other examinations in Europe had practically the same results.

In 1876 Professor R. W. Nichols dug a pit six feet square and five and a half feet deep in Back Bay district in Boston. He filled it within a foot and a half of the top with semi-liquid from a sewer man-hole and covered it with earth. Some of the matter which was buried was rich in sulphuretted hydrogen. Two weeks later he examined the air in and around this pit, and repeated the trials at intervals for about five months. "Sulphuretted hydrogen was not detected, even in the *air*, fourteen inches from the ground;" that is, four inches above the top of the deposit. "Ammonia was not found in any appreciable amount." Of what is called the "mephitic" gases there was none. He then examined the air one foot and a half from the base of the old Roxbury sewer, the bottom of which was not water-tight. "The examination failed to detect sulphuretted hydrogen or marsh-gas." Oxygen, nitrogen, and carbonic acid were the only gases he found. In an appendix, Professor Nichols gives a *résumé* of Fleck's investigations, which proved that air directly over a dead body was contaminated with no other than carbonic acid gas; that in none of his examinations in earth purposely arranged to favor the formation of so-called mephitic gases could more than mere traces of sulphuretted hydrogen and ammonia be obtained. The process of decomposition was one of oxidation, forming carbonic acid.

The population of the town of Kendall was fourteen thousand in 1878. One million gallons daily * of sewage

---

* Massachusetts Board of Health Report, 1879.

from this town are poured on five and a half acres of land for three hundred and thirty-five days in each year. This process has been going on for five years; the effluent water is entirely clear. The soil from this farm, taken six inches below the surface, has been shown by the repeated experiments of Dr. A. Dupré to be purer than soil taken from unsewaged surface of the earth.

	Soil of unsewaged surface.	Soil just below the sewage.	Surface of sewage farm. Soil six inches below.
Ammonia	0.0026	0.0043	0.0039
Nitric acid	0.0077	0.0076	0.0115
Organic nitrogen	0.1178	0.1983	0.0715
Total nitrogen	0.1218	0.2046	0.0785
Organic matter	6.81	5.56	5.50
Phosphoric acid	0.0214	0.0268	0.0214

The few pages that make the reports of these scientific men nullify the tons of imagery and verbiage about our contaminated soil that the sanitarians have put forth for nearly forty years. Yet they have continued to keep the public mind in a panic, as if these examinations had never been made.

Confounded by their own reports, which were full of evidence proving that a soil which received organic matter was incapable of producing disease through the agency of foul gases, the sanitarians again accused the germ; this, they said, sought the filthy soil and there multiplied. They made no investigations, no experiments, but they detailed with painful accuracy the way in which the germ propagated itself in the foul soil and was transported from thence to our bodies. They described the mode of generation,—the birth, life, and death of those disease-germs which had never yet been discovered, but only guessed at.

The same ridiculous result awaited them here as with their distracted ravings about the germ in the air and the water. M. Miquel * proved by repeated experiments that air which passes through the earth not only carries with it no germ, but is completely purified by the passage. He says, "Ainsi donc l'air qui filtre à travers le sol non seulement n'enlève pas avec lui les microbes-bactéries qu'on y rencontre toujours en quantités innombrables, mais encore se purifie complètement." He then, as he says, "exaggerated" his experiments by mixing earth with putrid meat and directing currents of air through the mixture; these currents of air were proved to be free from germs: "qui se montrèrent aussi vierges de germes que l'air soigneusement filtré à travers une longue bourre de coton stérilisé."

Before these facts, he says, it is impossible to believe for a moment that air which passes through the soil is capable of raising a single germ.

Flügge † says that in unsterilized earth pathogenic bacteria are destroyed by the ordinary putrefactive bacteria, and the acidity of such earth prevents the reproduction of pathogenic germs. That when organic matter is poured on the ground it is not fixed; it is decomposed, and the organic molecules are oxidized; organic substances are made inorganic (*vollständig mineralisirt*). This is brought about not only by chemical action, but by the putrid bacteria in the superficial layers. Sterilize the earth, he says, and this is not so complete. Natural ground always has micro-organisms to aid the chemical activity. Micro-organisms in the ground-air are seldom found. (*Micro-organismen werden in der Bodenluft ausnahmlos vermisst.*) He is no less positive that air containing them is not drawn into the houses. (*Auch in die Wohnhäuser werden mit der Bodenluft niemals Bakterien eingeführt.*)

---

* Les Organismes Vivants, 1883.
† Grundriss von Hygiene, 1889.

And again,—

Flügge * says the uniform result has been obtained that even strong currents of air are unable to carry a single bacterial germ through a layer of earth a few centimetres in thickness. He adds, "From our present knowledge it is improbable that pathogenic bacteria multiply in the soil. Hence, it appears to be of relatively little importance for the question of pathogenic bacteria in the soil whether the latter is more or less contaminated,—that is to say, impregnated with excreta." Lehman,† Professor of Hygiene in the University of Würzburg, is no less explicit in his declarations respecting the ground-air. He says it contains no micro-organisms. ("Die Grundluft ist pilzfrei.")

In *Annales d'Hygiène* ‡ is recorded the story of an investigation made by a committee of fifteen hygienists, to inquire if the spread of sewage at Gennevilliers was prejudicial to the public health. The report is made by M. Ogier. Experiments made with the typhoid bacilli proved that these retain their vitality in passing through sterilized earth three feet deep, while they are destroyed in transit through natural or unsterilized ground at a depth of from eight to sixteen inches. The committee concludes that the spread of the sewage of Paris, even with the system of "*tout à l'égout*," over the soil of Gennevilliers did not imperil the public health.

Dr. Cornil § reports the visit of a French commission to the sewage-farms about Berlin. On their arrival, Professors Koch and Virchow met and accompanied them on the tour of inspection. Professor Koch showed the commission that the farms took the fecal matter of one million five hundred thousand people; that there was no danger of progressive saturation of the soil; that the water of the drains had been

---

* Micro-Organisms, 1890. † Methoden der Hygiene, 1890.
‡ 1889. § Revue d'Hygiène, 1888.

drunk for years ("*tout le monde la boit et la trouve bonne*"); and that they had begun to establish asylums there for convalescents. Virchow told the commission that the earth there was a complete purifier; that "the pathogenic microbes are destroyed on the surface and in the superficial layers of the earth by their rivals, the putrefactive bacteria, which greatly outnumber them."

At the *Congrès Internationale d'Hygiène*, 1889, MM. Grancher and Richard reported on the "*action du sol sur les germes pathogènes.*" They say that according to Hoffman it requires two or three years for a solution of marine salt to reach a depth in the earth of nine feet; that the bacteria could not travel as fast as the solution; that they are in the superficial layers; on the surface they number one hundred and twenty thousand to the centimetre, while twenty inches below there is an abrupt fall to two thousand. They showed that the typhoid bacilli did not penetrate the earth deeper than fifteen to twenty inches. They declared it was now proved that air passing through the earth had not the power to take with it the least germ. *En résumé*, they say, "Disease-germs are found, if at all, only in the superficial layers of the earth; they multiply there with difficulty; are destroyed by putrid bacteria and by solar light; and the water-level six to nine feet below the surface is safe from them."

At the same congress M. Wurtz showed by experiment that the typhoid bacillus penetrated the earth in loose soil not more than sixty centimetres, even where the wetting (*arrosage*) with the bacilli was kept up for ten days at a temperature of 72° F. In *Revue d'Hygiène*, 1889, are recorded the experiments of MM. Grancher and Deschamps with the typhoid bacillus. They filled a tube eight feet long with earth and kept a continuous flow, for sixty days, of a fluid which held the typhoid bacilli. Not only was no trace of these found at the inferior end of the tube, but when this was tapped sixteen inches from its upper surface, not a typhoid bacillus appeared.

As we have seen in previous chapters, it did not need these physical proofs to convince us of the harmlessness of a soil encumbered with organic matter. The histories and mortality-tables of numberless small localities, entire cities, and two vast empires, which the sanitarians have unconsciously collected in their health-reports,—those sarcophagi of Sanitary Science,—evinced that what they called soil-saturation, instead of being accompanied by an increase of zymotic disease, was invariably attended with its diminution.

Through these reports they have put themselves in a contemptible dilemma. Either alternative confers on them disgrace. Either there is no truth whatever in what they have laid down as the fundamental principles of Sanitary Science, or with no apparent purpose except the acquisition of power and pelf, they have been guilty of the despicable fraud of exciting panics in the public mind by representing falsely the condition of the air we breathe, the water we drink, and the soil on which we tread.

Imagine the dismay of the enlightened man who has nourished a delicacy of sentiment regarding a pure air, a sweet water, and a clean soil, in the belief that thereby he promoted bodily health and longevity, and finds the illusion rudely swept away by the crushing evidence that the sanitarians have garnered! That henceforth material benefits are not to be looked for; that the sole reward for living cleanly is a sense of comfort, decency, and self-respect; and that, except this, as he meditates on the watchwords "pure air, pure water, pure soil," he must view them only as the exuviæ of Sanitary Science.

CHAPTER VII.

The Sewer-Gas.

The cerebral fecundity that evolved Sanitary Science out of the triune problem of pure air, pure water, and pure soil was no less active and fertile in providing accessory creations to sustain and adorn the new science.

That which transcended all others in importance was the invention of sewer-gas. This imaginary substance has afforded material for flights of sanitary fancy such as no element or combination of elements has permitted it to reach. We had got the air purified; sentinels were posted at every avenue to prevent its reinfection; the water had undergone thorough ebullition; the sleepless eye of the sanitary inspector was fixed on the ice; the soil, overcharged with filth, had been purged of its impurities through the sewers, and our minds were at peace, when it was suddenly discovered that the very thing we had been doing had augmented our dangers a thousand-fold.

The sanitarians told us that although we had in a measure relieved the filthy soil, we had actually prepared something whereby the essence of that filth was concentrated in the sewers as a gas, and practically had so arranged things that this gas was "laid on" in our houses. The sewers were vast laboratories in which it was generated; in the New York sewers alone [*] eight hundred million cubic inches of sewer-gas were emitted daily, and the larger the sewer [†] the greater the quantity of gas there was generated.

One woe following another so quickly, we came near cursing the day that Sanitary Science was born. The

---

[*] *Medical Record*, vol xxxi.   [†] London *Lancet*, 1882.

apocalyptic ardor which had pushed and firmly grounded the elementary principles of the new science received a fresh glow and a fiercer impulse. An extravagant fervor seized the amateurs of both sexes; they outdid the professionals in predictions of disaster from the new danger.

The triple alliance, which the reformers had made with the ladies and the clergy, was now to be reinforced by a liaison with the plumber for offensive and defensive operations against the sewer-gas. Elated with his new accession to the dignity of being tolerated in the company of scientists, he became in a manner transfigured; and as he lifted up his voice for sanitary reform he was no longer a plumber; he was a "Sanitary Plumber;" and he bore this device on his escutcheon in his future reprisals on the community.

This guide, philosopher, and friend proceeded without delay to fabricate and set traps for us, which he said would shield us from the deadly vapor. He had no sooner put one in than it was shown that the gas was generated in such quantities in the sewer that it was forced past the trap. The next one he placed went through that inscrutable process of "siphoning out," and we were worse off than if we had had no traps at all. We must now ventilate them; when this was done, the joints began to leak, and he said the material of the pipes was so weak that it could not stand the peppermint test; and in some way that future improvisation alone would explain, it was proclaimed that sewer-gas escaping from a pin-hole would cause disease much more surely than if it were passing out in volumes a foot in diameter. There was no safety but in tearing out all of the old fixtures and replacing them with new ones. After their renewal we were no better off, for not a day passed that the sanitary dervishes did not relate the poisoning of whole families by sewer-gas.

Just when, where, and by whom sewer-gas was invented

is shrouded in the mists of doubt and obscurity. Unlike other great contrivances, there seems to be no competition for the honor of having invented this. It was soon after 1850 when the gases of sewers first began to be talked about; but it was not until about the year 1857 that it was decided, not by chemical experiment or by any other investigation, but by a whim of the sanitarians, that there should be a distinct substance known as sewer-gas.

The most discordant and contradictory properties were at once imputed to it. Sometimes its gravity caused it to descend into the bowels of the earth; again, by its surpassing levity, it ascended to heaven. Its powers of lateral diffusion were illimitable: it would permeate masonry eight feet thick; its backward pressure was enormous; then, unlike other gases, instead of finding vent at the man-holes and large openings of the sewers, it had such affinity for the human system, to poison and destroy it, that it remained pent up until it could find egress through some crack or pin-hole and escape into our dwellings. Sometimes it had a vile odor; again it had a faint mawkish smell; but the climax of danger was reached when it was odorless. "Poisoning by sewer-gas [*] which has no smell is the cause of many maladies; we destroy the warning odor without destroying the poison." "We take the rattle off the tail of the snake that he may the better bite us with impunity." "Better let the atmosphere of a house be nauseating from the fumes of recent *fæces* or pestilential from the fumes of a cesspool, than poison its inhabitants with the demon sewer-gas skilfully laid on by a system of closed drains." Ventilating them by gratings in the street "is Machiavelian in its refinement of folly and wickedness."

It was a painfully well-known fact [†] that a whiff of sewer-gas produced a peculiarly marked effect; it often happened

---

[*] London *Lancet*, vol. i., 1882.   [†] London *Lancet*, vol. ii., 1884.

that *the victim suffered most who had been the least exposed to it.* "It makes its way out probably in gusts."

Dr. F. H. Hamilton* says that the united skill of the specialists has not succeeded in keeping our houses free from sewer-gas. "A generation has come and gone, thousands upon thousands have died, and looking at our decimated households we may well ask, How many more must be sacrificed to this terrible experiment?"

Its composition was almost as variously described as its properties. "Sewer-gas † is a comprehensive term used to designate a greater number of gases and vapors, very complex in composition. This complexity of composition is not due to the great number of elements represented in the gas, but to the variety of ways in which these elements may combine with each other. The elements present in sewer-gas, and indeed certain of their compounds, are practically without any injurious effect. There are, however, products intermediate between these elementary bodies and the more ultimate products which are deleterious, and in the possibility of the formation of just these products lies the danger of sewer-gas." "It is unfortunate ‡ that we do not know the exact components of sewer-gas, and that there are no means of ascertaining how it acts upon the system." The risk from sewer-gas § is probably not so great as many suppose; it is a slight risk, but "a slight risk of terrible danger."

"Not much is known ‖ of the noxious gases and vapors contained in sewers, or how they are generated," and "the quantity of gas in a sewer is of no consequence; it is the degree of concentration which is important." "Sewer-gas ¶

---

* *Popular Science Monthly*, vol. xxii.
† *Journal Franklin Institute*, 1887.
‡ Eighth Report of Massachusetts Board of Health.
§ E. C. Clark, Massachusetts Board of Health Report, 1879.
‖ Massachusetts Board of Health Report, 1880.
¶ Ziemssen, vol. xviii.

is a continually varying mixture of the gases which make up the atmosphere and a relatively small proportion of other gases." "Poisonous sewer-gas [*] cannot be clearly defined; it is known chiefly by its effect; even its odor is rarely a marked one." Dr. F. H. Hamilton [†] says that sewer-gas is a compound of air, vapor, and gases in constantly varying proportions. "Sewer-gas [‡] is of a singularly light character, and has a tendency to ascend or draw towards any heated part of the house. Also, it is so penetrating that I have known this gas to pass through floors and through chinks in two-foot walls. It will find out the smallest opening in any pipe." "The weight of sewer-gas [§] depends on circumstances, so that the specific gravity is always varying." Sewer-gas is a product of fermentation, and is heated by this fermentation, and so is rendered lighter than when it is cold. Dr. Griffin [‖] quotes Professor Lindsley as saying, "Sewer-gas is so subtle that its presence is many times not detected, and yet so laden with the germs of disease that diphtheria, scarlet fever, typhoid fever, and other fatal maladies are the sure event to those who dwell in such air-poisoned houses." Sewer-gas [¶] "is something *sui generis*, and is the distinct product of the modern system of sewage disposal;" and in the same article we read, "Concerning the nature and properties of sewer-gas proper, the sanitarians and sanitary engineers are ignorant."

Professor Kerr, in an address before the British Civil and Mechanical Engineers' Society,[**] said, "We know this gas has two qualities which are extremely obnoxious:" one of

---

[*] Waring, Sanitary Drainage, etc.
[†] *Popular Science Monthly*, vol. xxii.
[‡] Seventh Massachusetts Board of Health Report.
[§] *Builder*, 1881.
[‖] *Chicago Medical and Surgical Examiner*.
[¶] London *Lancet*, vol. i., 1882.
[**] *Popular Science Monthly*, vol. xxii.

these qualities was that "it ascended to the highest level by reason of deficient specific gravity; and the second quality was that when it reached the highest level it exercised a pressure, being an extremely elastic gas. When the sewer-gas had reached the highest level it exercised a powerfully-elastic pressure to force its way out, and succeeded in forcing its way. It got into the houses, and if there was no other, there was this grievance to complain of,—that this pestiferous and poisonous gas forced its way from the sewers into our houses, and of course reached the vital organs of those who occupied them." It would be difficult to devise a more perfect way to secure the entrance of sewer-gas into our houses * than the present system. We are practically laying on the gas of decomposition as we do illuminating gas. The closer and better the pipes, the worse we are off; with the perfect drains all the gas is retained and connected with the houses. "Sewer-gas is a special product of our refined and scientific system of sewage; we have now a perfect apparatus for treasuring it up and laying it on into our houses."

Surely, the reader will say, it is not possible that these men in the livery of science could put forth these statements without any experiment to support them. Such is the fact. Not an experiment, or an attempt at an experiment, to prove the truth of this incoherent balderdash had ever been made by these so-called scientists. Their work and that of Lemuel Gulliver rest on the same basis,—an exuberant imagination.

If we were a little confused by the diseases that water caused, those that resulted from sewer-gas bewildered and confounded us.

They were contradictory in their symptoms, and as various in their nature as were the properties and composition of the gas. Ague,† all forms of malarial fever, typhoid,

---

* London *Lancet*, 1882. † De Verona, Sewer Gas, 1880.

typhus, and spotted fevers, scarlatina, abscess, thrush, boils, headache, chicken-pox, cholera, diphtheria, dysentery, eczema, enteritis, diarrhœa, conjunctivitis, convulsions, croup, glanders, erysipelas, general debility, neuralgia, rheumatism, whooping-cough, broken breast, measles, rash, phthisis, plague, childbed fever, vertigo, apoplexy, yellow fever, small-pox, and *perhaps* myelitis of the anterior horns!* It must be remembered that all of these diseases, except "myelitis of the anterior horns," had prevailed for generations prior to the invention of sewer-gas.

We see that pretty nearly every ill that afflicts the human race is caused by it, except alopecia, strabismus, and ingrowing toe-nail.

We have said that no experiments had been made with sewer-gas.

In a great many—almost all—of the treatises published on that subject by the sanitarians, professional or *dilettanti*, Dr. Barker's experiments with sewer-gas are cited. Dr. Barker confined animals in a chamber and subjected them to breathing the air of a cesspool. Either the sanitarians who so profusely quote him never read the account of his experiments, or they do not consider the opening sentences which precede their detail of any importance. In the *Sanitary Review*, vol. iv., 1858–59, Dr. Barker says, "For the purpose of experiment I selected a large cesspool which received, together with the animal excreta, the liquid refuse of an inhabited house. The cesspool was full, and had at all times so bad a smell that during the hot weather the vicinity was scarcely tolerable. *The inhabitants of the house, however, had not for many years suffered from any epidemic, nor did the near presence of the cesspool seem to affect their general health.* This fact is but fair to state in connection with what follows."

---

* *Medical Record*, 1887.

Yet these experiments are constantly exhibited by the writers on sewer-gas to illustrate its poisonous effect on the human system under ordinary circumstances, when Dr. Barker expressly declares that it had no effect on the family which had been subject to it for many years. In a small chamber—capacity five thousand eight hundred and thirty-two inches—made of wood and glass, he introduced dogs, and exposed them from five to twelve hours to the cesspool air drawn into the box. The dogs had diarrhœa and vomiting, but soon recovered. He let a mouse into the cesspool to within three inches of its contents and kept it there four days. The animal was well and lively on removal, but the next day died. He kept one dog in the chamber for two days, giving him fresh air only when the chamber was rapidly cleansed. The dog was thin and weak for six weeks after, but recovered. By pouring sulphuretted hydrogen into this box he destroyed the life of some animals; but these experiments amount to nothing at all as tests of the pernicious effects of sewer-air on the human body.

In 1822, Parent du Chatelet[*] says that he visited every sewer in Paris; was present at all kinds of work that the employés were engaged in; questioned them separately and *en masse;* visited them at their houses to ascertain the diseases they were subject to; tried to get from them contradictory statements to compare them, and to invent new questions to correct errors. The sewers of Paris were then without means of ventilation, so that sometimes men who took no precaution on entering them were asphyxiated, as men often are on descending into wells and beer-vats. The record showed that between 1783 and 1823 thirty men had been asphyxiated in these sewers, and that many of them died. The diseases of the sewer-men were few in number and not of a grave character. Ophthalmia and rheumatism

---

[*] Hygiène publique.

were the only ones that prevailed among these workmen more than with other people. Ulcers, wounds, and cutaneous eruptions were not only not aggravated, but the workmen considered sewer-water an efficacious remedy for them. He found no exception, even when the sewers had been long neglected. The only disease that he thought was aggravated by this work was syphilis. In winter the temperature in the sewers was so soft that it favored the growth on their walls of fungi-like mushrooms; these were gathered with care by the workmen and eaten; they made "*un des meilleurs plats de leur modest repas.*" He says that in no way was the health and longevity of these workmen impaired. ("*Leur santé peut être considérée parfaite et fort rarement dérangée.*") Parent says he had often heard that this occupation was the cause of putrid fevers. He declares he had never seen one of these workmen who had such fever, and that "*la véritable observation*" had destroyed all of his previous opinions and prejudices.

If this sewage is so innocent when new, is it made noxious by time? Observation alone, he says, could solve this question, and a unique occasion presented itself in the Amelot sewer, which had been stopped for years. It was a terror to workmen; the authorities decided to clear it, if it could be done with safety. Parent du Châtelet was one of the commissioners to superintend the work. Thirty-two men were selected, sixteen of whom had never worked in sewers. Their ages varied from twenty to seventy years. It required six months to complete the work; two thousand loads of solid and six thousand loads of semi-solid sewage were removed. When the work was done, the men were assembled and one after another examined. They displayed the best of health,—"*la santé la plus florissante.*" Some of the more delicate ones had gained in *embonpoint* and vigor. During the work the air in the sewer was often examined; at no time was the oxygen above 19; it was generally 18;

once it was below 14. The nitrogen was from 80 to 82, and sometimes the carbonic acid was above 2.

In 1871 the Metropolitan Board of Works in London ordered an investigation to ascertain to what extent sewermen were subject to typhoid fever. The report was made by J. W. Bazalgette, C.E., on January 18, 1872. Five inspectors had been employed from twenty-three to forty-nine years; none of these had ever had fever. One hundred and sixteen men had been employed in cleaning sewers; their ages varied from nineteen to sixty-eight years. They had been employed from one month to thirty-four years. None had had typhoid fever. Many years before, four of these had suffered from typhus fever. Forty-two had been employed from one to fifty years as flap-keepers; their ages varied from twenty-five to seventy-five years. One of these had had typhoid fever. Twenty-three had been at the pumping-station; their ages varied from twenty-one to fifty years, and they had been employed from one to twenty-one years. None had had fever. At another pumping-station was a group of seventy-eight men, whose ages varied from nineteen to seventy-eight years, with a length of service of from one to nine years. Among these there had been one case of typhoid fever and eight cases of intermittent fever. Fourteen men were employed in cleaning ventilators; their ages varied from twenty-two to sixty-eight years, and their length of service from one to twenty-three years. No fever had occurred in this group. Ten were employed as chainmen; their ages varied from forty-two to sixty-nine years, and their length of service from four to twenty-six years; in this group there had been no fever.

Here is an astonishing report, an official one; it contains sixteen pages. Of two hundred and eighty-seven men, a large number just in the typhoid age, only two had ever had typhoid fever. This document does not say if these men were or were not subject to any other class of diseases.

The surveyor of Chelsea,* in his report to the vestry, says that, contrary to the dicta of many sanitarians, the sewermen there show vigorous health and vitality, though they spend seven hours a day in the sewers, in cramped positions, dealing with offensive and dangerous matter. One who is now pensioned off is eighty-six years old, and was a sewerman for twenty-eight years. Another has been at work over thirty years in the sewer; another is seventy-five years old, and has been at work thirty-eight years; another, fifty-five years old, has been at work in the sewer for thirty-six years, and all the other sewermen enjoy equally good health.

In 1878 Professor Bartholow read a paper before the American Social Science Association at Cincinnati. After speaking of the complications of sewers and sewer-gas with politics, he said, "We now know that no amount of *fecal* accumulation can cause typhoid, unless, indeed, its germ be present;" that specific diseases were caused by specific germs; that the germs cannot rise from a moist surface; that they must be dried and carried by the wind,—a process which could hardly take place in a sewer. He was followed by Colonel Anderson, the city engineer, who said that if sewers were properly constructed and ventilated, no sewer-gas could form to endanger health. Assistant Engineer Hobbie said that he spent many hours a day in the sewers; that the air in them was superior to much of the air in the streets and alleys of Cincinnati; that he and his men were exceptionally free from disease.

In the *Medical Record*, vol. xxii., 1882, Dr. Burral writes that during 1879 and 1880 there died in New York sixty-four plumbers; none of these died of zymotic disease, except one of cholera morbus. Yet here is a large body of men exposed to sewer-gas in that most dangerous of all condi-

---

* *Building News*, October 1, 1886.

tions, issuing from a pin-hole,—as well as where it escapes in huge volumes,—who are found to be specially exempt from just those diseases that sewer-gas is said to cause.

During the last three years, as occasion has presented itself, the author has inquired, in different cities, of engineers, superintendents of sewers, workmen in the sewers, and of plumbers, to ascertain the effect of sewer-air on the health of those exposed to it. He has asked if sewer-air in any way damaged their health, or if, after long exposure to it, whereby their clothing became saturated with it and befouled by sewer-contents, they took any precaution against exposing their families; or if they ever mistrusted that their families were in danger; or if they ever demanded or received extra pay on account of exposure to sewer-air; or if blood-poisoning ever followed the reception of a wound; or if they objected at any time to enter sewers lest disease should be contracted. Some to whom these questions have been put have smiled contemptuously; some have laughed outright; others have listened to them soberly; all, without exception, have answered "No" to each of the questions. Many of these men had been employed in the sewers for twenty years. Some of the superintendents declared that they had had two, three, four, and five hundred men under their observation, and one gave the number as over a thousand who had been subject to his orders.

In 1882 and 1883 there was a severe epidemic of typhoid fever in Paris.* Of eight hundred and fifty men employed in the sewers, seven had typhoid fever in 1883, twelve had typhoid fever in 1884, and in 1885 two more had the same disease.

This is the only instance which the author can find of sewermen having ever been affected by zymotic disease of any kind. Every year thousands of visitors make the tour

---

* Annales d'Hygiène.

of the Paris sewers. It requires nearly half a day to make the circuit. None of those who take this excursion—all are unaccustomed to sewers—ever use any precaution on entering them or any antidote on emerging from them.

In the Connecticut Board of Health Report for 1885-86 are accounts of inspections of the county jails in Connecticut. At the New London jail, one of the conspicuous defects was the imperfect plumbing, the sole intent of the plumber being to insure the passage of the sewage by gravity, without any attempt " to prevent the return of the sewer or cesspool gases into the building." At one point a free opening existed, through which exhalations passed directly into the kitchen. There were two large leeching cesspools on the premises; one was only fifteen feet from the jail; on the other side, a little farther off, was another which received excreta from the prisoners' pails. No sickness is reported here.

At Bridgeport jail there are no receptacles for the prisoners' pails, and no arrangements to carry off the effluvia; the water-closets are insufficiently flushed. From two points the flow of sewer-gases into the kitchens and corridors is unrestrained, which " must necessarily pollute" kitchens and corridors. This jail was then fifteen years old. " The officers of the jail stated, upon inquiry for the hospital accommodations, that there was no use for a hospital, it was so seldom that any one was sick." The warden of this jail, in an interview with the author, declared there had not been a case of zymotic disease there for fifteen years, and but one death, and that was caused by injury; a man with delirium tremens had dashed his head against the wall of his cell, and died of inflammation of the brain.

Litchfield's jail was in the same condition respecting ventilation and sewer-gases, and the disposal of the excreta of the prisoners. " The health * of the prisoners was good.

---

* Connecticut Board of Health Report, 1885.

It was said to be quite rarely that any sickness occurred among them." At Danbury* matters were still worse; sewage from the prison part is received in a cesspool only eight feet distant, which was full and overflowing on the surface, and fifty or seventy-five feet farther the excreta of the prisoners settled on a little surface of swampy ground. Other cesspools were near the building, one only ten feet from the entrance to the jailer's office. In reply to an inquiry by the author, the warden of this jail said that since its foundation, twelve years before, there had been no sickness among the prisoners except such as they brought in, and they speedily recovered after admission. The Tolland County jail was "by far the dirtiest jail" the inspector had ever visited. Here "there were no traps to sinks or urinals, and no protection against the admission of sewer-gas through them." The inspector—the secretary of the Connecticut Board—regrets that there is occasion to speak so plainly, but the law makes it his duty. The sheriff assured the author by letter that he never knew or heard of any sickness occurring in Tolland County jail, except such as the prisoners brought in.

During the last twenty years there have not been less than fifty thousand,—probably one hundred thousand,—people of both sexes and all ages above that of childhood admitted to the different county jails in Connecticut, who were sentenced for a period varying from two or three days to eighteen months; and—incredible circumstance!—there is no record that there was ever,—not an epidemic, but a single case of zymotic disease occurring in any of these prisons. In some of these towns numerous epidemics of diseases of this class have prevailed, sometimes close to the prisons; but for some reason the prisons escaped the visitation.

The New Jersey Board of Health, 1880, reports on the condition of the jail at Warner. It has no drainage except

---

* Connecticut Board of Health Report, 1886.

into a cesspool, which has no ventilation for the sewer-gas, and a great deal passes into the soil; there is also no ventilation for the cells. No sickness is reported as occurring in this jail. The sewerage at Morris County jail is by cesspools; in spite of the traps, foul gases are forced back into the building. "The health of the jail is reported excellent."

At the Barry County jail,* Michigan, "there is not the slightest attempt at ventilation, and when the air comes up from the dark and unventilated basement, laden with the fumes of vaporized tobacco-quids and the expectoration of diseased lungs, and mingles with the gases arising from the privy, the stench must be intolerable." No sickness is reported at Barry County jail. At Washtenaw jail a foul odor was noticed as soon as the door was opened. The report declares that a more ingenious arrangement for producing unsanitary conditions of the worst sort could hardly be devised than the water-closet arrangements. Sewer emanations are pouring into the building; "the only provision for air-supply was through the sewer-pipe by way of the filthy vault into the corridors." "A strong current of air was constantly ascending into the women's quarters, being drawn through the privy-seats of the men's closet through the vault, thus furnishing the female prisoners a doubly-contaminated air as their fresh-air supply." Rheumatism, bronchitis,—diseases of the urino-genital organs, are reported here. Typhoid fever and diphtheria are not mentioned. At Jackson County jail is the same shameful uncleanliness, but no sickness is reported. At the House of Correction in Ionia,† the sewerage, plumbing, and ventilation are in the worst possible condition; the arrangement here is "an admirable one for the equal distribution of poisonous gases through all the cells." No sickness is reported. At the Montcalm

---

* Twelfth Michigan Board of Health Report.
† Thirteenth Michigan Board of Health Report.

County jail * the defective condition of the sewer was such that offensive gases were constantly given off, making "the stench absolutely intolerable." The foul odors were driven into the female corridors and *vice versa*. Here was a case of diphtheria, and report says that the arrangements were such that a more ingenious device for spreading contagion could not be contrived. The cesspool is so arranged that it is "the best calculated to secure a wholesale poisoning with sewer-gas that could possibly be devised." Here seemed a test case; the disease and all of the conditions for its spread. As no further account was given of this diphtheria, the author wrote to the sheriff of Montcalm County jail for information. He replied that one young man died from this disease, but that it did not spread, and had not been here at any time previous to 1886.

We have seen what little effect sewer-gas had on the inhabitants of Newport.

The seventh report of the New Hampshire Board of Health records an inspection of Rockingham almshouse. The building is old and filthy; the water-closets are in an intolerable condition; the several sinks are all untrapped, and "there is nothing to prevent the building from being constantly filled with sewer-gas and other odors from cesspools. In fact, so far as is known, the entire sewer system of the institution does not possess a trap." This building had been complained of five years before as being in the state here represented. On the day of inspection there were one hundred and ninety-two inmates. Not a single case of disease is reported, nor is there any hint that there had ever been any disease in this institution.

The health of towns report for Massachusetts † says : "An inspection at Springfield disclosed that of two thousand nine

---

* Fourteenth Michigan Board of Health Report.
† Eighteenth Report, Massachusetts Board of Health.

hundred and thirty water-closets in use, more than one-half were not properly trapped or ventilated;" "of five thousand sinks, eight hundred and thirty-two discharged into cesspools and three hundred and thirty-three on the surface of the ground;" "of three thousand six hundred and thirty privies, a majority were in bad condition." That year the general death-rate of Springfield was 18.51 per thousand. Cases of zymotic disease were reported as follows: Scarlatina, twenty-six; diphtheria, twenty-five. No typhoid fever is mentioned.

As we have seen, the city of New Haven in 1885 * was reported to have five thousand untrapped sinks, out of which sewer-gas was pouring. This would imply that not less than about twenty thousand people of that city were continually exposed to this gas; but, instead of having a large proportion of zymotic diseases, this class in that city had never been large, and, besides, had been steadily diminishing for ten years. Those cities, on the other hand, like New York, Brooklyn, and Boston, where the plumbing is under the most rigid supervision, and sewer-gas excluded, show the largest amount of zymotic disease!

From the first the influence of sewer-air in producing disease was contemptuously rejected by the Germans. Dr. Soyka † says the sewer-gas theory of disease is now taking the place of the worn-out (*abgenutzen*) theory of drinking-water. Dr. Soyka shows by tables that in Hamburg, Dantzic, Frankfort, and Munich, as sewers and sewer-gas have been introduced into those cities, typhoid fever has been steadily on the decline; and that this disease is most plentiful in the portions of those cities where there is the least sewer-gas, and *vice versa;* and he declares that there is no proof whatever of any connection between this and

---

* New Haven Board of Health Report.
† *Viert. für Oeff. Gesund.*, vol. xiv., 1882.

the extension of epidemic disease. Professor Rohé* says that some physicians and sanitarians believe that sewer-air is the direct cause of typhoid and scarlet fevers, diphtheria, etc., while others believe that it is the breeding-place of infectious germs. "There is no absolutely trustworthy evidence in favor of either of these doctrines."

Every experiment to ascertain the chemical constituents of the air in modern sewers has proved that it does not differ materially from the atmosphere outside. In 1879 Professor W. R. Nichols examined the air in the sewers of Boston. His report comprises twenty pages. He purposely chose the Berkeley Street sewer, which "is an example of the worst type of construction." He says, "It was only extremely seldom that sulphuretted hydrogen could be detected, by employing for the purpose even a considerable amount of the sewer-air." While it may exist "in this sewer, I have never found enough to determine and to express it in figures." He gives the following examples of examinations of sewer-air previous to his own:

Year.	Name.	Place.	Oxy.	Nitr.	Carb. Acid.	Sulph. Hyd.	Marsh Gas.	Ammon.
1829.	Gautier de Claubry.	Paris.	13.79	81.21	2.01	2.99		
1829.	" " "	"	17.4	...	3.4	0.81		
1858.	Dr. Letheby	London.	19.51	79.96	0.53	trace.	trace.	trace.
1867.	Dr. Miller	"	20.71	.	0.11			
	" "	"	20.79	78.81	0.13			
1870.	Dr. Russell	"	20.79	...	0.40	....	....	trace.
1871.	Dr. Nicholson	"	18.44	81.10	0.55			
	" "	"	19.33	80.35	0.23 highest.			
	Berkeley Street Sewer	Boston.	20.48	...	0.40			

He found the oxygen in this sewer as high as 20.90, normal air being 20.96; and the carbonic acid was as low as .05. A glance at this table shows the immense advantage which the modern sewer has over those built before 1830,

---

* *Hygiene.*

in respect to the amounts of oxygen and carbonic acid which they contain. Neither Professor Nichols nor any previous observer found any of the "mephitic gases" which the sanitarians say abound in the sewers. Professor Nichols modestly suggests a query, "whether sometimes the sewer (or the water) is not made to bear the burden of charges for which there is no sufficient proof." He says that if a decomposing mass is so situated as only partly to fill a receptacle which contains in addition a limited amount of air and to which fresh air does not have access, a very large proportion of the oxygen of the confined air disappears, and the space above becomes filled with the gases which have been produced, as a result of decay, in addition to what nitrogen of the air remains. "But," he adds, "no one of these substances is of a character to account for the filth-diseases which are believed to be caused or favored by breathing the emanations from drains or sewers, nor indeed do we know of any gaseous substance which is capable of producing these effects." "Sewer-gas is sometimes alluded to as a homogeneous mixture of light specific gravity, with immense diffusive power, acting as a distinct body, with its own individual characteristics, and referred to as soluble in water to such a degree, etc.; this idea is utterly erroneous."

In the Proceedings of the Royal Society, London, 1887, are recorded the experiments of Professor J. S. Haldane and Thomas Carnally, who were appointed to ascertain the condition of the sewers under the Parliament Houses, and the cause of the bad odors that were said to proceed therefrom. After giving a brief *résumé* of the analyses of sewer-air which had been made previously by various observers, they say that the air in the sewers examined was much better than might have been expected; that the carbonic acid was about twice, and the organic matter about three times, what it was in outside air; whereas the number of micro-organisms was less. They assert that the air was much better than that of naturally ventilated schools, and, with the exception of the amount of organic matter, it had likewise the advantage of mechanically ventilated schools; that the sewer-air contained a smaller amount of micro-organisms than any class of houses; that the carbonic acid in the sewer-air was rather greater than in houses of four rooms and upward, but less than in two- or one-roomed houses. As regards organic matter, the sewer-air was only slightly better than the air of one-roomed houses and much worse than that of other classes of houses. The carbonic acid may be due to diffusion from neighboring soil, but probably to the oxidation of organic matter in the sewers. The average number of micro-organisms was always less in the sewer-air than in outside air. The state of filthiness in a sewer had no effect on the number of micro-organisms. They say, "In view of the fact that sewer-air is to all appearances comparatively innocent as regards its micro-organisms, experiments were also made to see whether it contained any poisonous volatile base of the nature of a ptomaine. These experiments, so far as they went, had negative results." Experiments by the same observers on other sewers led to the same conclusions. The Bristol sewers * are completely shut off from outside air; the number of micro-organisms here was exceptionally small; the amount of carbonic acid was less than was found in many schools. So far as micro-organisms are concerned, sewer-air "is twice as pure as outside air,—in summer, at any rate." They say † that the connection of sewer-air and typhoid fever rests not on satisfactory evidence, but largely on *à priori* reasoning; and "in the present state of our knowledge we should cease to attribute blindly to sewage emanations cases of disease of which we do not know the

---

* *Sanitary Record*, 1887.
† *Chemical News*, 1887.

cause, and patiently seek for convincing evidence as to the real cause."

Mr. Haldane says * that the result of their researches will "tend to mitigate some of the terror with which we have come to regard sewer-air." That while it has "been supposed to be loaded with micro-organisms, it turns out to be some of the freest air from micro-organisms that can be found." "What is," he asks, "the supposed evidence for the causation of typhoid fever and other diseases by the inhalation of sewer-air? We may dismiss as absolutely worthless collections of cases in which something has been found wrong with the drains in a house where typhoid fever has occurred."

Dr. Eben Duncan,† in summing up the evidence in regard to sewer-air, says, "The air of a sewer, when it is properly ventilated, is much better from a chemical point of view than the air of crowded houses and schools, or even the air of houses in which one-half of the population of Glasgow lives. To many of these people it would be of great advantage to be permitted to live in an average sewer atmosphere rather than in the air of their own houses." As for the germs in sewers, Dr. Duncan says, "The number in sewer-air is actually less than the number in the outside air." The conclusion of the whole matter, therefore, is that it is scarcely possible that the germs of such diseases as typhoid fever, diphtheria, or phthisis can be carried back into our houses through moist drains and soil-pipes.

Dr. Carmichael ‡ reports a series of experiments to ascertain the pressure of air in sewers, of which so much had been said by the sanitarians to terrify the people. Some of his experiments were made with a water-closet which had

---

* *Sanitary Record*, 1887.
† *Sanitary Journal*, Glasgow, 1891.
‡ Ibid., 1880.

been used many years. Its soil-pipe was very foul. Into the same soil-pipe, which passed untrapped into the drain, three water-closets above were discharged. The mouth of the sewer, three hundred yards away, was submerged at high tide. He examined the air which came through two water-closets in twenty-four hours. In the worst possible condition, when the outlet was closed by the tide-water, and the roof-ventilator was closed also, only thirty-two grains of carbonic acid and $\frac{1}{100}$ of a grain of sulphuretted hydrogen passed in twenty-four hours, and the amount of ammonia varied from $\frac{1}{400}$ to $\frac{1}{200}$ of a grain. "*These are the quantities of the only sewage-gases existing in the soil-pipe, in estimable quantities, which pass through an ordinary water-closet in twenty-four hours.*" These quantities, he says, are perfectly harmless. "Thirty-two grains, the largest quantity of carbonic acid that passed, is less than the quantity of the same gas given off when a bottle of lemonade is opened. A man exhales in the same period four hundred times the amount which passes through the trap from an unventilated soil-pipe."

The specimens of germs which he collected, "though kept from two to five months at cultivation temperature, have remained perfectly clean; and, even though examined with a lens multiplying nine hundred diameters, exhibited no trace of life." He adds, "These experiments seem to me crucial, and to warrant the conclusion that germs do not pass through a sound water-trap." But he asks, "Do the fetid organic vapors of which we hear so much indefinite *horror* expressed, but of which so little is known, come through in appreciable amount?" He admits that there may be traces of these vapors; but they are organic. If they pass through the trap they are included in the ammonia, very much less than $\frac{1}{100}$ of a grain in twenty-four hours. "This, I need scarcely say, must be harmless."

Experiments made by Parkes and Burdon-Sanderson * have shown that the tension of air in sewers is seldom very different from that of the atmosphere; or, if there be any difference, equilibrium is quickly restored. Of twenty-three observations made on four different days by these observers, the tension was less in the sewers than in air outside, in eight it was the reverse, but on the average there was a slight indraught.

In the *Builder*, vol. lvii., are recorded the experiments of Mr. Santo Cramp, C.E., who studied the pressure of air in sewers; very frequently the movement of the air was so feeble that the anemometer would not work. His experiments showed that where the action of air in sewers was strong enough to affect this instrument, down-hill currents were recorded two hundred and seventy-three times, as against up-hill currents on ninety-seven days.

In 1881 Dr. Rosahegyi,† after repeated trials with an anemometer, the fumes of ammonia, the fumes of cigars saturated with tincture of benzoin and with sulphuretted hydrogen, found that the movement of air in the Munich sewers is at all times very feeble, often none at all; that when it does occur, it is generally downward towards the mouth of the sewers, and that, too, independent of the wind blowing into them; that it follows the fall of the sewers and the downward flow of the water; that the air-current is strongest in the lower part of the sewers; that it happens only very seldom that there is a reversal of the current of air, and then it is only temporary and for short distances, and is dependent on accidental circumstances.

Professor C. D. Chandler,‡ in a lecture on the atmosphere, said, "The common idea that gas in our sewers exerts a

---

* Report on Sanitary Condition of Liverpool.
† *Zeit. für Biolog.*
‡ *Sanitary Engineer*, February 16, 1882.

pressure to get out is a great fallacy." He had yet to find a case to prove it. Some time before he had a lot of pressure-gauges made, which he distributed among his students, telling them to apply them wherever they could, but there has yet been no case of pressure reported. He had a vent taken from the street-sewer to his laboratory, thinking he could get sewer-gas in plenty whenever he wanted it by simply turning the stopcock, but no pressure has yet been indicated.

Again, March 2, 1882,* Professor Chandler said that sewer-gas itself was not dangerous, but the danger was from the bacteria for which it was the vehicle. The professor cites no experiments to prove that sewer-gas is "the vehicle" of disease-germs. May it not be that the reiteration of the charge by the sanitarians, that disease-germs are propagated in sewer-air, had imposed such a belief on him as it had on the medical profession at large and on the community?

In the *Medical Record* † Dr. H. W. Mitchell says that the effects of sewer-air have been greatly over-estimated; that experiments recently made by the Board of Health of New York City in connection with the authorities of Columbia College have shown that the pressure of air in the sewers is inconsiderable, even when a strong wind is blowing into their mouths. For twenty-five years this backward and lateral pressure of sewer-gas had been a terror to the people of New York: to save themselves from it they had submitted to most tyrannical laws, had suffered domiciliary visits from insolent officials, and they had been burdened with useless taxation. It might be thought that so soon as it was ascertained by experiments that there was no such pressure, those who made them would hasten to compose the fears of the people and relieve them from the useless

---

* *Sanitary Engineer*, March 2, 1882.   † July 11, 1891.

officials and burdensome taxation. Not at all. The author has sought in vain for a report of these trials by the New York Board of Health and the Columbia College authorities. A prominent official of that board writes to him, under date of September 21, 1891, "I am not aware that there has ever been published the results of the tests made in this office some years ago relative to the alleged pressure of sewer-gas from the public sewers, but it is true that a delicate gauge was placed upon the sewer connections in the office of the department, and careful observation made from day to day of the gauge, resulting in a failure to discover any appreciable pressure."

These experiments of Drs. Carmichael, Rozahegyi, Parkes and Burdon-Sanderson, Mr. Cramp, and Professor Chandler disperse completely the imaginary terrors of the sanitarians respecting the backward pressure of sewer-air. But, except by a few enlightened men like themselves, no heed was paid to their researches, or to those of Nichols, Haldane, Carnally, and other scientific philosophers, who had preceded them, and who, with balance, test-tube, and microscope, had demonstrated physically—what all previous observations had shown—that there *was* no sewer-gas, and that sewer-air was free from disease-germs.

The terrified public listened to the sanitarians. These made no counter-investigations to impeach the observations of the scientific men; they did not even debate them; sometimes they assailed them with vituperation; but, for the most part, snowed them under with wild tales of wholesale poisoning by sewer-gas, revised sanitary codes, and malicious prosecutions. Plumbing laws were passed, and plumbing inspectors appointed in every city; and statutes were made forbidding a man to introduce plumbing into his dwelling without the consent of the sanitarians organized as boards of health.

Baffled and entangled by the evidence collected in their

own reports, the flexible sanitarians again shifted their ground and took a position behind the germ; this, they said, got into the sewers, was borne upward on the gas, and forced past the traps into our houses. Light, they said, was now being shed on the pin-hole mystery; a single germ, endowed as it was with hermaphroditic power and energy, exuding from this orifice, could multiply itself hundreds of thousands of millions of times in forty-eight hours.*

As before with the air, the water, and the soil, they made no investigations, but their imaginations actually ran riot, and they again mounted the witness-stand and caused damages to be assessed against innocent and respectable citizens by solemnly swearing that the sewers were loaded with gas, and the gas was loaded with the germs of disease. Scientific men had hardly begun their researches into the composition of sewer-air when this evidence of the sanitarians was proved to be false. M. Miquel,† in 1882, declared, as the result of his experiments, that the danger from the spores which escape from sewers is not greater than that from the spores in the open air; that often the confined and humid air of the sewers is purer than the air in those streets which are reported to be healthy. In the summer of 1880 the air in the wards of Hôtel Dieu, although the hospital had been lately renewed and was carefully ventilated, contained six times as many bacteria as the air of the sewer beneath Rue Rivoli, and in one week, from the 1st to the 7th of November, the sewermen in this sewer were breathing only one-half the number of microbes that were inhaled by the passengers in the street above; and in summer the atmosphere of this street was always five or six times more impure from micro-organisms than that of the sewer.

As we have seen on a previous page, experiments both

---

* *Popular Science Monthly*, vol. xxii.
† Les Organismes Vivants.

with the cholera and typhoid bacilli showed that these germs lost their vitality and disappeared in sewers, just in proportion as these contained putridity.

In "Comptes Rendus de la Société de Biologie," 1889, is recorded in a few lines the experiments of M. Olivier with the water from a Havre sewer. Those specimens of this water which were sterilized, when sown with broth containing typhoid bacilli, were swarming with these germs the next day. Those specimens which were not sterilized, but which retained the putrid bacteria, remained unaltered.

Flügge * says that the views respecting the influence of sewer-gas in producing typhoid fever and like diseases, which are in such a drastic manner brought out in English books, are entirely unauthorized (*völlig unberechtigt*). The infective action of sewer-gas is the less likely to take place, as this, by repeated examinations, is shown to be free or nearly so from germs which propagate these diseases, and because the continually wet sides of the sewers make it almost impossible for them to be set free. If by exception the germs of disease are present, all of the chances are against the possibility of their being the cause of infection. The action of sewer-gas, he adds, is no other than that of bad-smelling air; and by ventilating the sewers this can be avoided. Professor Lehman † devotes, in his book of six hundred pages, only seventeen lines to the consideration of sewer-gas. He cannot conceive, he says, how typhoid fever can be produced by it, as it is always free, or nearly so, from bacteria (*stets pilzarm oder pilzfrei ist*).

As we have seen, the romance concerning the air, the water, and the soil, with which the sanitarians had cajoled the people, was only the prelude to a more exalted reverie. It is doubtful if the pages of the world's history can furnish a

---

* Grundriss von Hygiene.
† Methoden der Hygiene, 1890.

more striking illustration of the credulity of mankind than the sewer-gas dream. A pure creation, begotten in and floated from the sanitary brain without any investigation, it was, without examination, accepted and devoutly cherished by almost the entire people, wise and simple, of Great Britain and America,—a creation that, from the first, was viewed with contempt by scientific men of other countries. Pettenkofer said that it was as easy to show that infectious diseases had the same relation to lines of illuminating gas-tubes and telegraph-wires as to lines of sewers. Renk,* who made a special study of sewer-gas, declared that it was unpardonable nonsense (*unverzeihlicher Leichtsinn*) to claim that the contents of sewers had any relation to the causes of infectious disease. He could not conceive how men who work in, and continually breathe the air of, sewers should seem to be specially exempt from such disease, and people be made sick by the same air hundreds of thousands of times diluted. The French smiled at the new conceit and began early to train their wit upon it. Dr. Zuber † said the sewer-gas doctrine, more noisy and plausible than rational, was a transitory theory; it would soon be forgotten; and he asserted that there was not in epidemiological science a doctrine built on such a fragile basis; that the rich collection of anecdotes of sewer-gas ailments which the English displayed depended on their absence of method in etiological studies.

There were in reality as many reasons and facts to show lunar influence in the causation of infectious disease as that it could be caused by sewer-gas. Notwithstanding that in every instance where the effect of sewer-gas on the human system has been studied on an enormous scale, it has been proved to be innocent; notwithstanding that in every instance—even in Great Britain and America—in which

---

* Kanalgase. † Revue d'Hygiène, 1882.

scientific men have searched for sewer-gas with the most delicate apparatus, they have been unable to detect it, and have declared that it does not exist; the sewer-gas nightmare has so oppressed the people of these two countries that they have submitted to the most offensive legislation that was ever invented; they have endured odious domiciliary visits; they have suffered persecutions and prosecutions without number; panics have been fomented, arrests have been made, fines imposed, damages assessed, and imprisonments threatened, to save the people from sewer-gas. Under the pretence of peril from this deadly gas, the citizens of nearly every large town in our country are compelled by most despotic laws to submit the plans of every prospective dwelling to a body of men,—sanitarians,—organized as boards of health, whose technical and mechanical knowledge is usually of the lowest order, and millions of dollars are spent annually to protect the people from a danger as imaginary as the dragon of ancient and mediæval story.

CHAPTER VIII.

Cemeteries.

CHRONOLOGICALLY, the graveyard ghost was the first which the sanitarians erected to affright and torment the people about their health; it preceded by nearly twenty years the sewer-gas bugaboo.

In 1721 an anonymous London author issued a small book entitled, "Seasonable Considerations on the Indecent and Dangerous Custom of Burying in Church-yards." He says, "It is an undoubted truth" that the corruption of dead bodies in churches may be communicated to the living, and that many fatal distempers may be received from the effluvia of the dead, although the stench be not perceived. That a healthy man may impart strength to one that is feeble, is a matter of daily experience; "nor is it a new discovery, for certainly this was the case with David, who, being decrepit and feeble, was cherished by a virgin that lay in his bosom." In the same way, dead bodies can bestow weakness and decay. The power of human bodies, our author says, is exercised not only on their own species, but on other bodies; "for daily experience tells us that coral or amber, worn on a woman's breast, looks more bright or dull, more vivid or languid, according to the different degrees of her health." This sanitarian gives no more facts to sustain his argument, except to ask why it is that so many women faint in churches, if it is not from the stenches which arise from the graves. He demands that the practice of burial in vaults be forbidden.

In 1839, G. A. Walker, a surgeon in London, published "Gatherings from Graveyards," which professed to inform the people of the dangerous and fatal results of burying the dead in the midst of the living. Mr. Walker says that in China two men began to dig a grave where a few months before a human body had been buried. The spade pierced the coffin, when both men fell down nearly lifeless. Though resuscitated, they had severe heat and pains in the liver, and died in four or five days after. He relates a story from the New York *Gazette of Health*, which was supplied by Rev. Dr. Render. In the month of July, 17—, a very corpulent lady died at ———; she was buried in the church. The next Sunday more than sixty people were taken sick at the communion service; it was thought that the wine was poisoned and some arrests followed. The next Sunday a chalice of wine was exposed on the altar and soon "became filled with myriads of insects," and the rays of the sun denoted that they came from the grave of the corpulent lady. Four men were employed to open it, when two of them fell dead on the spot. "It was now clearly perceived" that this had caused the pestilence.

A very fat man was buried in a church in Burgundy. Of one hundred and seventy persons who attended the burial and who entered the church, one hundred and forty-nine were struck with a malignant fever. The nature of the symptoms left no doubt that the malignity was owing to the infection of the cathedral. Mr. Walker tells us, "it was said" that in Paris in 1765 the air about the cemeteries was so infected that necessary aliments became tainted in their vicinity.

In 1775 a curate breathed the infected air from a dead body while performing funeral rites, and contracted a putrid disease. Mr. Walker says, "Ramazzini assures us that those who dig graves do not live long;" and Haller relates that the exhalations from a single body, eleven years after burial, caused a dangerous disease in a whole convent. The soil of the cemetery in Portugal Street "is saturated, absolutely saturated, with human putrescence." The graves are desecrated; bones and heaps of coffin-wood lie on the surface; the effluvia make people who live in the vicinity close their windows. Mr. Walker gives a list of forty cemeteries which are in bad condition. St. Giles's has the melancholy notoriety of originating the plague in 1665. From official documents he shows that from 1741 to 1837 there had been buried inside the metropolis two million one hundred and five thousand one hundred and twelve bodies; and ten years later[*] Mr. Walker says that twelve-thirteenths of every dead body must dissipate and mix with the air we breathe. Some dreadful facts have now come to his notice; the bodies of four persons dead of cholera "were put into the ground against the floor-wall of a house, without an inch of earth

---

[*] Journal of Public Health, 1849.

between them and the wall." He does not say that any bad results followed.

The most remarkable feature of Mr. Walker's book is, that with the enormous number of bodies buried in these forty cemeteries, he should be forced to seek in China, Paris, and Burgundy for evidence to establish his belief that burial in towns is prejudicial to health. He gives no instance of any epidemic or any single case of infectious disease having been caused by any of the cemeteries in London; and he adds, " Our freedom from pestilence can only be ascribed to the natural or acquired power of resistance of its (London's) inhabitants, to favorable seasons, or to diminished temperature."

As the sanitary awakening had now begun, Mr. Walker's book excited a good deal of attention. A reviewer of it in the London *Lancet* * says, " There can be no doubt but that putrid exhalations from dead bodies in a concentrated state produce highly injurious and even fatal effects on the living subject." The *Lancet* tells "the frightful truth that in London alone a mass of human corruption, formed by nearly fifty thousand dead bodies, is annually deposited in the graveyards, there by its putrefaction to contaminate the air which the living breathe, spreading fever and disease." A committee of the House of Commons collected a large amount of " the most revolting but conclusive evidence," to show that these burials spread " a noxious pestilential poison over the metropolis." Of one of these cemeteries the *Lancet* says, " What a focus of disease such a burying-ground must be! thick masses of remains, lying in a state of putrefaction, within a few inches of the surface of the ground." † " The facts which substantiate the pernicious effect of putrid exhalations are too numerous and too well authenticated to admit doubt; for on two occasions the plague is said to have lingered in

---

* London *Lancet*, vol. i., 1839–40. † Ibid., vol. ii., 1845.

Paris longest near the cemeteries; and there are numerous instances where the health of grave-diggers has been disordered." "Not only are the emanations of the graveyards disseminated in the atmosphere, but the products of decomposition percolate the soil so as to pollute it to an immense distance."

Petitions began to pour into Parliament to suppress burial in towns, it being "in many instances productive of injury to the public health, owing to the noxious emanations from the soil."

In 1847[*] is a report that the rector of Minchinhampton removed a thousand cartloads of earth from a church-yard and strewed it on grass-lands in the vicinity. "By a kind of poetical justice, typhus fever invaded the rector's family and destroyed his wife, daughter, and gardener."

Through the inquiry made by the House of Commons, it was brought out that the laboring-people, though often living in one room, kept the corpses of their friends until putrefaction was far advanced. One undertaker testified that he had known bodies held for three weeks, and that every week he saw them retained until nearly putrid, although men, women, and children were living in the same room; that the smell from the coffin is often extremely offensive; and that it is not uncommon "for fluid decomposed matter to escape from it into the street and to run down over the shoulders of the bearers." "In the metropolis, of the fifty thousand deaths that take place annually, twenty thousand occur in single rooms, each occupied by one entire family." A clergyman writes that the dead body with the poorer classes "is never absent from their sight. Eating, drinking, or sleeping, it is still by their side, mixed up with all the ordinary functions of daily life." It "is pulled about by the children, and is not seldom the hiding-place for

---

[*] London *Lancet*, vol. i.

the beer-bottle or the gin, if any visitor arrives inopportunely." In a report made by Mr. Chadwick on the results of a special inquiry into the practice of interments in towns, great importance is attached to the fact that a number of deaths had followed the descent of men into vaults. Mr. Chadwick alludes to "the cadaverous appearance of grave-diggers and all the other signs of a slow poison;" and the evidence is such that "there can no longer be any doubt of the effects of air from church-yards on the inmates of neighboring houses." "In nine cases out of ten, the undertaker who has much to do with the corpse is a person of cadaverous hue." He says that Sir Benjamin Brodie mentions cases where students had taken small-pox in the dissecting-room. He quotes Dr. Copeland, who relates a case where a man went to church on Sunday; the next Tuesday the doctor was called to see him sick with a fever. The patient told him that when he was in the church he felt a rush of foul air. He died eight days later. Some years ago a vault was opened (he does not say where) in a church-yard, and a coffin burst. "So intense was the poisonous nature of the effluvia" that a great many people were seized with sickness. An undertaker had told him that he had known coffins to explode like the report of a small gun. "On the whole," he says, "the evidence tends to establish the general conclusion that offensive smells are true warnings of sanitary evils," and that these putrid emanations "furnish the principal cause of the most developed form of typhus; that is to say, the plague."

A chemist testified that gases from bodies would easily penetrate the soil and escape into the atmosphere, as was shown by coal-gas. Dr. Lyon Playfair affirmed that the "slightest inspection" showed that the putrid gases were not absorbed by the soil. He estimated that three million cubic feet of noxious vapors were yearly emitted from graveyards, and must pollute either the air or the water in the neighborhoods; and they have been known to pass through the sides of sewers, though these were thirty feet away.

Some describe the odor of the graveyards as the "dead man's smell."

Mr. Chadwick, however, says that *all undertakers* that he had seen "*state that neither specific disease nor the propagation of any disease was known to occur amongst them from their employment. Neither the men who handle nor those who coffin the remains, nor the barbers who are called in to shave the corpses of the adult males, nor the bearers of the coffin are observed to catch any specific disease from it, either in their novitiate or at any other time.*" "*It is not known that by their infected clothes they ever propagate specific disease in their families or elsewhere.*" "*Neither does this appear to be observed among medical men themselves.*" Mr. Chadwick passes by this direct evidence as if it were of no consequence whatever, and as if he could not see that, unexplained, it contradicted and overturned all of the other testimony which he had offered. The condition of the public mind, however, was such that this direct and undisputed evidence availed nothing against the anecdotes of the "corpulent lady," the "very fat man," and the "dead man's smell." One clergyman, Rev. H. Milman, had the candor to testify that the sanitary part of the question was most dubious, and rested on less satisfactory evidence than other considerations,—viz., the decency, solemnity, and Christian impressiveness of burials.

In 1849[*] we are told that Sanitary Science has made a "gigantic stride;" "churches are now considered under a double aspect,—a place of worship and a site of pollution by the festering bodies of the dead in the vaults beneath." As late as 1885 the same journal [†] declares that old burial-grounds are dangerous for children to play in; that the

---

[*] London *Lancet*, vol. ii.   [†] Ibid., vol. ii., 1885.

germs of plague, cholera, and fever are lying there only dormant, but not extinct. It is most strange that with all this profusion of talk about "noxious effluvia," "pestiferous gases," deadly exhalations infecting air, water, and soil, not a single instance of disease or epidemic that can properly be ascribed to cemeteries is cited, except that in 1665 St. Giles's Cemetery had the notoriety of originating the plague. There is no proof that the two million one hundred and five thousand one hundred and twelve bodies interred between 1741 and 1837 did any harm to the people of London.

Parkes* says, "It is a matter of notoriety that the vicinity of graveyards is unhealthy." "The air over cemeteries is constantly contaminated (he does not say with what), and water which may be used for drinking is often highly impure."

Tardieu says that Ramazzini deplores the fate of gravediggers, their livid aspect, their sad countenance, and that he never knew one to reach old age.

The sanitarians of our own country delight in conjuring up graveyard spectres. In a report on the Nyack Cemetery, the New York State Board of Health, 1885, p. 289, says, "It has long been a matter of experience that low fevers and various forms of filth diseases are apt to prevail in the neighborhood of old burying-grounds." Professor Lindsley, secretary of the Connecticut Board of Health,† says that the subject of cemeteries is one closely connected with the public health; that disregard of them had produced much evil. "The evidence on this point is overwhelming and unquestioned." "The emanations from the graves of the dead after hundreds of years of burial have communicated to the living fatal maladies of which many have died." The secretary offers no evidence. Professor Kedzie‡ says that,

---

* Hygiene.   † Eleventh Report.
‡ Fourth Report Michigan Board of Health.

unfortunately, we have not to go abroad to learn the effect of graveyard water on the public health; and he relates that a family at Grand Rapids had typhoid fever after drinking from a well near a cemetery. These "poor people," the professor says, "were drinking a cold infusion of death." In the Twelfth Michigan Board of Health Report are the replies of twenty sanitarians to the question, "Do you regard the presence of a large cemetery in a city as detrimental to the public health?" The board says it sought the opinions of "eminent sanitarians" on this subject. One replies, "Unquestionably detrimental." Another says, "I unhesitatingly answer in the affirmative; for although the evidence is, to a large extent, indirect and inferential, yet it seems to me conclusive." A third answers, "As a rule, I would say, extra-hazardous." A fourth replies, "There are no circumstances under which proximity to a decomposing mass of animal matter can be devoid of danger to the living." A fifth says, "If one thing in Sanitary Science is better settled than another, it is that decomposing human bodies pollute both the air above the ground, the ground itself, and the water that percolates through the ground." Still another replies, "The universal judgment of sanitarians is that cemeteries should not be located" within a populous area. A seventh says, "Yes, emphatically, yes; all experience teaches that it is." Another writes, "I do regard the presence of a cemetery, large or small, in a city, as detrimental,—yea, very dangerous to the public health." Though some of the answers were qualified, all to whom the question was addressed agreed as to the dangers of cemeteries. We have here some energetic vociferation, but not one of these authoritative replies is fortified by any evidence.

Dr. David Warman,[*] in discussing cemeteries, rehearses the tale of the corpulent lady and other like stories. In

---

[*] New Jersey Board of Health Report, 1883.

addition, Doddsley's Annual Register for 1773 reveals to him that of one hundred and twenty young people who were receiving their first communion in the Church of St. Eustace, in Paris, all but six fell ill, also the curé and grave-digger. The illness was a putrid fever. But he says we need not search the pages of ancient history for facts to establish the noxiousness of cemeteries; for during the Revolutionary War Trinity Church-yard emitted pestilential vapors, and in 1814, from a lot on Broadway where was a potter's field, "most deadly effluvia arose." Still later, the Weehauken Cemetery, which had been in use for sixteen years, is said to be offensive; Jacob Hausche, who resides on the north side of it, testifies that a very bad smell pervades the whole neighborhood. Jacob "describes the smell as that of rotten carrion, extremely offensive." Dr. Warman says that both air and water are contaminated by cemeteries.

Erichsen, in "Cremation of the Dead," says that "These are times that are trying men's bodies quite as much as their souls." "Zymotic diseases breaking out in what were once healthy villages may set even the blindest to seek for causes, and the most prejudiced may finally be forced to admit that the multitudinous graveyards are contaminating our water. New England villages, once so free from ills, are now taking on the air of invalids. People forget how they drink well-water the springs of which percolated through graveyards." Dr. Erichsen says, "Church-yard emanations can penetrate almost everything; they have a remarkable force." He quotes Dr. Marble, of Worcester, who says, "The monstrous delusion that the mere contact of the corpse with the fresh earth renders it innocuous is dissipated by overwhelming evidence." He also quotes Colonel Whitman, who says that the people in New England who deplore the advent of malaria, "the unseen vampire that sucks the red blood of the present generation," would do well to look about them and "see how the graveyards, old and new, have grown in two centuries."

The Eighth Report of the New York State Board of Health says the greatest danger has been noticed in exhumations from old cemeteries, but specifies none. It relates that the Board of Education at Port Jervis bought the old St. John's graveyard on which to erect a school-house. The Board of Health interfered to prevent it. Dr. Carroll reported adversely to its use on account of danger to the public health, but the Port Jervis people proceeded to exhume the bodies, two hundred and thirty-five in all. The work occupied four weeks; none of the persons engaged in it suffered any inconvenience. The Board of Health declared it was a source of danger to the children, and prophesied great disasters. They said that the soil here was of the nature to allow of the freest passage of gases, and that samples of the earth were swarming with bacteria, "the invariable concomitants of putrefactive processes."

So far we have numerous graveyard legends, an abundant supply of words about the mephitic gases, pestiferous, deadly gases, destructive effluvia, and the "dead man's smell;" we have the poor people who were "drinking a cold infusion of death," an unlimited supply of the opinions of the sanitarians, "the matters of notoriety," the "unquestioned and overwhelming" evidence, the "settled principles of sanitary science," the "universal judgment of sanitarians," etc., etc., but not one particle of proof of a scientific character. Indeed, not one of these sanitarians who so magisterially offer their opinions and their averments about these "gases" claims to have made any experiments of his own; neither does he offer the experiments of others to show that these mephitic, pestilential, deadly gases have any existence.

During the prevalence of cholera in London in 1849, the *Times* overflowed with articles on the dangers of burial in towns. Disgusting stories were told of grave-diggers boring through the interred bodies to find out if there was more room in the earth; of mutilated female corpses partly decomposed; of coffins being tapped, which would "let out a jet of gas that burns from ten minutes to a half-hour." Old graveyards were said to be so saturated with these gases that they could hold no more; and so, of course, they were emitted from the surface to be breathed. The graves were called "consecrated cesspools." One of these articles set forth the "dismal duty" of displaying the dangers of burial in metallic coffins, and directed "the public gaze on the shrouded slumberer within." The writer says, "Gaseous poisons are found so intense that their mere contact with a mucous surface of the body may occasion sudden death;" that they burst the coffins, and the fumes emitted will destroy those who happen to be in the immediate vicinity; and that "We have really only two alternatives as to the disposal of the resulting gases,—they must either be burned or breathed." Surely, the reader will say the *Times'* writer must have made some investigation, or, at least, have known of investigations by others, before he presented that communication to the public. But the fact remains that there had been no investigation.

The article met the eye of Waller Lewis,[*] who says he hopes to see entombment prohibited in the cities and churches, but deems it important to lay before the public the result of his inquiries, which extended over many months in 1849–50. He was impelled to make these investigations by the statements in the *Times*. He says that some of the allegations that cyanogen, hydrocyanic acid, sulphuretted, phosphoretted, and carburetted hydrogen gases were formed were so consonant with theory that one of the most eminent chemists in Europe, with whom he conferred, told him that he "was not prepared to deny any of the as-

---

[*] London *Lancet*, vol. ii., 1851.

sertions in the *Times*, though he believed they were largely theoretical." Unwilling to accept Dr. Playfair's "slightest inspection" as a proof of these gases, Mr. Lewis examined carefully between fifty and sixty vaults in the different churches in London; he noted the external condition of more than twenty-two thousand coffins, and examined the contents of nearly one hundred. The results were so different from what he expected, or from what is generally believed, that he was desirous of laying them before the profession. He says, "I have never succeeded in obtaining any traces of the presence of cyanogen, hydrocyanic acid, sulphuretted, phosphoretted, or carburetted hydrogen, even in the smallest quantity." In the air of one vault was a trace of sulphuretted hydrogen, but there was no certainty that it proceeded from the bodies. "I examined gases formed by bodies of all ages, from the still-born infant to those which had survived the age of ninety-two." "Those which had been there for a week were examined, as well as those which had been there for a century and a half." Death had been caused by accident, age, and disease; typhus, phthisis, small-pox, cholera. "Not one of the above circumstances seemed to influence, in the slightest degree, the composition or character of the gases." "All I analyzed or otherwise examined were composed of nitrogen and carbonic acid, mixed with atmospheric air, and holding decaying animal matter in suspension." The coffins do not burst. All of them do not even bulge. From the most searching inquiries of sextons and others, he obtained no evidence of the rupture of a coffin. He says that during this work, carried on in the stagnant atmosphere of these vaults, he sometimes had nausea and vomiting, throbbing pain in the head, and loss of appetite. A long time after he had boils and erysipelas. He does not say that he thinks the vaults were the cause of these disorders.

In 1785 the bodies in the *Cimetière des Innocents* in Paris were exhumed.* This cemetery was old in the twelfth century; it had been complained of for two hundred years as a source of disease and danger. It was said that the mephitic gases from it had penetrated the adjoining cellars in the vicinity, had collected on the walls, and had been condensed there into a most subtle and deadly poison. By simply touching the wall it acted on the human system like the venom from the poisoned arrow of the savage. The gorged earth of this cemetery was swollen eight to ten feet higher than the level of the street. The number of bodies surpassed all calculation: more than ninety thousand had been placed there during the thirty years anterior to 1785. M. Thouret says, here was an opportunity such as never before had occurred to observe changes in bodies, whether they were mouldering in heaps in a humble bed of earth, or proudly rotting apart in metallic coffins and stone vaults. All changes were noted, from the corpse which had dissolved in putrefaction to the more privileged one that was converted into a dried and fibrous mummy. The most distinguished families were confounded with the most inferior. "*Cette espèce d'hommage rendu au principe d'égalité que la nature établit parmi les hommes devait flatter la multitude.*" Between fifteen thousand and twenty thousand bodies, in all stages of decomposition, were exhumed. Through fear, the labor was begun with great precaution, but afterwards it was carried on without any. The work lasted many months, being continued sometimes in the heat of summer. Not an accident happened to workmen or people, nor anything that could disturb the public.

Parent du Châtelet † says that every year there are more than two hundred exhumations at Père-la-Chaise, made at

---

* Sur les Exhumations du Cimetière et de l'Église des Saints Innocents, par M. Thouret.

† Hygiène Publique.

all seasons, and often two, three, and four months after death, when putrefaction is at its highest point; and there is no record of any accident having occurred from these disinterments. Pellieux * expected to find "mephitic gases" in the cemeteries and vaults; he found none but carbonic acid and small amounts of ammonia. He thinks that other gases do exist in cemeteries, and he hopes in future to demonstrate the truth of this theory. If he afterwards found them, he neglected to record his observations.

Gaultier de Claubry reports † on the exhumation of the victims of July, 1830. Notwithstanding the difficulty of the work, the unfavorable temperature, the fatigue and bad odor, no one engaged in it suffered *serious inconvenience*. Five hundred and seventy-four bodies were exhumed. They were in all conditions, from dry bones, in some of the places, to the most absolute putrefaction in others, varying according to the soil and the quantity of bodies massed together.

Dr. O. du Mesnil says, for speculative purposes, the public health had been invoked to have a single cemetery established at Mery-sur-Oise, and to have the cemeteries in Paris closed. This involved an outlay of nearly thirty million francs. Dr. Du Mesnil ‡ says that our present knowledge assures us that the contamination of air, water, and soil from cemeteries is pure hypothesis. The experiments of M Schultzenberger with three specimens of earth—one from a virgin soil, one just over and one just under a coffin from the *fosses communes*—proved that the combustion of a body was complete five years after burial; and that from the present mode of burial there is no danger of saturation of the soil. In none of the examinations of air, either on the surface or from one to two and a half feet below, could a trace of sulphuretted hydrogen, ammonia, or oxide of carbon be found; and it was proved that bodies buried four and a half

---

* Annales d'Hygiène, 1849.   † Ibid., 1843.   ‡ Ibid., 1866.

feet in the ground are consumed in five years, without disengaging any gas that can affect the public health. In 1879 M. Schultzenberger * took air from the cemetery at Montparnasse at a depth of from one to two and a half feet, over graves many years old, and over graves six months and a year old. At no time could he find any gases except nitrogen, oxygen, and carbonic acid.

M. Carnot examined twelve specimens of water from six different cemeteries in Paris. He found a strong proportion of sulphate and carbonate of lime, with slight amount of salts of magnesia and alkaline chlorides ; *but no traces of organic matter, with one exception*, and this was where the well had been long unused. The usage of these wells, says Dr. Du Mesnil, by the inhabitants around, by the workmen, and by visitors, for centuries, furnishes the best answer to the accusations against them. G. Robinet,† in a thesis sustained before l'École de Médecine, on the pretended dangers of cemeteries, says that the decomposition of bodies in the earth is an organic combustion; the final products are carbonic acid, ammonia, and nitrates. He estimates the weight of bodies buried in Paris annually at about three million pounds; fifty-six per cent. of these is water, fifteen albuminous matters, twenty-one fatty matters, and eight per cent. ashes. According to his investigations, the transformation of all the carbon contained in these bodies into carbonic acid would make nearly three million pounds of this substance. As the daily production of carbonic acid in Paris by men, animals, and different sorts of combustion is more than fifteen million pounds, it would take five years for the bodies buried in Paris to produce as much carbonic acid as is formed in one day by the living in that city; and the talk about "*certains gaz*," "*certains produits volatils*," is imaginary. That ammonia may be found, he admits; but it is found everywhere in

---

* Revue d'Hygiène, 1884.   † Paris Theses, 1880.

nature, and is harmless. What are called *miasmes* do not exist in cemeteries, except as we can apply the term to microorganisms. All experiments in exhumation prove that the soil does not become saturated with either solids, liquids, or deleterious gases; that the water of wells near cemeteries has been proved by experience and by chemical analysis to be innoxious; that analyses of soil in and near graveyards show the same result as arable soil elsewhere that has never been infected.

In *Annales d'Hygiène*, vol. ii., 1884, is a report by Dr. O. du Mesnil on soil containing organic and decomposing matters, particularly the soil of cemeteries. The 21st of August, 1883, on the occasion of a burial at Montparnasse, a workman descended into a *caveau* and was suffocated. In the attempt to rescue him, three other laborers came near losing their lives. An architect was ordered to examine and report. He advised that energetic measures be taken to disinfect this soil, from which fetid emanations proceeded, endangering the health of workmen and even of that quarter which surrounds the cemetery.

Dr. Du Mesnil says it might be supposed that, in view of the counsel of such radical measures, some examination, some analyses of soil and air had been made. Not at all; not a single experiment with either had been undertaken. The government appointed a commission to investigate. In that part of the cemetery where the accident had occurred two *caveaux* were dug, about fifteen feet deep, separated by a space of twelve feet. They were bricked up in the ordinary way with cement, and covered. In less than twenty-four hours they were opened and a laborer descended. In a few moments he called to come out; bad air, he said, was in the *caveau*. A lighted candle let down was extinguished eight feet from the surface. The pit was filled with carbonic acid gas. In none of the analyses of the air was there a trace of sulphuretted hydrogen, nor of carburetted hydrogen, nor of oxide of carbon. The proportion of oxygen was always very feeble; it and carbonic acid oscillated, according to the temperature.

The commission concludes: First, that in any excavations in a soil where are decomposing organic matters, whatever may be their origin, two phenomena occur to put life in danger,—the rapid and free disengagement of carbonic acid, and a notable impoverishment of oxygen. Second, that these modifications of the air are limited strictly to the excavation, and have no influence on the neighboring inhabitants. The commission say that between 1863 and 1883 there were three hundred and sixty-seven thousand eight hundred and eighty-four descents into the different *caveaux*, without incommoding a single workman.

In 1883 M. Miquel published the result of his investigations of the air of cemeteries, and of the germs which he found in them. He proved that the number of microbes in the air at the cemetery of Montparnasse, even in times of drought, was very much less than in the streets of Paris, and he adds, "If it be proved that the soil to which is confided the vast numbers of the dead is incapable of emitting noxious germs, these vast fields of mourning, on which rest so many unjustifiable accusations, will be not only innoxious but will be a sanitary agent (*une cause d'assainissement*) for the great cities, like public gardens, wide passages, and spacious courts, which permit the winds, the purifying agents, to accomplish their mission." M. Miquel made most careful and repeated, but unsuccessful, experiments to ascertain if a single germ passed out of the soil of cemeteries.

Dr. Reimer,[*] after investigating the Jena church-yard, concluded that the buried body exercises no important influence on the number of bacteria in ground near it; that bacteria disappear almost entirely at a depth of from three to six feet. Neither near nor under the coffins in the Jena grave-

---

[*] Zeit. für Hygiene, 1889.

yard was the number of bacteria greater than at corresponding depths elsewhere, and it made no difference whether the body had been buried thirty-five years or only eighteen months.

In the sixth report of the Massachusetts Board of Health is a paper by Dr. J. F. Adams, on "Cremation." He sent a circular, to which he received one hundred and seventy-one replies—one hundred and thirty-one from Massachusetts—to the question whether cemeteries were a cause of ill health. This time the query was directed, not to professional sanitarians and health officers whose standing as scientists and whose subsistence depended on perpetuating a panic among their fellow-citizens, but to physicians engaged in the daily practice of medicine, who had neither inclination nor interest to excite alarm in the communities where they were located. Of the one hundred and thirty-one replies from Massachusetts, five said "Yes;" one hundred and twenty-six said "No." Thirty-two replies were received from other States. Three said "Yes," twenty-nine said "No." Dr. Adams says, "In searching for cases of recent date, of disease resulting from graveyard infection, we find that such are almost unknown to medical literature." After citing a number of wells (before alluded to, p. 119), he says, "After a diligent inquiry, we have been unsuccessful in obtaining a single example of disease presumably induced by water contaminated by the proximity of burial-grounds." He adds that the danger to health from cemeteries is "utterly insignificant. A living man in sound health is far more to be dreaded than is a dead man buried with ordinary care." Professor Rohé* says, "An unprejudiced consideration of the subject (cemeteries) shows, however, that there is no trustworthy evidence that any of the gases exhaled by decaying or putrefying bodies are injurious

---

* Hygiene, 1890.

to health." He asserts that the dangers from pollution of water by cemeteries have been over-estimated, and that there are no facts on record which show that infectious diseases are propagated from interred bodies. Fleck,* in his examination of the water of wells in Dresden, found no proof of ill effects of the use of such water. Neither the period of interment nor the closeness of the graves had any influence on the contents of the wells.

At the International Hygienic Congress at Vienna, Professor Franz Hoffmann † offered a paper on the sanitary dangers of graveyards. He said that not only the laity but great numbers of experts and writers maintain that emanations from buried bodies breed disease among the living. That men have been asphyxiated on account of descending into tombs, he admits. The same accident has attended descents into wells; but nobody is buried in wells. He declares that there are no gases in graveyards perceptible to the senses. The bacteria of putrefaction work with great energy; he had seen thirty-five per cent. of the soft parts of a child disappear in less than two months after burial. The objection that the germs of typhoid fever and other diseases are retained in graves is altogether unfounded.

Dr. Sigel said that numerous (*zahlreichen*) experiments which he had made with Dr. Hoffman in the exhumation of bodies during the last eight years had entirely changed his views respecting cemeteries. Before that he had stood on the ground of those traditions which might be called the legends of graveyards. He had seen large numbers of exhumations, two, three, four, and five years after burial; he never knew any accident to happen to assistants or to people in the vicinity. In Saxony, Dr. Sigel said, an investigation had been held to determine whether the laws could be mod-

---

* Appendix to Prof. Nichols's report on ground-air.
† Viert. für Oefft. Gesund., 1882.

ified with safety, which demanded a protracted interval of time between the burial of one body and that of another in the same ground. Twenty-eight physicians, to whom the investigation was assigned, reported, without any concert, that no danger to the public health would arise from a relaxation of those laws.

Dr. Rozahegyi said that his investigations had been made in the cemetery of Budapesth, where were from one hundred and eighty thousand to one hundred and ninety thousand bodies. The earth here is astonishingly active in oxidation, and the bodies defile neither the earth nor the water. Of fourteen specimens of water in the cemetery, compared with fifty-nine in other parts of the town, the former were the purer; and that the water from wells in the cemetery did no harm was confirmed by the comparative freedom from zymotic disease of those who drank it. These papers produced a profound impression on the conference. Dr. Kuby considered them as a blow in the face (*einen Faustschlag in das Gesicht*) of sanitary authorities, who had fostered the cemetery bugbear for a thousand years. Flügge* says that it is impossible in a well-regulated cemetery that the ordinary process of the decay of bodies should do any harm; that no specific gases, no so-called corpse-gases, are formed. He admits that a bad smell may arise from large masses of imperfectly-interred bodies, but declares that the absorption of these odors in a regular cemetery is complete; that contagious diseases never arise from a buried body; that it is impossible for the infective bacteria to escape from the ground; that according to many experiments the water of wells in graveyards is purer than that of wells in the same towns; and that it is not possible for the contagious microbes to escape into the wells of church-yards.

---

* Grundriss von Hygiene.

There are thirty-five cemeteries in New Orleans;* nineteen are within the city. Many of these are "situated in the heart of populous districts." In many places in New Orleans water is struck eighteen inches below the surface, so that burial in vaults is largely the practice there. For the fifty years ending in 1883 there had been 272,649 burials in the thirty-five cemeteries; 47,622 died from infectious diseases. Dr. Holt,† formerly president of the Louisiana Board of Health, says that the stench from human bodies in these graveyards pervades the houses. In Locust Grove Cemetery the pauper graves receive the bodies of many dead. A grave is opened, a coffin broken into, the bones are raked out, the old coffin broken up, pried out, and a new one put in place; the lid of the box only two inches below the surface of the soil. Flies swarm back and forth from the graves into adjoining houses and upon tables. Dr. Holt does not say that he ever knew or heard of a case of disease or an epidemic arising from these cemeteries.

Dr. Joseph Jones,‡ on intramural burial, says that in 1801 a Capuchin priest petitioned for the removal of one of the cemeteries, because it was a menace to the public health. This is the only evidence that the doctor offers in his short paper to show that any one in New Orleans ever suspected that these cemeteries were a source of disease. He has spent a large part of his life in that city, and has had better opportunities for obtaining information respecting its salubrity than almost any other observer; yet he does not say that he ever knew or heard of a case of disease or an epidemic arising from these cemeteries. He gives no opinion, although he may perhaps have the opinions of the twenty sanitarians before him which are recorded on p. 189. As a

---

* Louisiana Board of Health Report, 1880–83.
† *Sanitarian*, vol. vii.
‡ Louisiana Board of Health Report, 1882–83.

preliminary to the formation of an opinion, Dr. Jones suggests what? That a *thorough investigation* be made of intramural burial; and that the inquiry should embrace the nature of the soil of cemeteries; the effects of seasons; the chemical and physical properties of the gases exhaled; the nature of the organisms developed; the effects of putrefaction, etc., etc. If it be permitted to judge Dr. Jones's character from his printed works, it is not hazardous to say that when the conditions of his proposed investigation shall have been complied with, he will form an opinion and be ready to express it on proper occasions, so as not to be misunderstood.

The pensive wanderer through and about the crowded cities of the dead on Long Island seeks in vain from those employed there, or from the inhabitants near them, for any other than a negative reply to the question whether labor among these tombs or residence in their vicinity is prejudicial to health. Many of those to whom the query is put treat it with mild contempt. Nobody but the sanitarians seem to know that cemeteries or their surroundings are unhealthy, and their knowledge is not obtained by investigation. Did they acquire it by revelation or intuition?

In the old Cavalry Cemetery are two wells. One is located in a valley-like depression surrounded by graves, the nearest of which is only three feet away. Within six feet there are a dozen, and within twenty feet there are fifty tombs. The water of this well is drunk habitually by the workmen in the cemetery, and daily in the warm season by hundreds of people who visit the church-yard. The work men in this part of the yard praise the water highly. They say that of the other well, which is on the hill, somewhat remote from the graves, is likely to cause cramps, if taken too freely in summer. But this, in the valley, is a guileless water. Inquiries, however, about the well on the hill, which is fifty feet from a grave, bring out just as positive testimony as to its virtues. The workmen and the visitors who partake of this water deny the impeachment that it ever causes cramps. The clergyman at the church, which is near, casts no aspersions on the well in the valley, but affirms with energy that the water of that on the hill equals in every respect that of the one below, which is so close to the graves.

The well in the Lutheran Cemetery is about one hundred feet from the graves. Its water is so abundant and so precious that in times of drought the neighbors come for it from a long distance. Two wells are located between the Jewish and Cypress Hills Cemeteries, about one hundred feet from the graves. The water of both is sought after by the people around, when their own supplies fail, and is not only regarded as delicious, but is proved by long experience to be healthful.

Beneath the Church of Santa Maria della Concezione, in the city of Rome, are four mortuary chapels, adorned with the bones of more than four thousand (so the guide-book says) Capuchin monks, that have died in and been deposited beneath the monastery, since its foundation in 1624. In these chapels are seen uncoffined bodies in various stages of dissolution. When a new inhumation is made, the bones which have lain the longest are employed to decorate one of the chapels. The ghostly father who, three years ago, conducted the author through the gloomy cloisters and into the charnel of this abbey, when asked if such entombment of these bodies ever had any influence in impairing the health of those who ate, drank, worked, and slept over and around them, raised his brows in mild surprise and answered, "No." How many had here been entombed. He could not tell. As many as four thousand? "Many, many more."

CHAPTER IX.

Public Funerals.

ANALOGOUS to the Graveyard Ghost, but somewhat later, the sanitarians presented a new terror,—that of public funerals. There was no more direct, scientific proof to implicate the newly-dead body in the production of disease, sporadic or epidemic, than there was to so connect the one which was far advanced in putrescence. Indeed, during the research in England concerning intramural burials the question of contagion from the unburied dead was hardly alluded to. The whole scope of that investigation was only to establish the belief that bodies in their progressive decay, after burial, emitted noxious gases which excited disease. The positive testimony which Mr. Chadwick presented in his report (and which, unexplained, seems to overturn that of all other witnesses),—namely, that undertakers, sextons, bearers, barbers, all who had anything to do with the preparation of the corpse for burial, never either became infected themselves, nor by their bodies or clothing ever infected others,—these observations, too, extending for generations,—would seem to settle beyond a doubt the non-contagious property of the recently-dead body.

But the public mind, constantly kept in agitation by the sanitary terrorists, was always open for the reception of new apprehensions of evil, and the excitement about the cemeteries had hardly subsided when they seized the opportunity to widen their dominion by frightening the people respecting public funerals, and persuading them to yield to boards of health the power to prohibit such ceremonies.

The tender sentiments which had pervaded every people, barbarous and civilized, since the beginning of time, regarding care for the dead, must now in a great measure be suppressed. The sanitarians, as usual, without making any investigation, began with pontifical solemnity to exhort the people against the dangers of public funerals; and it soon became one of the "settled principles of Sanitary Science" that the body dead of any contagious or epidemic disease could communicate the same to the living.

The Mosaic hygiene, which the reformers had so reverently held up for us to imitate and admire, sustained them in this instance, for it was written that whosoever touched a dead body should remain apart and unclean for seven days.

Tardieu[*] says that there is no doubt that bodies dead from contagious diseases can communicate the same; notably those diseases which are transmitted from animals to men; though he gives no example. Arnould[†] says it is difficult to conceive a more unhealthy object than a dead body among the living. Were it not for the sentiment of religion, we would instinctively withdraw ourselves from it, even when there is no fear of contagion. Thousands of experiments have put beyond doubt the deadly energy of the "*poison septique.*" In fact, he says every organism that life has abandoned is immediately invaded by the formidable phenomena of putrefaction. He gives no example of the deadly energy of the "*poison septique.*" It is related in the Second Report of the New York State Board of Health that an undertaker laid out two children and carried home the infection to two members of his own family. In the Fourth Report of the New York State Board of Health is an account of diphtheria at Indian Lake; a public funeral was held and the disease spread. The same disease is reported at Arietta as having been brought there by two members of a family who had slept in a bed at Johnstown

---

[*] Dictionnaire d'Hygiène, article "Contagion."
[†] Nouveaux Elémens d'Hygiène.

in which a child had died of diphtheria two weeks before. In the Ninth Report of the Massachusetts Board of Health it is recorded that a man attended a public funeral of a child dead with diphtheria; after returning to his home "his two children were taken ill and died:" the account does not say what other exposure there was, or how long after the return of the father the children were taken. The same report says a mother laid out these children, took the disease, and died. Again, it is related that diphtheria broke out at Winchendon; "four children in one family, being the first cases that had occurred in the town for several years, died." "Out of forty who attended the funeral, thirty-seven took the disease." At Webster a child died of diphtheria. "At the end of three weeks sixteen of the children who attended the funeral were down with the disease." At Rockport "a corpse dead from scarlet fever, brought from abroad, was exposed in the church. There was then no case in the town; but a few days after (how many is not stated) one who attended the funeral and was near the coffin was struck down with the disease." The report does not say if any others who attended the funeral were seized with the disease. This report of the Massachusetts board, which records these cases, says, however, "It is an open question whether in the cases enumerated the contagion was received directly from the cadaver, or whether from some one or more of the accessories" of a funeral; those who are associated with the sick being "more active distributors of the poison than the corpse;" and it adds that some of these cases were obscure enough to lead physicians to the belief that the disease in many of them was started *de novo*. Rev. S. Bridenbaugh, in a paper read before the State Sanitary Convention at Norristown, Pennsylvania, cites three instances where disease was communicated through public funerals; but these, like all of the others, are so carelessly reported that they have no scientific value. In all of the places except Winchendon and Rockport epidemics were prevailing. At Winchendon diphtheria had been absent for "several years" anterior to the seizure of the four children, and these cases were probably the beginning of an epidemic that had no reference to the public funeral. The *Sanitarian*, November, 1890, reports that a child died in Wilmington, Delaware, the physician who attended the case certifying that the death was from pneumonia. The body was transferred to Prospect, Maryland, and the coffin was opened. A few days after, how many is not stated, diphtheria broke out in Prospect. To admit this as a case of contagion at a public funeral, we must assume that the physician made a false certificate, apparently without any motive. There is no record that diphtheria was prevailing at Wilmington. The author has inquired carefully into the circumstances of this case, and can affirm that there is no proof that the child died of this disease. The undertaker declares that it did not die of a contagious disorder, and, except that there was a "report" that the case was one of diphtheria, and that the family was "reticent" after the funeral, there is no evidence whatever. Yet this case has gone into sanitary literature, to be quoted by sanitarians in future as positive proof that a public funeral may communicate contagious disease.

It must be kept in mind that all of these cases are reported by those most interested in creating and sustaining panics; but, allowing all of the significance that the sanitarians wish to derive from them, it is plain that the prohibition of public funerals in such cases can accomplish nothing to prevent the spread of epidemics, unless all who have been connected in any way with the deceased while passing through the sickness are quarantined for a period of time which has not yet been defined. Except one or two apocryphal cases, to be afterwards alluded to, the author has found none reported in the medical journals that would indicate that there is any danger from the body dead of contagious disease. In reality, the conductor of a respectable medical journal would not admit to its columns, without qualification, these accounts of cases of contagion at funerals, which are welcomed and bruited by the sanitarians.

In 1831 Parent du Châtelet* and M. D'Arcet were appointed to examine the *salles de dissection* in Paris. They say that up to the beginning of the present century dissections were carried on privately in different parts of that city; about that time they were forbidden, except in certain institutions. In six of these Parent had studied anatomy for five years; a very large number of young men were under his observation; in spite of the crowding of bodies and of students, in spite of continued labor of three, four, or more hours daily over the bodies, he could find no proof that the emanations from them—dead of all sorts of diseases—had ever any influence on their health. He had seen young men indisposed, and even fall sick during their studies with dyspepsia, general *malaise*, colics, and diarrhœa; but they were spontaneously cured in a few days, and the same disorders prevailed among the students of other professions in Paris. Of the medical students not more than one in ten or twelve were ever even indisposed, and they were as often so in summer when dissections were not going on. Parent was one of four hundred students who dissected at one place in 1812; not one of these was sick that year.

Nothing, he says, could be more foul or more crowded than the amphitheatre of the Hôtel-Dieu; one hundred and fifty students worked here; he could find no proof that any illness occurred among them. Lallemand observed the same facts, and cited Desault, who said that those who frequented the hospitals were often taken sick with contagious disease, while those who frequented the amphitheatre were

---

* Hygiène publique.

exempt. Desault expressed his opinion—founded on observation—of the harmlessness of a body dead of contagious disease, in the axiom, "*morte la bête, mort le venin.*" Boyer, who directed crowds of students, Professors Dubois, Ribes, Dupuytren, Roux, Dumenil, Beauchene, Jadelot, Serres, Breschet, and Andral testified to the same effect. Serres had charge of La Pitié for sixteen years, where were dissected twelve hundred bodies annually by six hundred students, all or nearly all young men. He declared that, except a slight diarrhœa, he had never known any illness among them. Dr. J. C. Warren * says that he had seen, both in Paris and in Edinburgh, students in great numbers —some of them in delicate health—devoting long hours at a time to dissections, enduring great fatigue, opening bodies dead from typhus and yellow fevers in hot weather, and when epidemics were raging, without any harm having come to them from this work. He had made many inquiries of sextons and undertakers in Boston regarding contagion from dead bodies. They all agreed that they knew of no instance where fever had been communicated in this way. Dr. Warren quotes Professor Lawrence, who for ten years had observed the students at St. Bartholomew's Hospital, and who declares that he never knew a case of illness to occur from anatomical studies. Tardieu, who is so positive as to the communicability of sickness from the dead body when he discusses contagion, does not, in his article, Amphithéâtres de Dissection, try to controvert Parent du Châtelet's conclusions. He only claims that the dissecting-rooms should be under surveillance on account of the interests of neighboring proprietors. Dr. Gull † says, "Those who were engaged in making post-mortem examinations of cholera subjects seemed to incur no risk of thereby taking

---

* *Boston Medical and Surgical Journal*, 1829.
† Report of Royal College of Physicians on Cholera of 1849.

the disease." Dr. Allison * says, "It is certain that the dissecting-rooms in Edinburgh were supplied during the greater part of 1848 and 1849, as they were in 1852, almost exclusively by cholera subjects, and in neither was there a single case of the disease among the numerous students attending these rooms." In the *Lancet*, vol. i., 1871, Mr. Samuel Wilks, surgeon, calls the attention of the medical profession to the terror which is spread abroad regarding contagion from the dead body. He says, "There is no harm in the segregation of the dead, but do not let scientific men act solely on false popular fancies." He gives his "opinion framed from negative evidence," that all fear of contagion ceases with death. At his hospital he had seen students take typhus fever after a slight exposure to a case in the wards, while fifty students would pass a whole hour over the same body, in a small room, making a post-mortem examination, without harm; and he thinks there is no reason to believe that emanations from a corpse can produce disease. Dr. Ewens † writes that Dr. Wilks's letter has "completely upset" all his previous views; that for twenty years he had urged the early interment of persons dead of contagious disease, under the impression that the risk of infection was great, and that his labors had been ineffectual. " I am compelled honestly to admit that, in spite of solemn protestations against such practices, I cannot recall to mind a single instance in which there was any good ground for supposing that the dead body did the mischief anticipated." Messrs. Hawkins and Hovell ‡ believe that they have seen small-pox communicated by dissecting bodies dead of that disease. One correspondent in the *Lancet* offers no facts to disprove Dr. Wilks's views, but regrets their publication,— not because they are false, but because it will weaken the

---

* *British and Foreign Medico-Chirurgical Review*, 1854.
† London *Lancet*, vol. i., 1871. ‡ Ibid.

hands of those who advocate the establishment of mortuary houses in crowded cities; and he adds, with charming simplicity, that he fears he will "not have so good a case" in his efforts to establish one of them. A number of hearsay stories were brought out by Dr. Wilks's publication, among others that of a girl who died of enteric fever; the carpenter who made the coffin was seized with sickness *immediately* at the time of placing the body therein, and died of that fever. Like Dr. Ewens, the author has made many "solemn protestations" against public funerals, which were unheeded. He can call to mind numerous instances of severe disease of an acute inflammatory type, which probably had for exciting causes exposure to cold and wet at funerals. He can remember no case of contagious disease which was ever communicated by public burials. He has conferred with many physicians of wider experience than his own, and has found none who could remember a single case of contagion from a funeral. In conversing on the subject with an octogenarian in New York, who was a practitioner during the cholera of 1832, the veteran at first was impatient that any one should doubt the danger of public funerals, but when pressed for his experience said, "If you ask me to put my finger on a case, I cannot do it. The subject is worth looking up to see if there is danger from them." The author has inquired of many pathologists who have been engaged for many years in making autopsies on bodies dead of all kinds of contagious diseases; many times in these operations they have been surrounded by large numbers of young men who were more than ordinarily susceptible to contagion by reason of age, and who have often been permitted to handle the diseased organs. None of the pathologists offer an instance to show that infection can be spread by a dead body, but each affirms his belief that such contagion cannot occur. The author has the testimony of ten undertakers located in four cities of one hundred thousand, two hundred and fifty thousand, eight hundred thousand, and one million six hundred thousand inhabitants respectively, whose business experience varies from seventeen to fifty years. Some of them have known of the celebration of many wakes over the bodies of those dead of infectious diseases. None of them can remember that any ill effects ever followed these ceremonies. All say that they or their employés never take any precaution against contagion when called on to officiate; and all except three unite in expressing their belief that a body dead of any contagious disease is incapable of communicating such disorder to the living. Each of the three exceptions was an undertaker who had a case in mind where one of his employés was seized with variola a few days after burying a body dead of small-pox, though all of the three admitted that the disease was prevailing at the time. The remaining seven were as confident that small-pox could not be imparted by a body dead of that disease as they were that other diseases could not be so communicated.

The international conference which reported on the cholera of 1865 * says there is no proof that the body dead of cholera can transmit the disease; it is only prudent to consider it dangerous.

At a meeting of the Suffolk District Medical Society, April 29, 1876,† a committee, to whom the subject had been referred, reported on the Dissemination of Diphtheria at Funerals. The committee regret that the evidence presented has been so slight that they have been unable to come to as positive conclusion as would be desirable. Four hundred circulars were sent to as many physicians. There were two hundred and thirty-nine replies. "Of these, one hundred and forty-three report a belief in the possible danger of contagion at the funeral of those who have died of

---

* Annales d'Hygiène, tome xxvi., 1866.
† *Boston Medical and Surgical Journal*, 1876.

diphtheria." "Seventeen indicate an opinion that there is danger from funerals in the houses of the deceased, but none in the churches." None of these one hundred and sixty physicians names a single case to support his "belief" or "opinion." "Twenty-nine correspondents consider that physicians are not justified, in the present state of knowledge, in prohibiting public funerals." "Eight report instances where the transmission of the disease from a cadaver seems probable." Of these the committee cite the four most striking, but in none of them was it possible to conclude that the source of infection was the funeral; for, in each instance, diphtheria was present in the town where the funeral occurred. Two doctors recollect cases where an apparent spread of the disease followed funerals, but the committee say of these physicians, "They are unable, however, to give full particulars." The committee say, "An examination of the text-books, and of the French, German, English, and American journals of the last forty years, has given no evidence of the contagion of diphtheria from a corpse, or even a hint of the possibility of it," except in the *British Medical Journal* of February 5, 1876. The particulars in this instance, however, are taken from a local newspaper; and "the committee cannot consider the testimony as very valuable." The committee conclude, from the answers received, that the "opinion" is tolerably current among the profession that there may be possible danger from public funerals, but that the evidence is insufficient to establish that there is any danger of contagion from a body dead from diphtheria. It is not unfair to conclude that the remaining two hundred and three physicians, who make no reply to the circular, had no facts to communicate which could bear on the subject, or they would certainly have furnished them. This report would indicate that the danger from public funerals of the dead with diphtheria is so slight as not to be worth considering.

CHAPTER X.

The Meat.

THE rising orb of Sanitary Science had poured its beams through the atmosphere; its light had been diffused through the water, and dispersed over the soil; it had pierced the gloom of the sewers, illuminated the graveyards, and kindled the funereal torch of public burials; but as yet had failed to shed its benignant rays on what we should eat. Since the beginning of time, the appetites of the various races and tribes of mankind had stimulated them to extort a pitiless tribute from almost the entire vegetable and animal world for their subsistence. Instinct and experience had taught them that everything which pleased the taste and could be digested and assimilated to the human body was good for food.

There were tribes even that ate the crude soil and found nutriment therein. The aborigines of South America and New Caledonia appeased their hunger with a fat, ferruginous clay and a friable earth. The native of the polar regions revelled in the blubbery fat of the walrus and whale. Roots, fruit, and fish, with an occasional roast of the enemies or friends of their own species, nourished the savages of the tropics. Semi-civilized and civilized people had discovered and contrived an endless variety of gastronomical pleasures. The Hindoo was content with his rice; to pacify his craving for food, the gentle and meditative Brahmin shuddered at the thought of taking the life of even the most insignificant of his fellow-mortals, which, he religiously believed, partook with him of a common origin, and were to share with him a common destiny.

> "Kill not—for Pity's sake—and lest ye slay
> The meanest thing upon its upward way."

The timid Chinese shrank from the destruction of life; yet he did not hesitate to eat the flesh of beasts which had been slaughtered by others. The nomadic Tartar on the steppes of Asia fed on the diseased sheep and oxen of his own flocks and herds. Though he gave to the stranger and sold to the alien the flesh of animals dead of disease, the lofty and disdainful Jew would himself eat none that had not been struck down in the flush of health and animal enjoyment. And when Israel's people were led captive by the armies of Babylon, they became the strictest vegetarians rather than defile themselves with the king's meat. "Give us pulse to eat and water to drink," they said; "prove us for ten days." And at the end of ten days they were fairer and fatter in flesh than those who had eaten the king's meat. And when brought before the monarch, it is recorded of this stubborn race, ever triumphant in moral and intellectual contests, that "in all matters of wisdom and understanding that the king inquired of them, he found them ten times better than all the magicians and astrologers that were in all his realm."

That man should kill and eat the domestic animals about him which have been his companions and friends, and which have requited his caresses and love, has deeply affected many a pitying heart. Burns, in his "unco mournfu' tale" of poor Mailies's lament, says,—

> "Oh! bid him save their harmless lives
> Frae dogs and tods and butchers' knives."

And Goldsmith says,—

> "No flocks that range the valley free
> To slaughter I condemn;
> Taught by that power that pities me,
> I learn to pity them."

Shelley exploded in a passionate protest against destroying animals for food; and counselled those who persisted in eating flesh to do like the carnivora, who seize their victims, tear and devour them while the flesh is still fluttering with life. (Note to "Queen Mab.")

> "No longer now
> He slays the lamb that looks him in the face
> And horribly devours his mangled flesh."

Though these were the sentiments of three of the most tender and contemplative cranks of their day and generation, they have doubtless been shared by many gentle souls who never gave utterance to their thoughts. Against the practice of slaying our domestic animals for food is a gentle protest in "Jane Eyre," and a hidden rebuke in Manzoni.

That man can exist and enjoy robust health without animal food, that man can live by bread alone, is proved by individual examples, as well as by many nations which are so scantily supplied with flesh that it may be said not to enter into their diet. The thirty-six millions of Japanese are "essentially vegetarians."* They are restrained from eating flesh by poverty and by religious prejudice. There are in Japan only two head of cattle to each one hundred inhabitants, while in the United States there are seventy-three head for the same number of people. Fully one-half of the cattle slain in Japan are consumed by the foreigners. One-half of the Japanese eat fish every day, one-quarter two or three times a week, the remainder once or twice a month. The food of the masses is more than ninety per cent. vegetable. Nourished by this food, Mr. Van Buren says, they show endurance of body, power of intellect, and cheerfulness of disposition.

---

* Van Buren, Consular Report, 1881 : Food of Japanese.

The whole history of man, however, and the close but melancholy resemblance in the anatomy of his digestive organs with that of the swine, show that, like the latter, he is omnivorous; that since the Creation both have tasted and enjoyed every fruit and every animated thing, and every conceivable combination of the same which have been placed before them; that their instincts and experience have never failed to point out what is good, and, except the sanitarians, nobody has doubted that each individual is as capable of selecting what he requires to eat as any board that may be constituted by law.

In some countries, the cat destined for alimentation is not hid under the guise of a rabbit, but is flaunted in the stalls side by side with the dog; neither dog nor cat is wanted to kill the rats, for the disposition of these rodents is a special and important industry. The luscious toad is roasted and made into a hash; spiders are a *comestible recherché;* so are maggots and caterpillars; the silk- and earth-worms are held in high repute; and birds'-nests and an egg ready to hatch are the choicest of foods. Maggoty and putrid cheese is greatly esteemed in Germany and Italy; its motive power is often so great that a glass cover is used to prevent its escape altogether.

But the reign of Sanitary Terror would not have been complete if it had not assumed in some way the control of our supplies of food. As early as 1857[*] it was discovered that "enormous quantities of diseased and half-putrid meat" are on sale in London, and "the soup or gravy" is described as "an extract of pathological products of pleuro-pneumonia and typhoid fever." Mr. Gamgee [†] says that people both in England and in Scotland are consuming enormous quantities of diseased meat. Much that he saw at Newgate Market

---

[*] London *Lancet*, vol. i.
[†] Letter to Sir George Gray, 1857.

was unfit for food; yet it was being sold, scattering "the seeds of disease and possibly the stroke of death." Mr. Gamgee's periods are startling; but he does not offer a single instance of any disease having arisen in England or Scotland from eating these "enormous quantities" of diseased meat. In 1863 the flesh of cattle dead with malignant pustule is common in the market, and at all times the flesh of splenic-fever animals is for sale. No investigation was undertaken to show that this meat was harmful, no charge was brought that any person had ever been made sick through eating it; but a great outcry was raised that if such meat got into the system * it would "cause either low fever or nausea and vomiting." In 1876 it was declared that "the fattened beasts which it is the practice to provide about Christmas" were unhealthy. Laws were passed against the sale of diseased meat, inspectors were appointed, and prosecutions followed without number. Health-officers condemned meat which people were desirous to consume; the dealers resented the interference, and trials followed. Some of them were interesting, not to say ridiculous. On one occasion † the health-officer testified that the meat he had condemned would cause "various diseases." Dr. Stewart, pathologist of the infirmary, and Dr. Wood declared under oath that this meat was entirely wholesome; that under the microscope it showed as well as any other meat. Professor Dick testified that he would eat it; when asked if he would give it to his friends, he replied, "Yes; and they would lick their lips after it." At another trial the evidence was plentiful that the condemned meat was "splendid." Later still ‡ we learn that large numbers of cows in the last stages of tuberculosis are sold for food; and in 1889 great numbers of cows are

---

* London *Lancet*, 1862.
† Ibid., vol. ii., 1864.
‡ Ibid., vol. ii., 1885.

afflicted with phthisis; yet as long as they can walk their milk is consumed.

Most astonishing of all is, that no case of illness, during these forty years since the outcry against this meat, was ever proved to have arisen from its use.

But the public mind was in such a ferment about the public health that in nearly every prosecution the defendant was convicted. Surgeon T. H. Simcocks * says, "From a physiological and theoretic stand-point of view, all meat from animals killed when in a febrile state is unwholesome; but having regard to the facts and experience of the last thirty-five years, I believe there is little or no evidence whatever to show that the flesh of pleuro-pneumonically affected animals has ever caused disease or even the most transient illness." He had eaten such meat himself; had found it excellent, and he called on the gentlemen who opposed its sale to state if they had ever known it "to produce any, even the slightest, ill effects in persons who have eaten it." Dr. Simcocks said, "The point at issue is not whether such meat is sold in England, but whether it is, or is not, fit for human food." This brought a rejoinder from one of the terrorists, who said it was reported that such meat had caused malignant anthrax in Africa; and that since it had been sold in Scotland it was said that boils were more common. This hearsay evidence of what happened in Africa and Scotland was the only testimony to show that the enormous quantities eaten in London caused disease. Dr. Rubridge † says his colonial experience was, that the natives of South Africa eat the flesh of animals dead of disease, even those which die of malignant pustule. When pleuro-pneumonia was introduced into the colony, the "natives universally ate the diseased cattle, even those which died, and I never heard of any

---

* London *Lancet*, vol. ii., 1877.
† Ibid vol. ii., 1864.

evil result from it." Tardieu * says it can be affirmed that there is not a single proved fact (*pas un seul fait avéré*) that the alimentary usage of flesh from animals dead with contagious disease has communicated sickness.

The fright about diseased meat was one of the first that was brought forth in our own country. The Massachusetts Board of Health took it up quite early, and in searching for evidence found that " tons and tons" of decayed and putrid meat were made into sausages and eaten ; and that hitherto no case of disease could be traced to this cause. The board quotes the special investigation of the Privy Council by Dr. Thorne, who visited thirteen towns in England, some of which had large populations. No instance could be found of disease having occurred from the use of flesh of animals sick with foot-and-mouth disease; and it was affirmed that some of the primest meat used in London was taken from beasts suffering from acute inflammations. The Massachusetts report for 1875 says the people of the Faroe Islands habitually eat meat in a high state of putrefaction ; that sausage is the grand receptacle of vile meat, but that through cooking it may be relied on not to cause disease ; that putrid meat was largely eaten in Paris during the siege of 1871 without ill effects ; that in many of our large towns great quantities of meat from cattle sick with Texas cattle-disease are eaten without harm ; that in Illinois was a fatal epidemic of splenic fever, and that there was not a single case of disease from eating the milk and flesh of these cattle. The board concludes, " We think that the fact that meat has been taken from diseased cattle should be of itself enough to condemn it; and no meat should be allowed to leave the shambles in any part of this State without thorough inspection and permission of sale being given by a properly-qualified person." Unaccountable conclusion ! The board has

---

* Dictionnaire d'Hygiène, article " Contagion."

not adduced a single case of disease produced by such meat; on the contrary, it has presented a vast amount of testimony to prove that it is harmless.

Long before the reign of Sanitary Terror, or even before the dawn of Sanitary Science, the question as to the noxious character of diseased and putrid meat had been studied, not only by individuals, but by government commissions. Dr. Chisholm, in the Edinburgh *Medical Review*, vol. vi., after referring to Spallanzani's experiments on animals with putrid meat, and to the testimony of numerous travellers who affirm that whole tribes and races consume, without hesitation, putrid flesh and that of animals dead of disease, concludes that if, by selection, accident, or necessity, putrid or diseased flesh becomes the food of men, it is harmless.

One of the earliest reports to the French government on this subject was made by M. Huzard on the " Phthisis of Milch-Cows in Paris in 1814." Large numbers were brought here by forced marches and huddled in low, unventilated stalls; they developed inflammation of the lungs (epizootic); many perished; the greater part, however, were consumed by the people, and no evil results followed.

In 1814, when the allies came to Paris, they brought with them thousands of cattle which they had seized *en route;* these soon became afflicted with contagious dysentery. None of those which died were lost; they were eaten by the allied troops, who had no other meat; this was consumed also in the hospitals. No inconvenience followed its use. In Annales d'Hygiène, vol. x., 1833, is a report made by M. Huzard (*fils*) to the *Conseil de Salubrité* of Paris on the sale of the meat of animals dead of disease; he concludes that *charbon* alone can give unwholesomeness to meat; that there are many examples to prove that people have eaten the flesh of beasts dead of *charbon* without any accident following; but that this disease has been communicated to butchers who have killed the animals.

In Annales d'Hygiène, vol. xxii., 1839, is a report of seven scientists, to the French government, on the epizootic which prevailed among cattle in Paris at that time. After minutely describing the symptoms of the disease, the commission declares that to the eye, nose, and taste (*à la vue, à l'odorat, au goût*) the milk had all the signs of the best milk; that it had been used partly in the country, but mostly in Paris. When it was given to calves, some were attacked with the disease, others were not; when given to pigs, dogs, and poultry, no accident happened. That the milk was harmless to man was proved by the fact that long before the investigation began, and when the disease was at its height, it had been largely used without any derangement to the public health. As to the flesh, the commission says that the facts are more pronounced. Large quantities had been consumed before the disease was generally recognized, and since then the consumption had continued, and no case of disease could be found to have occurred from its use.

In a report presented to the French government* is a reply to the question, What maladies of cattle render their flesh unwholesome? The commission reports that though *charbon*, when innoculated into butchers, veterinary surgeons, and physicians, has caused death, yet it is equally certain that men in great numbers have eaten the flesh of animals dead of that disease, and that the best-informed veterinary surgeons agree that such flesh is harmless. The reporter says we ought to declare loudly (*nous devons le dire très haut*) that it is the fear that a meat may be unhealthy, and not the certitude, that causes us to reject it, and that it is the disgust that this diseased and putrid meat inspires, and not its unwholesomeness, which has caused it to be proscribed by municipal administrations.

In 1851, M. Delafond, veterinary surgeon, issued a pamph-

---

* Annales d'Hygiène, 1848.

let entitled "De l'insalubrité et de l'innocuité des viandes, etc." He says that the butchers' monopoly in Paris made a great clamor that the public health was in danger from diseased meat, because at one of the markets flesh was sold at auction. M. Delafond keenly intimates that nobody knows as well as these same butchers, who are suddenly so solicitous for the public health, how much diseased meat is sold in Paris. A mass of imposing facts, he says, have been brought for the last one hundred and fifty years to prove that the consumption of flesh from animals dead of contagious diseases is harmless: by Ramazzini in 1711, Carcain in 1714, Camper in 1745, Dufau in 1775, Beaumont in 1796, Huzard and Deplas in 1795, Huzard and Mérat in 1814, Groguier and Cozé in 1814 and 1815. M. Delafond had often witnessed the slaughtering and sale of diseased animals for food, and he believed it positively demonstrated that such meat was harmless. Decroix, V. S.,* says that for ten years ending with the siege of Paris, 1871, he had taken every opportunity that presented to eat the flesh of horses dead of glanders, typhus, hydrophobia, etc., cooked in different modes,—sometimes well done, sometimes rare. In spite of the disgust which it excited he never felt the slightest illness. A good many people had partaken with him of this food,—sometimes knowing its character, at other times ignorant of its condition. He even ate raw flesh of horses dead of glanders. He concludes that the meat of animals dead of any disease is perfectly proper for alimentation.

In the reports of the imperial customs of China are recorded many accounts of epidemic diseases of cattle in that country. It is almost always stated that the flesh of these animals is consumed by natives and foreigners, and generally that it is proved to be harmless, and disease from such

---

* *La France Médicale*, 1871–72.

cause is never reported. Dr. Scott, of Swatow,* reports a grievous epidemic among cattle, destroying as many as ninety per cent. of cows and buffaloes in some districts, and so virulent that they die in a few hours after their seizure. "The Chinese ate the flesh of the diseased animals with impunity," and ascribed the good health of the people that season to the fact that all of the disease was among the cattle. At Chefoo, Dr. Carmichael says, the flesh of animals that died of cattle-plague was eaten, and was followed by no ill effects. At Wauchow the meat sold was for the most part from animals dead of disease. Dr. Somerville, of Foochow, says that his inquiries were searching as to the effects of eating the flesh of diseased animals. "It might be eaten by any one *audacious* enough to do it." "It was a case of pure choice and audacity." He "excludes meat infested by parasites." M. Zundel, *vétérinaire supérieur* of Alsace-Loraine, in a communication to the Society of Sciences and Agriculture of Basse-Alsace,† says that though the transmissibility of tubercle from man to certain domestic animals by the stomach has been shown by some, but denied by others equally competent, there is no reason to conclude that this action is reciprocal in man, and that there is no proof that the flesh of tuberculous animals has ever caused human phthisis, though the subject has been under close observation by medical men for fifteen years. In the same volume of this review is a communication from Berlin by Dr. Villaret, who shows that by far the greater number of the cattle and swine adjudged to have phthisis in that city are consumed for food, and that each inhabitant of Berlin has a chance of eating more than two pounds of tuberculous meat annually. In the same volume Dr. Vallin says Paris is a *Minotaure;* it devours everything which can be eaten: "*tout ce qui peut se manger il le dévore ;*" that from

---

* 1880. † *Revue d' Hygiène*, 1883.

p

one hundred and fifty leagues about the city meat is brought in that is undoubtedly diseased; that the inspection is an illusion; that if this meat were condemned the inspector would be stoned by the mob; and that the question whether such meat is unwholesome is yet undecided. The only case that the author can find of disease being caused by eating the flesh of sick animals is reported by Dr. Horsch,* who saw a patient, seventy years old, with an enlargement of the inguinal glands. He found that his patient had eaten meat from an ox that had a swelling near one of its ears. The doctor says, "This is a case where I found no other cause than that diseased meat." He does not say that his patient was the only one who partook of this beef, or if he ate the whole ox! Dr. Horsch also reports that a man died of fever after eating some salted flesh of an ox which had murrain. He does not say whether in this case the patient ate the whole ox. That tubercle can be imparted to man by the ingestion of flesh of animals which have had this disease is, so far, pure hypothesis. M. Trasbot † declares that no case of such transmission has yet been brought forward, yet the opinion has become a dogma. This "opinion," which has become a dogma, is sharply contested by the most acute French and German observers. Flügge ‡ says that raw tuberculous meat *may* give phthisis to man,—he does not say that this is proved,—but when cooked it is harmless, and that there is no proof that the ptomaines, which are sometimes developed in putrid meat, are caused by sick beasts.

A most important fact is this, that in all or nearly all of the cases of poisoning from meats, cheese, or other animal foods which have been reported during the last seventy-five

---

* Proceedings of the American Public Health Association.

† *Revue d'Hygiène*, 1890.

‡ *Grundriss von Hygiene*, 1888.

years, no investigations have determined the cause; and in almost all it has been proved that the aliments were of entirely good quality. In Kloten, canton Zurich, in 1878, at a musical fête, a large number of people fell ill after partaking of the banquet. None of the viands were distrusted except the flesh of a calf, which, "it was said," had evaded inspection. It was not proved that the calf was diseased, but it was so presumed; and as many persons were seized who had declined the veal and had eaten only beef, it was presumed that all of the meat had been contaminated by using the same knife which had sliced the calf! The disease was diagnosed gastric catarrh and typhoid fever, which autopsies confirmed. One hundred and twenty-one of these cases were pronounced by M. Walder to be true typhoid. It would seem that either the diagnosis of typhoid fever was not correct, or the meat was not at fault; for the day after the fête there was an abrupt explosion of the disease; the larger number of cases developed before the third and fourth day, and none later than the ninth, after partaking of the meat. There had been no case of typhoid fever at Kloten,* nor in the neighboring villages, for a great many years. On this occasion the public listened with avidity to all sorts of stories. A dog which had gnawed a bone that had been tossed from the banquet-table was seized with convulsions, and a crocodile that was in a menagerie in the town had misery in the bowels (*maux de ventre*), after having been fed with some of the beef, though the account does not say to whom the honor belongs of making the diagnosis.

It may be that this attack of so many people at Kloten is of that large number of mysterious seizures which are unexplained. It is stated that forty or fifty of these cases followed by propagation from the original ones. All efforts to inoculate calves or other animals by feeding them with the stools

---

* *Revue d'Hygiène*, 1879 (Zuber).

of the sick people failed. Another singular fact is, that these examples of what is supposed to be poisoning from meat, cheese, and ice cream almost always happen where a large number of people are gathered together, as at a fête. The true cause of these attacks is not yet explained.

One hundred and sixty-five pages of the report of the New York City Board of Health for 1868 record the appearance, course, symptoms, and morbid anatomy of the Texas cattle-disease in that city in July, 1868. There was widespread alarm, which was increased by a notice from the board that there was a sudden rise in mortality in that month from diarrhœal disorders; and it was bruited about that the meat of sick cattle was the cause. It was known that large quantities of it were eaten; but if the board made any inquiry to discover whether it was connected with the bowel troubles, they made no record of the fact. The fact is ignored entirely in the report, but is recorded elsewhere in the board's general report, that the mean temperature for the three months ending October 3, 1868, was five degrees higher than it was the year before in the same months. The fact is ignored also, but recorded elsewhere in the report, that the great rise in mortality that year was among children under one year. Not a single case of disease from the use of the meat is offered; and although it was known precisely where the great increase in mortality lay,—namely, in the Nineteenth, Twentieth, and Twenty-second Wards,—if it bore any relation to the use of the meat and the board knew it, they were careful not to mention it. The Registrar of Vital Statistics that year says, "The phenomena of rapid increase and obstinate fatality of diarrhœal disorders" were more marked in the Twentieth and Twenty-second Wards, and later in the Nineteenth, "than in the infamously ill-housed and crowded Fourth and Sixth Wards." The quantity of fatal disease in the Twenty-second Ward is "something enormous, though the social and domestic condition of the people and the natural advantages of locality are far superior to those of the Fourth and Sixth Wards." Was the diseased beef sold and consumed in the well-to-do Nineteenth, Twentieth, and Twenty-second Wards? The board does not say whether any inquiry was made to ascertain the fact. In an authoritative way the reporter, who presumably speaks for the board, says, "It is idle to talk of such meats being safe and wholesome for food." We respectfully submit whether it is not "idle to talk" about it at all until positive information is obtained. The board made no effort to still the great alarm of the people; on the contrary, their words and actions all tended to exaggerate it; and this report is not unfrequently quoted by sanitary writers to prove the noxious character of diseased meat.

Parkes,* after summing up the testimony on diseased meat, directs the British army surgeon, in time of war, if healthy beasts are not attainable, to "allow the issue of all animals ill with inflammatory and contagious diseases, with the exception of small-pox, and perhaps splenic apoplexy in sheep." But he says, if necessity compels the use of animals sick with small-pox, great care in cooking should be used. The untutored man in Sanitary Science is here again puzzled. Why a food which is suitable and healthful for a soldier in time of war, while he is defending his own or invading another's country—just at the period when it is most important to preserve his vigor—should, as soon as peace is declared, become so perilous to the public health that those who buy and sell it are to be fined and imprisoned, must forever remain one of the impenetrable mysteries of Sanitary Science.

Notwithstanding that the sanitarians had no evidence to implicate in the causation of disease any of the foods which they had proscribed, the public mind, which had been so

---

* Hygiene, 1887.

susceptible to fright about the air, the water, the soil, and the sewers, was now in a condition to be panic-stricken about diseased and putrid meat.

That the flesh of animals begins to resolve itself, under the influence of heat and moisture, into its original elements as soon as life is extinct is a self-evident proposition. The sanitarians now began to excite the fears of the people that the public health was in danger, unless they were endowed with power to determine through their inspectors the precise moment when this decomposition was a source of peril. They professed to see great danger in the consumption of immature veal; and in many instances procured the passage of ordinances forbidding its sale, if slaughtered under four weeks old. What diseases such food would cause they never deigned to tell us. It is, in fact, as innocent as a new-laid egg or a raw oyster. It was known that in many countries *fœtal* veal was largely eaten; in some it was esteemed as a delicacy. Why it should be wholesome and desirable in Italy and the Argentine Republic, and impair the public health in England and the United States, we were never informed. Regarding roast pig, we have the uncontradicted testimony of Charles Lamb,—" It must be under a moon old; guiltless as yet of the sty." The fat,—" Oh! call it not fat, but an indefinable sweetness growing up to it,—the tender blossoming of fat." " The lean, no lean, but a kind of animal manna, or rather fat and lean (if it must be so), so blended and running into each other that both together make but one ambrosial result."

The sanitarians never explained why feathered game, and venison and mutton, in various degrees of offensive putrefaction, were to be eaten with impunity and even with gusto, and beef and poultry in the same condition should be shunned as dangerous. Not one of them raised his voice against artificially-diseased geese-livers, which the gourmand so highly prizes under the name of *pâté de foie gras*, and which lately, according to M. Mouillé, have been found to be a pure culture of tubercle. But they succeeded in getting their inspectors appointed, and for a while the public tranquillity was restored.

CHAPTER XI.

The Milk.

AFTER the installation of the meat inspectors a new sense of security pervaded the community, and the outlook for the public health grew brighter. The air, the water, the soil, the public improvements, the cemeteries, and the meat were under careful surveillance.

Once in a while we were somewhat dejected, and our faith in Sanitary Science wavered a little, as we surveyed the new dignitaries and reflected on their antecedents. They were chiefly either party mendicants or the poor relations of more ambitious politicians,—men who had never before in life been noted for anything but a general inefficiency. Not one of them made any pretence of possessing scientific attainments. We tried to think that maybe by some process of laying on of hands by the apostles of the new doctrine, a kind of sanitary grace or unction had been imparted to them; but in spite of these reflections we had to confess to ourselves that if, before their appointment, we had needed any counsel regarding the supply of our table, we would not have sought for it from the new inspectors.

But the sanitarians, in their journals and on the platform in convention assembled, cheered us with the news that Sanitary Science was making "gigantic strides;" that yet a little while and the enemies of the public health would be subdued, and the mockers at the founders of the new truths would be silenced. All at once, like a thunderbolt out of a clear sky, an alarm was sounded from a most unexpected quarter. There was poison in the MILK!

If there was anything on the earth's surface which had hitherto escaped the charge of original or acquired sin, it was this bland and innocent fluid. Ever since we had recourse to the maternal bosom, our respect for its purity had been maintained. The sanitarians now commenced to smut and defame it as if it were a vile drab. It was the messenger of typhoid fever, diphtheria, and scarlatina; and from thirty to fifty per cent. of all milch cows poured the tubercular bacilli fresh from their udders. The blow was all the more stunning, because for twenty-five years there had been a singular concord of opinion among practising physicians that milk was a most appropriate diet during the progress, and especially in the convalescence, of the first three diseases; while in consumption it was specially adapted, and unless some idiocrasy forbade it, it had been unconditionally recommended to be used as soon as possible after it was received from the cow. But now we were told that the animals whose milk had been esteemed the highest for consumptives, on account of its richness in cream, were precisely those that furnished the tubercular bacilli in the greatest abundance.

For a quarter of a century the doctors had been drenching us with these bacilli, and had been adding fuel to the flames which were destroying us, and at the same time had been reporting improvement of their patients, and an occasional cure, under the use of milk. There had been this consoling thought about the meat: it was consumed by adolescents and adults, who possessed a certain amount of resisting power to morbid influences; but the new danger menaced us at a point which we were least prepared to defend. It shook the foundations of society, and threatened the annihilation of the race by a wholesale destruction of infantile life. Our very hopes of posterity were now to be blasted by the Herods of the milk-can.

The dispensers of this fluid, the sanitarians said, not only sold diseased milk, but they were careless in the delivery of that which was sound. Milk was quick to undergo changes, and these variations in its character were fruitful sources of disease. It was of no use for some to mildly protest that for six thousand years men, women, and children had taken milk in all sorts of conditions; fresh from the mother cow; acid from commencing decomposition; still further resolved, it had been prized under the appetizing name of bonny clabber; children had begged for buttermilk; after it had parted with its caseine and fat, sick, well, and convalescents had swallowed the whey with delight.

The sanitarians said that all this violated "the settled principles of Sanitary Science;" that milk passing through these changes was filth, and would cause filth-diseases; and that the only way to protect the public, save the infants, and restore the invalids, was to have a proper number of milk-inspectors appointed. As soon as these were invested in office, the public tranquillity was, in a measure, again restored.

Those who knew that any examination of milk, to be worth anything, called for expert knowledge in the operator and a familiarity with the technicalities of organic chemistry which required years to obtain, saw through the limpid humbug; to the mass of the people, in the excitement, it was enough to be told that *somebody* was inspecting the milk.

Twenty years after the era of sanitary reform, typhoid fever was continually breaking out in England, just as it had done previously to 1850. Some new creation was necessary to continue the excitement and extend the panic. It often happened in that country, as elsewhere all over the world, that a typhoid fever epidemic was circumscribed in its extent by a few acres; that sometimes it would be confined to one side of a single street, or to a certain number of houses in a single block. The London *Lancet** reports a typhoid epidemic limited to an area of forty square rods, where were twenty-four cases. Another at Greenock was confined to a spot of five hundred yards, in the most airy, best-drained, and reputedly healthy portion of the town. One epidemic in Massachusetts was confined to a space fifteen rods square. It happened sometimes that these circumscribed places received the same or nearly the same milk-supply.

In none of the so-called epidemics from the use of milk which the author has studied is there any truly scientific evidence to support the belief that this was at fault. In all of them there is a varying proportion of cases of the disease which are admitted to have some other—some unknown—origin. That the admitted unknown origin was competent also to cause the other cases is denied by none. It is ten years since the typhoid bacillus has been discovered; the sanitarians avow that without this germ typhoid fever cannot occur. In no case of so-called milk epidemics of typhoid fever has the bacillus been found in the milk. A singular fact is that almost all of these epidemics have occurred in England, and, it might be said, close to the ear of the London *Lancet*. More than seventy had been reported as early as 1880. As soon as it appeared that this cause of disease was to become the mode, and that stories of milk-poisoning would be acceptable to that journal, the compliant English physicians poured in accounts of epidemics caused by milk, thicker than were the tales of Popish plots two hundred years before.

At first it was necessary that a case of typhoid fever should have occurred at some time, near or remote, somewhere in the vicinity of the suspected dairy; later, a case of rheumatism answered the purpose; and in one extensive epidemic † it was proved by the testimony of two surgeons

---

* Vol. i., 1874.   † *Chemical News*, vol. xxxii.

that the person who was accused of spreading typhoid fever through milk actually had heart-disease. Later still, no disease at all among people of the dairy was required to make out an epidemic from milk; the affidavit of somebody that a cow had been dallying with a pool of dirty water or had drunk from a well which was located near a stable or privy was enough.

It was asserted that typhoid fever had been caused by the milk of a sick cow. One epidemic was ascribed to the milk of a cow suffering from metritis. In another, a dairyman admitted that three of his cows were affected with a "stiffness" a short time before typhoid fever broke out in his family. It sometimes happened that the disease had existed days and weeks on the milk route before it broke out at the dairy; but these passed for epidemics from the use of milk, just as if the occurrence of the fever had been reversed. One epidemic is reported as arising from milk which had been contaminated by a pool of stagnant water from which the cows had drunk. One surgeon writes that he is convinced that the milk of cows which have drunk dirty water can produce typhoid fever; in India he had seen them licking manure, which he thought made the milk impure, and caused typhoid fever among the soldiers! This was a little too much for even the London *Lancet;* it was not quite ready to accept the stagnant-water theory, "since it would be subversive of all that is known and admitted concerning the etiology of the disease." One epidemic of typhoid fever caused by milk is reported at Ascot,* which lasted four and a half years. No epidemic had been known in Ascot for forty years; only now and then a case had occurred there during these forty years.

Dr. Beveridge † reports an epidemic caused by milk from the Oldham Reformatory. An official investigation showed

---

* *Sanitary Record*, 1879.   † *Sanitary Journal*, Glasgow.

that a cistern at the dairy had a loose cover, and it was decided that the water could absorb exhalations from the barn, and corrupt the milk through washing the cans with it. Analyses of the water were made by two different chemists; one pronounced it pure, the other disclosed albuminoid ammonia. The investigator chose the latter analysis, and so convicted the milk of causing the disease. There was no sickness of any kind at the Reformatory; it is expressly stated that all of the boys were well, although they partook freely of the milk. This is explained by the fact that the boys did not use the milk "exclusively," and that "what the boys got was evening milk, skimmed; and various circumstances show that the evening milk, especially if skimmed, was the safest thing about the place to use." Nothing is said whether any one at the dairy drank of the water which infected the cans, which infected the milk, which infected the people.

In 1883 * there was an outbreak of typhoid fever from the use of milk at St. Pancras; but there was a remarkable immunity from disease at and near the farms that supplied the milk, and now it is hinted that maybe the cows had this fever in a mild form, although they appeared in good health. The editor of the *Lancet* says, "It is a debated question whether the milk is infected on account of the cows being affected with a mild attack of the fever," and being, "so to speak, the wet-nurses that carry the infection of enteric fever." "Exact observations," he says, are much needed to elucidate this subject.

Dr. Louis Parkes † relates an epidemic at St. Albans among the customers of a particular dairy. There was complete absence of any evidence that the milk had become infected. Dr. Parkes says that an outbreak of scarlatina

---

* London *Lancet*, vol. ii.
† Transactions of the Sanitary Institute of Great Britain, 1879.

was traced to milk from a farm at Farnham. Mr. Power reported that it was impossible that the milk could be infected "in any of the commonly-believed ways." However, the investigation was not allowed to rest there, and the astute inquirer, who showed great sanitary discernment, was rewarded by the discovery that the cow had "here and there lost a portion of her coat, and that her buttocks and posterior udder were soiled and stained by excrementitious matter." This revelation satisfied the investigator that the milk caused the scarlet fever.

In another scarlatina epidemic which was caused by the milk, some vesicles were found on the teats and udders of the cows. The animals showed no other signs of illness; they took their food, and gave their usual supply of milk, and their bodily temperature was normal. In fact, these vesicles would very likely affect the health of the cow much as a pimple on a man's face would affect his general health. The English investigators, however, seem to have been perfectly contented with such explanations of these epidemics. Dr. Parkes says that in epidemics of diphtheria caused by milk, "It has not been possible in a large percentage of the cases to trace the source from which the milk derived its infective quality." One outbreak of this disease at Hendon was attributed to a "ropiness" in the milk due to garget in the cows, although no evidence was obtained of this or any other disease being prevalent at the farm among the cows prior to the outbreak." It may be thought that Dr. Parkes was indulging in a little pleasantry in the presentation of this paper on "milk and disease." However, his hearers do not seem to have perceived it, for it is related that "a vote of thanks was accorded" to him for his discourse.

Diphtheria at Camberly is reported * as being caused by milk. The sanitary condition of the dairy which furnished

---

* *Sanitary Record*, 1887.

it is exceptionally good; no disease is there, either among the cows or the people. The undaunted sanitarian who investigated this epidemic pushed his inquiries still further and found that "two of the animals showed some signs of chaps on their teats," and this accounts for the diphtheria at Camberly.

The *Lancet* of August 13, 1892, tells us of outbreaks of scarlet fever in London which were "traced to the distribution of an infected milk-supply," but "careful inquiry failed to show" that there had been any disease at any of the dairies. The investigators did not cease their labors until they had found a cow which was said "to have a cough" and to be "off her feed." This cow and four others had some abrasions on their udders. The six days that the cow suffered with the cold were "the six days of maximum infective power of the milk," and the very day on which treatment of the cow was commenced "the infective property became less marked." In spite of the solemnity of the editor of the *Lancet*, who records and comments on these remarkable facts, we are sure he is jesting, and his imperturbable gravity heightens our amusement as we listen to him.

In 1889,[*] Dr. Anderson reports at Dundee an epidemic of typhoid fever occurring on a particular milk-route. "The most searching inquiries failed to find any trace of the disease among those handling the milk, or in their households." On one of the cows, however, which was in perfect health, an eruption like a ringworm was found on one of the teats. Dr. Smith had the temerity—no one else shared his rashness—to say that he saw no connection between this eruption and the epidemic of typhoid fever at Dundee.

The excitement reached such a pitch in England that the

---

[*] *Medical Record*, September 7.

most stringent regulations were made; inspectors were appointed, and, of course, prosecutions were commenced. For the most part the dairymen were glad to flee the wrath of the implacable sanitarians, but here and there was one determined not to succumb without a struggle. He was always worsted. There was one who attempted to defend himself on the ground that there was no history of any disease at or near his dairy, and he refused to go out of business. His testimony was unshaken, but the *Lancet* told him sharply that "those dairymen were likely to fare best who swim with the stream, instead of attempting to stem it." The dairyman seems to have taken the hint, for we hear no more of him afterwards.

It hardly ever happened that there was any accusation of a direct infection of the milk, but a favorite method of accounting for the epidemics was that the stools of typhoid patients, thrown on the ground or into a privy, retained their infecting power for months and even years, and that the germs percolated through the ground and infected water-supplies which were used to cleanse the milk utensils. The germs of typhoid fever were said to hold their vitality in the ground * for an indefinite period; "for several years at least." Dr. Frankland reported an instance in Switzerland where the typhoid poison filtered through earth a mile away.

There was something very bewitching to the sanitary fancy in this roundabout way of accounting for an epidemic of typhoid fever from milk. It looked to the reformers like science, and they seemed to imagine that they were unravelling some plot against the public health. It must be remembered that no one claimed to have any proof that water sources were thus polluted. The sanitary Gullivers never pretended that they had made any investigations to show

---

* Connecticut Board of Health Report, 1882.

the possibility of so defiling water-supplies. They relied on their imaginations and the credulity of the public. By the light of later experience on a great many sewage-farms, and the still more positive evidence of experiments with the typhoid bacillus in ordinary earth and in the presence of putridity, every one of these epidemics which have been ascribed to water polluted in this way must be thrown out altogether.

Towards 1883 the English mind seems to have been surfeited with these tales, for Dr. Saunders * warns people against "being too ready to credit milk with more than its due;" and he shares "with many others the opinion that in a number of cases in which milk-pollution has been asserted to have produced disease the proof has by no means been clear," but in 1884 the excitement is renewed, and the *Lancet* says one can hardly take up a morning newspaper "without discovering that disease and death have been disseminated through the milk-can."

No attempt seems to have been made to transplant epidemics from the use of milk to the continent. That they arose nowhere but in England attracted the notice and excited the merriment of the French, who described the milk pathology of typhoid fever in England,† as "*l'espèce d'entrainement*" which had reigned there for the last few years. Jules Arnould ‡ says that the merit of having made the sagacious observation of surprising milk *en flagrant délit de colportage* of typhoid fever belongs to Dr. Edward Ballard, and that this high treason (*haute trahison*) on the part of the milk was first committed at Islington. Arnould says a little calm is now § restored, although there is no reason for thinking that any more trust should be put in a milk-dealer. He thinks that these milk stories are

---

* London *Lancet*, vol. ii., 1883.
† Annales d'Hygiène, 1882.
‡ Nouveaux Élémens d'Hygiène.   § 1881.

wanting in precision, and do not show much *perspicacité* on the part of the historian. Dr. Vallin * says that there is not a week which does not bring forth some story of an epidemic caused by milk in England, and that an English physician who would throw any doubt on the reality of such a mode of propagation would be considered *malaisément.* Dr. Vallin says that what justifies our scepticism regarding these epidemics is that the English have been content to *présumer* that the milk is infected. He is not aware of a single microscopic examination. No one there, he says, has even sought the "*corps du délit*" in the milk. He adds that a serious observer in France would not relate such an epidemic until he had at least searched the milk. Bouchardat † says that in France, where the use of milk is much more general than in England, we rest in complete security respecting its being a conveyer of typhoid fever.

The sanitary Anglomaniacs of this country have made some desperate efforts to introduce the milk theory of epidemics here. One of the most brilliant achievements of Sanitary Science in this direction is recorded in the report of the Connecticut Board of Health for 1890. Here is related what is called an investigation of the rise and progress of an epidemic in Waterbury, Connecticut, in June, 1890. The investigators in this example were true to their imitative instincts; they servilely followed as closely as possible the methods, and are entitled to the acclamations of the English explorers. For the previous five months typhoid fever had prevailed in that city, but in June there was a sudden increase of the disease. Fifty cases happened in that month, and forty-one of them occurred on the route of one milk-dealer, who furnished six hundred quarts of milk to between twelve hundred and fifteen hundred people. Three cases broke out

---

* *Revue d'Hygiène*, 1881.
† Traité d'Hygiène.

before the 10th, eighteen before the 15th, and twenty-seven before the 20th of June. The farm-hand, who was the only one accused of infecting the milk, was taken sick on the 7th and removed to the hospital on the 9th of June. Before the 7th he had defecated in the barn-yard, thus says the report, "infecting the mass of material there," and portions of the infected mass being dried were blown as dust into the milk-cans, which were placed in a tank in the barn. *The fact is established that the farm-hand had no diarrhœa previous to his* removal to the hospital. If it was possible for him to thus infect the manure, so that this by the aid of sun and wind could corrupt the milk, the latter infection could not have happened before the 9th or 10th of the month. But by the 10th three (one occurred on the 8th) of the forty-one cases had typhoid fever, by the 15th eighteen, and by the 20th twenty-seven of the forty-one cases were down with the disease, eleven days after exposure.

The reporter of this epidemic gives from twelve to twenty days as the incubative period of the disease; but here we have twenty-one of the forty-one cases occurring within a week, and twenty-seven within eleven days after exposure. The experiments of Karlinski[*] showed that of twenty-one cases of typhoid fever that he examined, the typhoid bacillus could not be found before the ninth—in the larger number it was the eleventh and twelfth—day after the patient was seized with the disease. To admit that the milk was the cause of these twenty-seven cases, we must reconstruct entirely the theory of the incubative period of the disease, which the reporter of this epidemic recognizes as from twelve to twenty days. There were a number of facts which, if the Connecticut Board of Health obtained them, they were careful not to record. A cluster of twenty-five families living in the "Burnt Hill" district took this milk and

---

[*] *Centralblatt*, 1889.

no fever followed. Twenty-one people visited the farmer's house; all partook of the milk and none of these had any fever. Eleven men worked in the infected barn-yard during the month of June; nine of these drank of the milk and none of them took the fever.

Nevertheless, this tale has gone into sanitary literature as a genuine epidemic from the use of milk. Strange to say, in this instance, like all others of the sort, the bacillus was not searched for either in the manure which infected the milk or in the milk which infected the people. The water in which the cans stood, but which was not accused at all, was examined; no bacillus was found. The only really important work which the Connecticut board accomplished in this pretended investigation was to ruin the business and smirch the reputation of two unoffending citizens. Can it be possible that the board did not believe a word of their theory respecting the cause of the epidemic? that for the glory of Sanitary Science and to inflate their own vanity they were ready to sacrifice the reputation and prosperity of two innocent men? We are fain to think so; for we would absolve them from the cruel alternative, which is this, that if they believed what they said they were guilty of criminal negligence. There were at least sixty cart-loads of this manure the mass of which the reporter said was infected with typhoid-fever germs, yet the board of health allowed it all to go uncremated without a word of warning; they did not even recommend its disinfection. Those sixty cart-loads of manure were hauled out of the barn-yard,—so the farmer writes the author,—spread on different portions of the ground of a large farm, to be dried by the sun, blown by the wind, to be absorbed and filtered through the earth, infecting perhaps the air, water, and soil of the whole State of Connecticut!

The milk terror regarding tuberculosis is still hovering over us. Frightful tales are told of ten, twenty, thirty, fifty, even sixty per cent. of some herds being afflicted with phthisis. There is no proof that the ingestion of meat or milk of tuberculous animals has ever caused phthisis in man. Although cases of this mode of infection have been sought for twenty years, the terrorists are constantly challenged to produce a single instance. Bouchardat* says that, though there is not a single fact to establish the transmission of tuberculosis to man from milk, he thinks it is safer to boil that from a tuberculous cow. But, he asks, what inhabitant of Paris has not drunk tuberculous milk? The untutored man in Sanitary Science may —somewhat ostentatiously perhaps—apply the common-sense or ancestor argument here as he did about the ice. When he has recovered from his fright he will be likely to ask some questions. For example, if the bovine race is so thoroughly tainted with tuberculosis, and so prone to communicate it to man, why is it that those who are constantly employed about these animals, who are handling their milk and manufacturing it into butter and cheese, are less subject to consumption than those engaged in many other employments? Dr. Bryce, in a paper presented to the Conference of the State Boards of Health in Washington, May, 1891, says that phthisis is twice as prevalent in our large cities as in the country. The common-sense man will inquire, If the bovine race is one of the great sources of propagation of this disease, who is it that those who are in daily contact with these animals are not nearly so likely to have the disease as those why seldom or never come near them? He will ask, If tuberculous meat can convey phthisis to man, why is it that butchers who slay these animals, and who are daily handling the meat, are not more subject to phthisis than many other persons who are not exposed to it, except as it is served to them after it is cooked? He will ask, If one-third of the milk which is consumed is tuberculous

---

* Traité d'Hygiène.

and can impart tuberculosis to the consumer, why is it that those who take the most of this milk, at the most tender and susceptible age,—children under five years,—should only rarely have phthisis, and those beyond puberty, who take the least milk, should be the most obnoxious to the disease? The common-sense man would like to know if the rule in Sanitary Science, that two and two make five, and that the less contains the greater, never suffers deviation.

The common-sense man will ask why the people of those countries where neither flesh nor milk is eaten should be just as subject to phthisis as those who are large consumers of both. George Godet * says that the ox in Japan is of small size, and has always been reserved for agriculture and a beast of burden. It is only within the last ten years that he has served for food; and that now there is hardly beef enough in Japan to supply the foreigners. Phthisis, he says, is common at Hakodaté. Dr. Scheube † says that only very lately has beef been eaten in that country. In 1879 only twenty-four thousand cattle were slaughtered, and in 1880 only thirty-six thousand. Butter and cheese are not made at all in Japan, and milk is taken only as a medicine. Dr. M. Miura ‡ says that phthisis is a widespread disease in Japan (*ausserordentlich verbreitet*). In the five weeks ending August 2, 1890, three hundred and twenty-three deaths were reported as caused by phthisis in Tokio.

Cattle in Alaska (Bancroft) are very scarce. The natives, being accustomed to fish for a diet, do not relish milk or flesh, and regard animals as a nuisance. Among the fatal diseases that prevail there, Bancroft names consumption as the first; then fevers, syphilis, and scrofula. On the other

---

* Les Japonais Chez Eux, 1881.
† *Archiv für Hygiene*, 1883, " Die Nahrung der Ja aner."
‡ *Virchow's Archiv*, 1885.

hand, in Corea Dr. Koike * says that not less than three hundred thousand cattle are slaughtered annually, which are eaten by seven millions of people; while in Japan not more than thirty-two thousand are eaten by thirty-seven millions of people. Dr. Koike says that phthisis is rare in Corea, though he frequently met with scrofulous children. Dr. Fuji † says that one of the least common diseases in Corea is phthisis. In the monthly return of cases in one hospital, from 1879 to 1884, three cases only were found. A Japanese physician told Dr. Fuji that he examined only one case of consumption in Corea between 1886 and 1889. A Corean physician assured him that the disease is seldom found among old people, and its scarcity everywhere in Corea is an undeniable fact.

Every Friday in New York City, at the intersection of Ludlow with Hester Street, there is the centre of an open-air market which radiates west nearly to the Bowery. It almost touches Clinton Street on the east and south, and north it reaches close to Canal and Grand Streets. There is probably no such collection of people or goods to be found elsewhere in the world. Fish is for sale here at two cents a pound, eggs at six cents a dozen, and cheese at four cents a pound; pineapples at three cents a-piece, and strawberries at a cent a basket. Here are vegetables and fruits in such conditions of crudeness and decay that a man of sensibility would not feed them to his pig without first offering an apology. Here is poultry so emaciated that its bony anatomy can be studied without further dissection. Beef and mutton are in much the same condition. Nothing but phthisis in its last stages could produce such leanness.

Saturday nights on Ninth Avenue, between Thirty-sixth and Forty-second Streets, are offered the same kind and

---

* *Archiv für Ethnographie*, " Zwei Jahre in Japan."
† *Sei-i Kwai Medical Journal*, 1890.

variety of food, with the addition of pork. All day Saturday in the Italian quarter on Mulberry Street, between Chatham Square and Bayard Street, is a market superior to either of the others in neatness, but the merchandise here so tastefully displayed is the refuse of the better New York markets. Æsthetically considered, the food exposed for sale in these three markets should be swept into the sea or used on the land as a fertilizer. All of it is repulsive to the eye, and the most of it is disgusting to the nose. But, from an economical point of view, its destruction would be a crime. The fact is, that tens of thousands of people consume it, and to all appearance thrive on it as well as their more fortunate fellow-men who purchase their supplies from more luxurious markets. It is doubtful if the former suffer as much from digestive symptoms as the latter, and that the consumption of the goods from any of these places can cause infectious, zymotic disease, or in any way imperil the public health, is pure absurdity. Nearly every week, however, a troop of urchins can be seen at the heels of some insolent official of the New York Board of Health who swaggers through these markets and orders the destruction of the stock of some poor, cringing vender who can ill afford to bear the loss, and whose goods if they were consumed would harm no one. The next morning the papers record the deed with approbation, and the people of New York for the moment breathe freer, in the trust that if a great disaster to the public has not been entirely averted, it has at least been postponed.

CHAPTER XII.

Filth and Fecal Diseases—Typhoid Fever.

At a very early period of the sanitary excitement the reformers were seized with the notion that certain diseases originated in filth, in putrid emanations and in drains; and at length they grouped all epidemic and contagious disorders under the name of filth-diseases. At one time or another each disease of the zymotic class, no matter how specific might be its nature, was said to be not only propagated by filth, but actually to originate in and be created by filth. Could this be destroyed, the extinction of such diseases would necessarily follow.

This idea permeates the whole mass of sanitary literature which has deluged the public for fifty years. It has never failed as a theme for the sanitary orator. It has been the basis of all sanitary codes; has formed the groundwork for innumerable annoyances and prosecutions; it has been the business foundation of boards of health. These proclaimed that privies, cesspools, sewers, and drains were competent to excite any and all of the zymotic diseases. The absurdity of this doctrine was so apparent that, after a few years, the craziest sanitarian forsook a portion of it. To have yielded it entirely would have sent the whole body of reformers into obscurity, unless some new device had been hatched to terrify the people; so they proclaimed that diphtheria, typhoid fever, cholera, and yellow fever were dependent solely on filth, and that the last three were fecal diseases. No scientific inquiries were made previous to this manifesto to confirm its truth, and although the theory was as genuine a caprice as any which had preceded it, soon afterwards it became one of the "settled principles of Sanitary Science" that the source of these three diseases lay in human fæces.

A glance at the process of digestion shows nothing on which to build this opinion. The food which largely supplies the entire human family,—except a few tribes,—and which is almost the sole nutriment of the majority of mankind, undergoes decomposition as soon as it passes the lips and meets with the saliva. If nitrogenized aliment be taken also, both foods pass into the stomach, where they come in contact with a new fluid and ferment, which further disintegrate and decompose them; their original properties are lost. The result is, from the sanitarian's point of view, a mass of filth. It is giving forth gases which are identical with those which are evolved by putrefaction outside the body; but if these putrefactive and digestive changes were to cease we should die. They proceed until the contents of the stomach not directly absorbed into the circulation are poured into the small intestine, and the residuum of the contents of the small intestine are emptied into the large one. Here the deposit is retained for a variable period in different individuals, who may all enjoy the same measure of health. It may be evacuated once, twice, or thrice in twenty-four hours; or it may not be dismissed for two or three days. Through mechanical obstruction it may be held for many weeks without ever exciting infectious disease. In either event the act of defecation is as respectable as that of deglutition, or the intervening involuntary peristaltic movements. Because the final product suffers expulsion, it does not merit disgrace. It is just as precious as any that has preceded it; and instead of being rushed ignominiously into a sewer, and thence into the sea, should be carefully stored for use in the soil. There a new career is prepared for it, where it can participate in new growths,—for therein lie the resurrection and the life,—and reascend in leaf and bud and flower and fruit, to be again appropriated by man and bird and beast.

That human fæces ever produced specific disease is a pure phantasy of the sanitarians; they never offered any evidence to sustain the assertion; there is nothing in analogy to support it. The tree that sheds its fruit and leaves does not sicken and die when these decay through heat and frost and rain, and the resulting products are reabsorbed in trunk and limbs preparatory to another fruition. On the contrary, it will wither and fade if they are taken away and their place is not supplied by other putrefied matter. None of our domestic animals are made sick by contact with their own fæces, whether these are lately or remotely passed.

But human excrement is the most dangerous and the vilest of all filth, say the sanitarians; they neglect to furnish the proof. They never even made an investigation to obtain such proof. Long ago, Baron Liebig told us that human fæces were, in chemical constitution, almost exactly like guano; and the practical mind of that philosopher was grossly offended that the English should pour their sewage into the ocean, and then send ships fifteen thousand miles away over the same ocean to bring back the fæces of birds, to repair the loss of the wasted fæces of men. He contrasted this prodigality with the economy of the Chinese, who, since the building of the pyramids, had made their soil not only to support the swarming millions in China, but to contribute widely to the food-supplies of other nations. He declared that the example of the coolie, who, after the sale of his produce in the town, invariably transported human excrement back to his home to renew the fertility of his land, should be followed by every European farmer who took his sack of grain to market. He warned the people that the supply of guano would some time be exhausted; that famine awaited the nation that did not return to the soil those elements which had been abstracted from it by vegetable growth; and that every scrap of animal refuse,—blood, bone, muscle, sinew, and fat,—and every pound of human fæces, and every ounce of human urine, should be saved, that they might perform the dignified office of restoring the equilibrium of the soil, to the end that the human race might not disappear from the face of the earth. That the shallow mind of the sanitarians could comprehend any of these considerations was not to be expected; besides, their standing as scientists and, to a great extent, their revenue depended on perpetuating the terror respecting the power of human fæces to produce disease; and we were told that there was nothing in Sanitary Science better established than the fecal origin of

TYPHOID FEVER.

If this now ubiquitous disease afflicted mankind before the beginning of the present century, its distinction from other forms of fever was not recognized by physicians. In 1813 Petit and Serres made known that in one form of continued fevers the glands of the mesentery and intestines suffered a lesion. They called this disease *la fièvre entero-mésentérique*, and said that it had been seen in Paris for eight years. "La fièvre entero-mésentérique," they ask, " est-elle distincte des autres ? Est-ce une maladie nouvelle dépendante d'une constitution atmosphérique ?" They remarked that it attacked, for the most part, young people lately arrived at the capital; that youth and change of abode were predisposing causes; they said nothing further respecting its etiology. In 1829, Louis published a voluminous work of nearly a thousand pages, wherein, through the most laborious and exact investigation, he seeks to illustrate the morbid anatomy, symptoms, diagnosis, causes, and treatment of the disease, which he calls *affection typhoide*, and which has since been known all over the world by physicians as typhoid fever. The keynote to the whole book lies in the motto, from Emile, which decorates its title-page. " Je sais que la vérité est dans les choses, et non dans mon esprit qui les juge, et que moins je mets du mien dans les jugements que j'en porte, plus je suis sûr d'approcher de la vérité."

It is not too much to say that, except for the aid the thermometer has bestowed in proving the diagnosis and assisting the prognosis of this disease, and except for the latter biological researches respecting its bacillus, nothing of a truly scientific character has been added to our knowledge of the causes, symptoms, pathology, or treatment of typhoid fever since Louis put forth his work more than sixty years ago. Except age, change of habitation,—perhaps sex,—which favor its development, the causes of this disease, he says, are unknown (" La plus profonde obscurité regne donc sur les causes de l'affection qui nous occupe").

Five years later (1834), Chomel (" Fièvre Typhoide") says that the causes of this disease are wrapped in the deepest obscurity; that we know some of the circumstances under which it develops itself with a marked preference; but that the determining cause has so far escaped investigation; that misery, famine, fatigue, or mental anxiety may have something to do in its production, but that none of these is the cause. Littré,* still later, declares that there are no known causes of this disease. Bartlett † says that the only causes of typhoid fever the influence of which he has positively ascertained are age, recent residence, and contagion. The actual producing, efficient cause, like that of most other diseases, is unknown to us. Dr. Jackson ‡ supposed it to be connected with the soil, " although not at all depending upon any filth or decomposing substance, since no such could be discovered, and since the houses were often new, clean, and in good situations."

Without making any new discovery, or even any new investigation whatever, the more noisy promoters of sanitary

---

* Dictionnaire de Médecine, article " Dothinenterite."
† 1847.   ‡ Bartlett, p. 90.

reform began, very early in the movement, to declare that this fever originated in drains, sewers, and putrid emanations; and by 1862 it was rechristened *pythogenic fever* by Dr. Charles Murchison, who said that it is "generated and probably propagated by certain forms of decomposing matter." Facts were tortured in every conceivable manner to fit the new theory. Sometimes the incubative period of the disease was only a few hours, as at Croydon in 1852, where it "immediately followed" exposure to fetid emanations; or it might be forty-eight hours, as at Clapham, or be extended to three, or four, or six weeks.

A favorite way of accounting for an epidemic of typhoid fever was through the defilement of drinking-water by the proximity of privies or cesspools. It did not matter if it was shown that such water had been drunk for generations without having caused disease; it did not matter if it was shown that, in an infinite number of localities where there was no fever, the water for domestic purposes was proved to be more impure than where the fever occurred.

It should be kept in mind that no experiments had ever been made to show that it was possible for typhoid infection to filter through the earth from a privy to a well. Dr. Franz Hoffman [*] showed that it required three months for a solution of common salt to pass a distance of twenty inches through the soil of the cemetery at Leipzic, and to pass less than ten feet it would require four hundred and eighty-four days. In ground like the streets of Leipzic the same solution would not pass ten feet in less than seven hundred and thirteen days. Unless positive proof is offered to show the power of typhoid-fever infection to pass through the soil, is it not an absurdity to suppose it possible?

At Trinity College [†] there was an outbreak of the disease

---

[*] *Archiv für Hygiene.*
[†] *British Medical Journal,* April 10, 1869.

where the water was found to contain a large percentage of organic matter; but this water had been used for thirty years, and the college had had immunity from this fever during that time. Besides, within a year new sanitary improvements had been added, and the fever was limited to that part of the college where the best sanitary conditions prevailed.

Sometimes the water absorbed sewer-gas, as at Croydon in 1876; sometimes mud or a dead turtle, or a dead mouse was found in the bottom of a well, or some cows had drunk polluted water. In the mind of the sanitarian, any one of these events sufficed to account for a wide-spread epidemic. Again, decomposing twigs or leaves in water were enough; and two cases of typhoid fever are reported, in all soberness,* where no other cause could be found than the evacuation of a Newfoundland dog; and the reporter thinks that this gives a clew to explain why this disease so often attacks the rich! Room is found in two or three numbers of the *Lancet* for the discussion of this canine etiology of typhoid fever. It breaks out in Dublin in 1891. Prince George is seized with it. Bad smells from water closets had caused it twenty and thirty years before, when his father and grandfather suffered with it. In the prince's case no effort was made to show that he had honored any disreputable privy with his visits. It was found out, however, that he had eaten oysters at Dublin, and it is now asserted that these had taken up the typhoid germ which had been transmitted to the oyster-beds through the sewers, and thus had infected the prince. It might be thought by those unfamiliar with the methods of the professors of Sanitary Science that the germ had been discovered in the oysters or in the beds, or at least in the sewers which led to them. Not at all. It had not even been looked for.

---

* London *Lancet*, vol. ii., 1875.

If an epidemic broke out in a poor portion of a town, it was ascribed to the privies in the yard. If a well-to-do part was attacked and the poorer people escaped, it was because the former used water-closets in their houses, and the latter were supplied with privies.

In reality, the history of the rise and progress of this fever in Great Britain is contemporaneous with the rise and progress there of sanitary reform, and from its first appearance it continued its ravages, as typhus and other diseases had before it, and as diphtheria did later, in spite of, and without any regard to, sanitary measures. Before its recognition in Great Britain the reputed causes of the disease were not absent. There was no lack of fecal deposits. Private and public water-supplies were polluted with twigs, trees, and stumps, and were just as unprotected from subterraneous connections with leaky privies and cesspools, and the water from them was used to cleanse milk-cans; untrapped house-drains were just as common; heedless turtles tumbled into wells; frolicsome dairy-cows wantoned in puddles of dirty water; Newfoundland dogs were petted by the rich; royal Georges had stuffed themselves with oysters; yet, with all of these multiplied agencies, any one of which, the sanitarians say, is competent to produce enteric fever, it was unknown and unsuspected in that country before the third decade of the present century.

This filth-doctrine of the etiology of typhoid fever was, from the first, sharply contested by such men as Professor Christison, W. P. Alison, Hughes Bennett, Stokes, and Graves, and by many medical men of lesser note. They declared that it was inconsistent and irreconcilable with facts. Dr. Smith* writes that he had observed thirteen thousand cases; that whatever may be said of the new fever (typhoid) being generated by overcrowding, filth, impure air, poverty,

---

* *Edinburgh Medical and Surgical Journal.*

and such like, facts seem still to be wanting to establish the theory, or to give it even a semblance of plausibility.

Dr. W. P. Alison * says that the opinion that putridity generates fever is not only speculative, but that ample experience shows it to be erroneous; that, though it has often been supposed by medical men to be the cause, yet the greater number who have carefully investigated it have rejected the theory. He had observed fever in Edinburgh for twenty-five years. Hundreds of times he had noted places with putrid effluvia, but the same places were free from fever for years together. In his own time the city had been much improved, so far as cleanliness was concerned; but there had been no corresponding improvement in the health of the city; and for the last three years fever had been more wide-spread and more fatal than ever before. When it does prevail, it is not among the filthiest, but among neater and better-ventilated, parts and people.

Dr. Christison said that it was not until 1847 that typhoid fever began to be studied with care in Edinburgh. By and by it became common, and is now (1863) encountered, not as typhus was, among the destitute, but among those in easy circumstances and in the best houses. Of all forms of fever, none had been so confidently ascribed by some writers to filth. "Has London," he asks, impatiently, "been worse drained, or the habits of its people alternately been more or less cleanly, as this fever has been alternately growing or diminishing in London? . . . Does it generally appear where the drainage is bad and the water-closets faulty? . . . Our drains in Edinburgh have been very much improved during the very period that typhoid has been increasing; besides, cases are occurring where it is impossible to discover foul air, any more than in thousands of houses where it does not occur. . . . The majority of deaths, so far this year, happened in

---

* Observations of the Generation of Fever.

localities to which no objection could be made. . . . Further, this fever does not break out where the streets are ill-drained and the habits of the people filthy. . . . In countless places of that sort in Edinburgh it is unknown." Dr. Christison concludes that we do not know the cause of this disease, as of other epidemic diseases.

While Dr. Graves* admits an advancement in public health from drainage, improved habits, and increased comforts, he believes that the increase or diminution of fever arises from some unknown cause, independent of locality or drainage.

Dr. Hughes Bennett says that we have emanations without epidemics, and epidemics without emanations. He sees no connection between the two conditions. When in Naples, in 1863, he found the city full of the vilest odors: for half a mile from the inner margin of the bay the water was like turbid pea-soup, and there was a constant smell of sulphuretted hydrogen. Here it might be supposed enteric fever would prevail. There were one thousand beds in one hospital that he visited; the ammoniacal exhalation from the water-closets here "was something astounding." "Yet the reply, on inquiry, was that typhoid fever was trifling." Again, in the military hospital, with similar arrangements and disgusting emanations, but one case of fever was all that could be discovered in eight hundred beds. The director of the hospital certified that no typhoid fever arose from the latrines. Naples was in a deplorable condition, but it was no more subject to typhoid fever than other cities. Foul smells and fetid exhalations have been escaping from the irrigated meadows at Edinburgh for two hundred years; "yet, notwithstanding the magnitude of the nuisance, enteric fever in that locality had never occurred." Dr. Stokes † says that the origin of this fever is one of the most difficult

---

* Clinical Medicine. † Continued Fever.

and obscure subjects in the whole range of physical inquiry.

Dr. Grabham * writes that, while he does not deny the possibility of typhoid fever arising from polluted water, yet "it is vain to seek to trace many well-marked cases of pure typhoid to sources of filth-contamination." In Madeira he has, again and again, seen such cases occur in the persons of young, healthy new-comers, in situations many hundred feet above the city, far away from all dwellings or sources of filth, and with the drinking-water flowing from springs beneath basaltic columns of rock.

Dr. Elliot Laine says that in Australia, where he has been for six years, he has cases exactly similar to typhoid fever, except for the rash, brought in from the bush, where the subjects have been exploring for new squatting-ground.

Dr. Bakewell † tells of an epidemic of typhoid fever that broke out on an emigrant-ship, after it had been a month at sea. He apologizes that he, an unknown practitioner, should presume to criticise Sir William Jenner as to its etiology, but thinks he may "be pardoned" if, after having treated typhoid fever in London, the rural towns, the Crimea, Australia, and the West Indies, for thirty-two years, he has "an opinion of his own."

Sir Joseph Fayrer ‡ says, "A form of fever exactly like European typhoid, except in its etiology, exists in India and other hot and malarious countries; and it is due to climatic causes, not to filth or specific causes." "Enteric cases are reported from nearly every station in the Bengal presidency, where the filth-element is at its minimum." He offers the testimony § of Surgeon Ryley, who, as the result of sixteen years' experience in New Zealand, Fiji, New South Wales, and Australia, says that a fever having all the characteristics

---

* London *Lancet*, vol. ii., 1879. † Ibid., vol. i., 1880.
‡ Ibid., vol. ii., 1879. § Ibid., vol. i., 1880.

of typhoid occurs in those countries, as in England; that in no case could he trace it to contagion; that in a few instances it seemed to have arisen from foul drinking-water, but in the majority of cases it has broken out among Europeans, "new-comers," first inhabiting a virgin soil, apparently from paludal emanations or other climatic causes.

Of this disease in India, Surgeon-General Ker Innes* says, "It cannot, except in some rare instances, be traced to importation from England, nor to fecal contamination of air, water," etc.; and "medical officers have utterly failed in India to trace out the ultimate connection of this disease with filth-causes or specific infection, with which, according to European authorities, it is invariably associated."

Surgeon Kavanagh says that he is convinced that it is an error to seek for the unsanitary starting of this fever *outside*, and he suggests that inside the body decomposing substances are capable of producing it. Surgeon-General Cumming says that typhoid fever in India is, in the main, a disease of the young, and occurs chiefly in Europeans new to the country. The importation and germ theories will not account for it. Enteric fever in India is the result of climate telling on constitutions unaccustomed to the strain, and is favored, as all diseases are, by local unsanitary conditions. In the report of Indian health statistics† it is stated that in the great majority of instances "no clew whatever can be found to the causation of the disease (typhoid) in any local unsanitary conditions."

Fleet-Surgeon Norbury‡ entirely agrees with those authorities who think that typhoid fever arises *de novo*. From January 23 to April 4, 1879, his command was practically cut off from the outer world. Before the march away from this base no case of this disease had been heard of, neither was it known at Elkowe, the termination of the

---

* London *Lancet*, vol. ii., 1879.   † Ibid., vol. i., 1886.   ‡ Ibid.

march. Yet many cases of typhoid fever broke out at Elkowe. Surgeon Norbury thinks it is as nearly proved as it can be that typhoid fever can arise *de novo*.

The number of English troops in India* is sixty-two thousand. Enteric fever is the most fatal of all diseases. In Bengal, "at none of the stations could the origin of the disease be traced to local conditions," and the conclusion arrived at was that "it was due to climatic effect in young or recently-arrived soldiers." In Madras, "at none of the stations could any definite cause be assigned for the prevalence of enteric fever." At Bombay it was the most fatal of all diseases; "at none of the stations could any local cause be traced."

Sir Joseph Fayrer† says that enteric fever occurs in India, as it does in England, but "fever in India with diarrhœa, peyerian ulceration, and typhoid symptoms is not necessarily caused by a specific contagion derived from fecal matter or from the intestines of another person." It was not, he says, until 1853 that it was suspected that persons died of enteric fever in India. In a few years it became a prevalent and fatal form of disease, especially among young and susceptible Europeans, and it is now the "chief fever death-cause among our young soldiers in that country." He quotes Dr. Gurdon, late chief of the medical service at Madras, who says that enteric fever occurs in India "that cannot be traced to anything pythogenic or otherwise specific."

Deputy Surgeon-General Alexander Smith "denies the specific origin of enteric fever, and attributes it to general and climatic causes." Dr. Alfred Clarke, of the Army Medical Department, says, "I have seen enteric fever in India where all filth-causes, in the ordinary sense of the term, are absent."

---

* London *Lancet*, vol. ii., 1888. † Croonian Lectures.

Dr. Bryden asks, if enteric fever is due in India to fecal emanations, "how can we explain the fact that, in the midst of the city of Dacca, containing, as it did in 1872, sixty-nine thousand inhabitants, densely-populated quarters are to be seen, in which the fecal deposits of generations are collected in unsightly heaps, or thrown into privy-wells within a few feet of the wells from which drinking-water is obtained, which causes diarrhœa when first used, but never any form of fever is observed? Toleration is soon established and comparative health enjoyed." The Dacca jail and lunatic asylum stand side by side, on the driest and most elevated piece of ground within the city. No water-lodges and no sewers exist. The dry-earth system is followed; night-soil is buried in the gardens and vegetables are planted therein. The drinking-water is pure; that for the lunatics is filtered through charcoal and sand. No outbreak of fever has occurred here; but isolated cases are met with, at intervals of weeks and months, which refute the idea that the seizures are due to any local defects within the walls.

Brigade-Surgeon Marston writes, "I came out here (India) imbued, rather than otherwise, with a belief in the truth of the views of European pathologists; but Indian experience has compelled me to recognize that those views as to the causes of enteric fever are too exclusive and quite inadequate to account for the facts."

Surgeon-Major Clark reports to Sir Joseph Fayrer from Natal and Zululand that enteric fever occurred there in healthy camps on virgin soil, and in bodies of picked men where the water-supply was good, no sewers, no drains, dry earth, and trenches carefully attended to. Staff-Surgeon McLean* writes that if enteric fever is always associated with defective sanitation, "then Ascension, of all places

---

* London *Lancet*, vol. i., 1880.

with which I am acquainted, ought to be exempt from it. There is no such thing as a sewage-drain or a cesspool on the island, all sewage and other filth being removed daily and thrown into the sea to the leeward of all dwelling-houses." But he adds, "Cases of this disease have occurred from time to time without any discoverable connection with each other. My investigations have utterly failed to connect the fever with any of the conditions commonly believed to be essential to its production."

Surgeon-General Marston * says that none of the theories that prevail in Europe can account for it in India,—neither filth nor contaminated water nor specific infection. He concludes that change of climate, surroundings, diet, acting on young, recently-arrived soldiers is sufficient to cause the disease.

In 1890 † enteric fever in India caused thirty-seven per cent. of the total mortality, against twenty-five per cent. for the previous year. In Bengal the mortality from this disease was forty-two per cent., against twenty-seven of the year before, and in Bombay it caused forty per cent. of the deaths. Dr. Ferguson ‡ relates epidemics of typhoid fever in his command at Bermuda; he "could not trace it to any defective sanitary arrangement." "The dry-earth system was adopted." "The water used was rain collected in tanks removed from any possible pollution." Dr. Hamilton § says of typhoid in India, "In no part of the world is sanitation so perfect as in India, and in no station could the dry-earth system be more thoroughly carried out than here. Yet year after year, with the advent of young soldiers from home, we see the same recurrence of the disease."

Dr. Thomas ‖ reports that the sanitary arrangements of

---

* Notes on Typhoid Fever in India.
† London *Lancet*, vol. i., 1891.
‡ Ibid., vol. i., 1880.
§ Ibid., vol. ii., 1888.
‖ Ibid., vol. i., 1880.

Algiers are bad, the system of water-closets imperfect, and bad smells only too common. Dr. Robertson says that at the principal hotel here the water-closets all open into one shaft without proper traps, and the closets and corridors are in direct communication with the sewers of the town. Dr. Thomas says that one thousand English visitors have been here the last winter, "not one of whom has suffered from typhoid fever."

In August, 1889, typhoid fever breaks out in London at West End. The *Lancet* says it is impossible that it can be due to unsanitary condition of the houses, and it suspects the milk. Further investigations fail to reveal any cause. It breaks out the same year in the stock exchange. The cause is not the general insanitation of the building, and there is no record that it is ever ascertained. It breaks out at Tullamore jail;* an investigation shows that "the sanitation of the building is perfect." The town is "devoid of sanitary arrangements," but no typhoid is reported in the town.

Dr. Tew † personally investigated one hundred and two cases of typhoid fever in 1890. Five per cent. were traced "without reasonable doubt" to specifically polluted water. He does not say that the bacillus was found in the water; while in twenty or twenty-five per cent. the water was of inferior quality, he does not say in what respect it was inferior. He thinks that although water-supply and drainage are important in considering the causes of this disease, inquiries into the fruit, vegetable, and meat supplies are of increasing importance.

Typhoid fever is reported at Dublin in 1891;‡ as regards the cause "no special light has been obtained." "Sir Charles Cameron, the medical officer of health at Dublin, reports the

---

* *Medical Press*, 1891.   † *Lancet*, vol. i., 1891.
‡ *British Medical Journal.*

sanitary condition of the city good. The epidemic, he says, is not "due to the defective state of the sewers and accommodations." "In seventy-five per cent. at least of the cases no sanitary defects were observed." "The disease seems to be more prevalent among the upper and middle classes." The only explanation he gives of the epidemic is that "it is a disease the germs of which exist in the soil, and ascend from it into the atmosphere at some seasons in greater numbers than at others." Dr. Cameron does not condescend to explain why these germs ascending at certain seasons of the year should have such malignant effect in 1891 on the upper and middle classes of Dublin. M. Rochard * insists on overcrowding as a cause of typhoid fever. The French naval vessels, he says, are extraordinarily (*exagérée*) cleanly; the men in this respect are irreproachable; water is taken from sources specially selected for purity; the excrements fall into the ocean; yet in our national as well as in our emigrant ships, typhoid fever appears when people are crowded together (*dans les milieux encombrés*).

Vienna, which for some years had been only lightly visited with typhoid fever, experienced in 1882 a serious increase in the number of cases.† The examination of water, well, and spring gave negative results. The reports of the physicians detailed the condition of the houses, habits, diet, and pursuits of patients, the quality of the water they had drunk, the situation of wells, privies, and cesspools, and the cleanliness of yards. No origin could be assigned for the increase of the fever. *L'étude de ces rapports n'a pas permis jusqu'ici d'assigner une origine nettement déterminée à la récrudescence de la maladie.*

The commission appointed to report on the severe epidemic of typhoid fever at Zurich ‡ say that poor and rich were alike affected; density of population had no influence

---

* Annales d'Hygiène.   † *Revue d'Hygiène*, 1890.   ‡ Ibid., 1885.

on it; *neglected and bad-conditioned privies seemed to be a protection against the disease rather than to favor its development;* it was improbable that the air of sewers had any influence to spread it; neither was it affected by meteorological changes. All sorts of bacteria were discovered except the typhoid. The commission declared it could not give the origin of the epidemic.

The filth theory of this disease, especially the polluted-water part of it, was received from the beginning by the Germans with derision and contempt. It was more kindly entertained in our own country; indeed, the Anglomaniacal sanitarians here adopted it without hesitation. In the *Sanitarian*[*] we are told that typhoid fever is traceable to filth with as much certainity as smoke is to fire, and is a disease wholly preventable by proper sanitary measures. During the Newport excitement it was declared that scarlet fever "originates in filth" and is fostered in filth; and that typhoid fever is "purely a filth-disease, and only spread through filth-exposure." The Massachusetts Board of Health, however, was a little doubtful after its first investigation; for it said [†] we must believe that English drinking-water is "exceptionally dirty, or that medical observers are unconsciously influenced by preconceived opinions, based upon the ingenious speculations of men of ability, who have directed their attention to this form of danger." The board said, if this be the cause, we should expect to find this disease most frequent and virulent where privies and wells are in closest proximity. "There are many towns in Massachusetts where the surface of the ground is dotted all over with these structures. There are a very large number of towns with a population of from five thousand to ten thousand, compactly built, with no water except from wells, and no means of disposing of their excrement except by privies; and we know from the

---

[*] 1880.   [†] Report, 1871.

registration returns that, in proportion to population, the people of these towns are more free from the pest of typhoid than the inhabitants of agricultural districts." "If we may deduce any conclusion from the mortality in Lowell for a single year, it would appear that, though filth, putrid air, and impure water are active agents in causing scrofulous, tuberculous, and bowel diseases, they have but little, if any, power to produce typhoid fever." For two years, 1869-70, in Lowell, a very much greater number of deaths had occurred from typhoid fever in good localities than elsewhere, and only a very few in places that were pronounced bad. In that city* the number of deaths from typhoid fever in the worst localities was five; in localities somewhat better, five; and in well-ordered sections the number of deaths was twenty-six. To the question that the Massachusetts Board sent to different towns, as to whether any variation had been observed in the prevalence of typhoid fever between families that used spring-water and those which used water from wells, there were ninety-four replies. Twenty-three said yes; seventy-one said no. In 1871 ten towns reported no typhoid fever; and in some it was said that the disease was almost unknown. Wherein these towns differed from others is not stated.

The disease prevailed at West Boylston, but the cause was not obvious. The physician from Watertown writes that he has "diligently searched for some local cause; but, except in one case, has never been able to discover any satisfactory one." The doctor at Worcester has "never been able to trace the disease to any particular cause." Another physician in Worcester can find no cause. Still another says, "The worst cases have been on high ground, and under apparently the best hygienic influences." At Chester and Cambridge no cause could be found. At Coleraine, "often the most negligent families seem to escape." To the physi-

---

* 1871.

cian who reports from Franklin it seems strange that he sees so many places where sink-water is deposited at the back door and no evil results follow. "On the other hand," he says, "I have had cases in families where the surroundings seem conducive to health." In Salisbury and Swampscott no cause could be found. In Orleans was an unusual amount, confined to the highest, driest part of the town. The people use cistern-water, and the cisterns were cleaned three months before fever broke out. There were no cesspools here, nor other filthy deposits; cellars were dry, and free from decaying vegetables.

Among the specified causes of this fever, which are given by some of the Massachusetts physicians, are swamps, drains, decaying vegetables, low water in ponds, heavy rains, pools of water, wet cellars, a cider-mill, a slaughter-house, swimming in a foul pond, repairing an old house, putrid fish as manure, cesspools, decaying wood, clearing up the soil, open drains, stopped-up drains, cleaning a cellar, bad wells, privies, etc., etc.

The Massachusetts Board says[*] that this fever does not prevail most in the filthiest towns, nor in the filthiest parts of towns, nor in those years in which filth is most accumulated; and in the sense of being an index of filthiness, it is very far from being a filth-disease. It goes on to declare[†] that there is an "unknown factor" in this disease, of which we know as little as we do of the germ. "If we ignore this 'unknown factor,' it must be admitted that the negative evidence enormously preponderates over the positive facts, which tend to show that filth, independently of a specific poison, produces typhoid fever." The board says, further, that the evidence seems to be that filth "plus an unknown factor" is sufficient. This is a practical abandonment of the case; for the board's own reports show that, in a great

---

[*] Report, 1879.   [†] P. xviii.

number of cases, the "unknown factor," without the filth, has produced the disease. Really, the cause, pure and simple, of typhoid fever in Massachusetts, is the "unknown factor."

In 1882 * a severe epidemic of typhoid fever broke out in New Haven. The untutored man in Sanitary Science will say, of course, that it started, at least, among those citizens who have been so carefully storing up their filth close to their houses. Not at all. It broke out in Yale College, and some weeks later spread from thence to houses in the town. Professor Brewer, a member of the Board of Health, certifies to President Noah Porter that New Haven is an unusually healthy town, and that the late typhoid fever in the college could be traced to no local cause. Four months before, the entire plumbing had been renewed, and Professor Brewer described it as "the most elaborate and expensive that has yet been devised."

In 1884 a circular was sent to the physicians in Connecticut, inquiring for any cases of typhoid whose origin was from foul drinking-water or other local conditions. The secretary of the Connecticut Board of Health says that the reports "will confirm the teachings of the best sanitarians in regard to the close and dependent relation of disease to surrounding conditions." There were one hundred responses. Seventy-three did not answer the question at all as to the *cause* of typhoid fever. Ten said plainly that there was no local cause. Dr. Hill, of Stepney, said that malaria had "squelched" typhoid. "It must have removed our filth and purified our drinking-water." Dr. May found the disease where the water was pure and the sanitary condition the very best. Dr. Pinney said that it was dependent on some cause too subtle for him to detect. "In fact," he adds, "the general sanitary condition is better, but the typhoid is

---

* Connecticut Board of Health Report, 1883.

more." In Suffield it occurred in one of the healthiest highest, and driest localities, with absolutely no unfavorable surroundings. Dr. Newton writes, "We find this fever on high ground and sanitary condition perfect." He believes it is due to climatic influences, and not to filth. Of the fifteen who mention a cause, three say drinking impure water, five lay it to cesspools, three to pouring slops on the ground, one to wading in a muck-pond, one to a damp cellar, one to a privy, and one to a place where was "formerly some filth!"

The Connecticut Board of Health Report for 1887 presents the results of another official inquiry among the physicians of the State concerning the cause of typhoid fever. One hundred reported more or less of the disease. Eighty-two of the reports did not allude to the cause. Eleven named some cause,—three, polluted water; one, disturbing the soil; one, a garbage-heap; two, imperfect sewerage; three, sink-drains; and one, "may have been a sink-drain." Seven said plainly that there was no cause. Dr. Goodwin, of Thomaston, said that it was impossible to trace the cause in any case. Dr. Smith was able to trace the real source in none. Dr. Worden could tell of no reason why it prevailed in Bridgeport with unusual severity. Dr. Lewis writes, "Careful search fails to find the cause." Drs. Young, Fox, and Brown could find no cause for the disease. It occurred in one locality in Middletown where there was no imported case and no other cases in the vicinity. Here it was ascribed to pouring slops on a raspberry-patch four months before!

An epidemic occurred in Madison, and the secretary of the State Board of Health investigated its source. He says, "There are no facts respecting garbage, sewage, cesspools, pig-styes, barn-yards, or privies, which have been discovered, to distinguish the houses in which the fever occurred from other houses." He ascribed it to low water in the wells.

But the water was just as low in the wells of adjoining towns where there was no fever, and it was never explained why low water in the wells should cause typhoid fever in Madison and prevent the same disease in the towns of Guilford and North Haven.

In 1890,* to the question whether typhoid fever could be traced to any special cause, the reporters from sixty-six of the towns which reported more or less typhoid made no reply. Thirty-two said distinctly that no cause for the disease could be traced. Dr. Goodwin, of Thomaston, said, "After careful search and inquiry no cause could be found." Some cause was assigned by twenty of those who reported on the disease. One physician writes that the cause in one case that he had was "a duck-pond," but he could not trace the cause of the other cases. Nine report bad wells, or suspected wells, though no report is made that there was any examination of the water. One was caused by a filthy cellar; bad locations caused it in three cases; low ground and bad water in another; six accused faulty drains; and in one instance there was an "unsanitary condition." Dr. Brownson, of New Canaan, writes that he has searched diligently for causes of typhoid fever, and believes that they are beyond our knowledge. In the Connecticut report for 1891 are recorded the replies of physicians to the question if typhoid fever could be traced in any case to a specific cause. Sixty-six said plainly no cause could be found. Many of these stated that they had carefully searched for the cause. Forty-eight made no reply to the question. The disease was unusually prevalent at Bridgeport. The county jail was the nucleus. Of course the reader will say it was due to the vile sanitary condition of the jail, decribed on a previous page. Not so. Dr. Werdon writes, "All the cases occurred in the new building, the sanitary

---

* Connecticut Board of Health Report.

FILTH AND FECAL DISEASES—TYPHOID FEVER. 271

outfit of which is thought to be complete. Careful investigation failed to reveal a cause for the disease."

The New Hampshire Board of Health * announces that the evidence is overwhelming that typhoid fever is a filth-disease. In the same report, however, fifty physicians who report on it say no cause could be found. Sixteen others make no allusion to the cause. Five are particular to say that the sanitary condition was good where it occurred. Twenty-five say bad water was the cause, though there is no record that the water in any case was examined. Between thirty and forty others give a variety of causes, as a foul cellar, filth, sewer-gas, overwork, a pond of water, a dry pond, poverty and filth, lying on the ground, smell from a carcass, contagion, etc. Dr. Blaisdel said his cases "were surrounded by unusually good hygienic conditions, and the water-supply was better than the average." Dr. Tucker said he had given very careful attention to learn the cause, but had "utterly failed" to find it.

Dr. Cook, of Concord, writes, "I have given special attention to cause, but have not discovered it." Dr. Boynton gave "special attention" to the cause, but could not find it. "All used pure spring-water." Dr. Mooar investigated for the cause, but could find none. Dr. Chase writes that the past few years Peterborough is exempt from typhoid fever, why, he cannot see; a few years ago there was a great deal, and the water-supply and the sanitary condition were the same then as now. Dr. Fleeman says he uses as good common sense as he can to find out the cause; he does not say, "This is the cause or that is the cause," and he adds, "It is quite doubtful if any of us know quite as much as we think we do in relation to what causes disease."

In the Fourth Report of the New Hampshire Board, thirteen physicians who report on typhoid fever make no

---

* Third Annual Report.

mention of cause. Forty others say they could not trace it. Thirty-two ascribe the disease to polluted water, but do not say that in any case there was any examination of the water. Forty physicians give a variety of causes,—a tannery, a newly-plastered house, a privy, low vitality, atmospheric causes, swampy land, foul air, bad meat, contagion, pig-sty, low water, foul sink, decomposing vegetables, malaria, overcrowding, etc. Dr. Jenking had given special attention as to cause, but was forced to say that "the wind bloweth where it listeth;" there were at least "no unsanitary surroundings." Dr. Junkins investigated every case: "Investigation revealed nothing." Dr. Frost gave "special attention to probable cause," but could not find it. Dr. Flanders investigated all cases and could not find the cause. Dr. Hill gave special attention to the cause and could not find it. Dr. Dearborn "investigated thoroughly for the cause, the sanitary surroundings being perfect." Dr. Parson gave special attention to the cause, but could not discover it. Dr. Lee made "strict inquiry," but no specific cause could be ascertained. Dr. Porter reports the sanitary condition good where his cases of typhoid occurred.

For three or four years the New Hampshire Board of Health investigated typhoid fever with results similar to these. Dr. Hutchinson writes to the New Hampshire Board * that there is no typhoid fever in Milford. He says he has examined several wells, and he is certain that these, as well as many others, are "*polluted* waters," but there is no typhoid fever. Dr. Jenkins writes as to the causes of typhoid fever, —"their name is Legion." Dr. Roberts never saw "a case where I thought contaminated water was the primary cause."

In the Twelfth Report of the Michigan Board of Health forty-nine of the health-officers who are asked for the cause

---

* Fifth Report.

of typhoid fever in that State make no reply. One hundred and nine stated that the cause was unknown. Five ascribed it to overwork; three, to decomposing vegetable matter; one, to animal decomposition; one, to heat; one, to hot and dry weather; one, to exposure to wet and cold; one, to a low temperature; one, to living in a swamp; one, to decaying slabs and sawdust, etc., etc.

The Thirteenth Report of the Michigan Board of Health records typhoid fever occurring in two hundred localities during the previous year. Forty-nine report the cause unknown. The board says there were "other indefinite replies." Among the causes named were overwork, exposure, impaired vitality, bad sanitation, etc. Twenty-three ascribed it to impure water, but none of them report any examination of the water. The Fourteenth Report of the Michigan Board of Health tells of typhoid fever in two hundred and ninety places the previous year. The board says, "The reports show that the health-officers, as a rule, do not yet succeed in tracing outbreaks of typhoid to their source." In one hundred and two replies nothing was said as to cause. Fifty said plainly that it was unknown. Various theories were advanced to account for special cases; for example, impure water, low water in the wells, contagion, exposure, sleeping with a filthy person, overwork, a dead cow, a lumber-camp, hot, dry weather, exposure to wet and cold, privy-vaults, etc.

The Board of Health of Ohio (Report, 1887–88) received one hundred and twenty-eight replies to questions as to the origin of typhoid fever. Thirty-nine say that nothing is known as to the cause. Forty-five say bad water, or "possibly" bad water. Forty-five attribute it to direct exposure to other cases by nursing, or by living in the house with cases. One cause is given as travelling around with a threshing-machine; one, to sleeping in a loft over a stable.

Dr. S. S. Turner,* of Fort Buford, says that he has seen typhoid fever on the prairies of Dakota, where the air is exceptionally pure; and it is observed at military posts, where the sanitary conditions are "seemingly the best." " I have seen the same in commands living in tents upon the dry prairie, where the natural conditions are perfect and the privy-sinks remote from the camp." Dr. Stillman † writes that between July 1, 1849, and January 1, 1850, ninety thousand people arrived in California. It was roughly estimated that one-fifth of this number died within six months after arrival. Here was an entirely new country, with pure air, pure water, and pure soil. Dr. Stillman says that typhoid fever was very common and very destructive among these emigrants.

The editor of the *Medical Record* (1883), in commenting on the typhoid fever in New York that year, says, " We do not remember when our city has been in a better condition as regards cleanliness than at present;" yet there had been two hundred and forty-four cases of typhoid in August, 1883, against ninety-six in the same period of time the year before.

Dr. G. W. Robinson ‡ says that in Trinidad, Colorado, a large number of cases of typhoid fever occurred in 1883. The people used water from three different sources, all pure; thirty-five miles away from these cases he had others, where the water-supply gushed from the mountain-side; forty miles in another direction he attended typhoid fever, where the water was from a spring at the bottom of a cañon. There were no other cases in the camp or neighborhood.

Dr. Robinson believes that, to develop typhoid fever, it is not necessary to have the germs from a previous case. Dr. Patterson § says, " I have observed the disease in my field-

---

* *Medical Record*, vol. xix. † *Edinburgh Medical Journal*.
‡ *Medical Record*, vol. xxxi.
§ California Board of Health Report, 1886.

service to affect men who were absent for months from permanent settlements, in the uninhabited mountain regions of Northern Idaho and Southeastern Nevada, where the presence of the specific germs of typhoid fever could hardly be suspected to exist."

Dr. E. Stone* says it often happens that a physician is called to typhoid fever where the patient has not been away from home for weeks, and there has been no case in the neighborhood for years. He may find ground soaked with slops under the kitchen-window, or a bed of muck, or a running-over privy; or he may not find these conditions. Of one thing he is sure, that his patient has not been poisoned by the fecal discharges of other patients.

Dr. Stone says that in the town where he has resided for twenty-six years quite a number of persons are employed in removing the contents of privies from Portland; their working-hours are constantly taken up in handling the disgusting material. Contrary to what might have been expected, if human excrement contains the typhoid poison, he says, "I have never yet had occasion to attend a solitary case of typhoid fever among this class of persons." Dr. Stone says the "excrement theory" of this disease "so essentially" conflicts with his experience of more than thirty-six years' practice that he is "compelled to reject it, notwithstanding the high authority by which it is sustained."

Dr. Fr. Hue† made a careful inquiry into the health of workers in night-soil at Rouen, and particularly as to their liability to typhoid fever. Sixteen are employed in emptying privies. Their ages vary from thirty to fifty years. These men have families of from eight to ten children, and one has eleven. Twenty-nine other men are employed in making poudrette of the fæces which the sixteen collect.

---

* Transactions Maine Medical Society, 1877–79.
† *La Normandie Médicale*, February 1, 1892.

The families of these men live in the depots where the poudrette is made. Dr. Hue says that, except wounds, alcoholic excesses, and ophthalmia, there is a remarkable immunity from disease of any kind among all of these workmen and their families. Only one case of typhoid fever had ever happened among the workmen or their families; this case occurred two years before, and was a young girl who attended a school. If the families of the poudrette-makers were one-half as large as those of the men who empty the vaults, there must be in Rouen not less than three hundred people exposed to these fæces. Dr. Hue says no hygienic precautions are taken, but they are "indemnified" from typhoid fever. *La Normandie Médicale* two weeks later reports eighty-three deaths from typhoid fever in Rouen in 1891, which indicate that there were about one thousand cases of the disease in that city in that year. Dr. Hue seems surprised at the result of his inquiry. He asks, May not the freedom from disease of this people be due to a kind of "slow vaccination?"

Dr. Bramlett * says that he has studied for many years the causation of typhoid fever in a mountainous country, supplied with pure spring-water, where there are neither ponds nor marshes. The elevation is twenty-five hundred feet above the sea. The cases that he narrates could not be traced to water, nor filth, nor "importation." "From whence," he asks, "did the contagion come? Obviously, I think, from the bodies of those having the disease." Dr. Bramlett is persuaded that it arises spontaneously, and that it is contagious.

Professor Cabell † concludes that "there are yet many unsolved mysteries" with regard to its development and spread; and it is to be regretted that the exclusive ascription

---

* *Virginia Medical Monthly*, 1877.
† Etiology of Typhoid Fever, 1877.

to putrescent matter, contaminating air or water, should divert attention from other less easily assigned, but possibly equally influential, agencies.

Dr. Hicks,* of Michigan, says that he has practised medicine in a new country, in a community which had sprung up in six months. "Two or three hundred people built their houses, and typhoid fever broke out before there was any kind of sewage. We had it on new farms, and we had it every year more or less; and he adds, "If the cause of typhoid is known, *I* don't know it."

We have seen on a previous page (p. 98) the condition of New Orleans regarding its exposure to fecal emanations. The people here, Dr. Holt says, live on a dung-heap, and have a privy in common. For the four years ending 1889 the number of deaths from typhoid fever in that city was forty-six, forty-one, thirty-four, and thirty-five. The city of Washington, with a population less by twelve thousand than New Orleans, furnished during the same years one hundred and twenty-eight, one hundred and sixteen, one hundred and sixty-eight, and one hundred and seventy deaths from the same disease. "But," says the sanitarian, " New Orleans drinks pure rain-water from cisterns, and therefore has immunity from typhoid fever. The Louisiana Board of Health Report for 1886-87, p. 169, says that for a limited time, almost yearly, New Orleans suffers a water-famine. "The roofs and gutters carrying this water accumulate quantities of impurities, such as dust, highly charged with organic matter; and these with each rain are washed into the cisterns."

Assistant Surgeon Charles Smart,† in his report on the water-supply of New Orleans, says the rain-water carries into the cisterns soot and condensed ammoniacal vapors; an

---

* Proceedings American Public Health Association, vol. xiv.
† National Board of Health Report, 1880.

infinity of débris, organic and inorganic, together with more massive fragments, as of dead insects and decaying leaves, all of which forms a soft, black, pultaceous sediment, rendered impure by being stirred up by every succeeding rain-fall; and the water is unfit for use after such rain-fall. Of a sediment from one cistern, one hundred parts yielded fifty-four parts of albuminoid ammonia. In October, 1891, Dr. Metz, chemist of the Louisiana Board of Health, pronounces the water of New Orleans as altogether bad. More than half of the samples which he examined were unfit for use, many of them being similar to "stagnant swamp-water."

This is just such water as has caused typhoid fever thousands of times, according to the sanitarians. Would Washington change its bounteous supply of the torrents of the Potomac for the scanty quantity of the foul water that the Louisiana Board of Health, Dr. Smart, and Dr. Metz say is supplied to New Orleans.

On account of the importance of this subject the author makes no excuse for prolixity in discussing it.

In surveying the evidence which the sanitarians themselves have collected, one is struck with astonishment at the agreement of the working members of the medical profession that the cause of typhoid fever is not ascertained. The silence of the hundreds of physicians in Massachusetts, New Hampshire, Connecticut, Ohio, and Michigan, when appealed to for the cause, is as full of meaning as the outspoken opinion of those who say that the cause is unknown. It cannot be that these hundreds of physicians, who are daily rushing about the cities, and jolting over the hills and plains of those States, watching the rise, progress, and termination of cases of this disease under all circumstances, know the cause and will not divulge it.

The evidence here exhibited shows that the acutest observers and profoundest intellects of the medical profession have recorded their observations of this omnipresent fever, on a colossal scale, over three continents and the islands of the sea. Is it not conclusive that, in respect to our knowledge of its etiology, we are where we were sixty years ago, when Louis and Chomel described the disease to us, and when they declared that, except age, recent residence, and perhaps sex, its causes were wrapped in the deepest obscurity?

The doctrine is taught in the schools that the typhoid dejecta are harmless when passed, but that under exposure to the air, or in the presence of filth, they develop a poison which is capable of giving rise to this fever. This dogma is accepted by so many truly learned, as well as renowned, medical men, that to escape the charge of disrespect, we will not openly combat it. We would, however, ask if there is any proof that the typhoid stools suffer any such change, either in filth or on exposure to air.

It seems as if this hypothesis had been evolved out of the previous one,—that typhoid fever is a pythogenic disease, or that the telling of one untruth had led to the telling of another. Typhoid fever was born of putridity; but it was shown that, in thousands of cases, exposure directly to typhoid excreta did not induce the disease (Murchison). Instead of questioning its pythogenic nature, and seeking elsewhere for its cause, it was easier to formulate a new hypothesis, which could be mortised into the original theory, —namely, that these excreta, so innocent when passed, acquired by time and change the power to spread the disease. Again, it was shown in the cruel epidemics at Croydon in 1876, at Paris in 1883, at Berlin in 1889, and in ordinary seasons at a great many other places where typhoid excreta were poured upon sewage farms, that those who were most exposed to the fecal matter did not take the disease. Therefore still another theory was necessary,—that the typhoid dejecta, in their transit through the sewer of, say, from twelve to twenty-four hours, suffered a relapse, but reacquired their innocence; that they not only found their depravity in the sewers, but their purgatory also.

The Massachusetts Board of Health* say the typhoid poison is destructible readily on free exposure to the air, as was shown at Croydon, where fourteen hundred cases occurred, and a great portion of the dejecta, not disinfected, passed into the sewers and was spread over an irrigated farm, to which large numbers were exposed. "The water of the surface-wells in that vicinity, really the purified sewer contents, constituted almost the sole supply for drinking and other domestic purposes for many people, and in no case was the fever propagated in that vicinity."

If there be no proof that such changes take place in the typhoid stools, are we not justified in saying that there is nothing in analogy to warrant the opinion that time or place can so transmute them? Is the poison of the cobra, or that of the rattlesnake, more virulent after it has become decomposed? When we vaccinate a child, do we seek for virus that has undergone a change and become something else than vaccine? or are we particular that it shall be as pure as possible? If it has undergone "certain changes," do we not know that it is inert?

CHAPTER XIII.

Yellow Fever.

The belief in the filth and fecal origin of this disease is firmly established in the minds of the sanitarians. It is unnecessary to go into its history here, any more than to say that, as it has disappeared from many localities in our

---

* 1879.

own country where it formerly prevailed, the opinion is pretty firmly grounded, in the minds of most medical men, that it is an imported disease. Though it is said that now its home is in the West India Islands, it was not recognized there until the middle of the seventeenth century. Neither Columbus * nor any of the early Spanish adventurers mention it. At the close of the last century and the beginning of the present it was not uncommon as an epidemic in the Northern States; and some acute observers, who made special investigations at that time, were persuaded that it was indigenous to the seaboard States in general.

Noah Webster made a careful inquiry into the causes of the epidemics of 1793 and 1794 in New York and Philadelphia. He became convinced of the fallacy of the opinion that it originated from abroad; and he thought the evidence was complete that yellow fever could arise in this country between the parallels of 41 and 44 north latitude. He found repeatedly that the stories of its source from infected clothing and from intercourse with ships were idle tales of the interested or the ignorant. On the other hand, the evidence of its origin in New York, Baltimore, Charlestown, Boston, and other places was to him clear and unquestionable. He says,† "Nothing is more common than for yellow fever to be imported into the West Indies in vessels directly from the United States." Mr. Webster thought there was not much doubt that it was a pestilence among the Indians before the landing of the Pilgrims. He did not believe that it was contagious.

In 1800, Dr. Ramsay ‡ wrote to Dr. Miller, of New York, "The disputes about the origin of yellow fever, which have agitated the Northern States, have never existed in Charleston. There is but one opinion among the physicians and

---

* Tytler, "Yellow Fever," 1799.   † Vol. ii., p. 447.
‡ *Edinburgh Medical and Surgical Journal*, vol. lxxviii.

inhabitants, and that is that the disease was neither imported nor contagious."

In 1804, Dr. Stubbins Pfirth published a thesis to prove that yellow fever not only had its origin here, but that it was non-contagious; and he records his experiments. Previous to the Philadelphia epidemic, he believed that the disease was imported and communicated from one patient to another; but he was led to change his mind. He fed a dog with bread soaked in black-vomit; in three days he became so fond of it that he took the black-vomit plain. It had no effect on the animal's health. He fed the same to a cat, with no ill results. He injected black-vomit into the cellular tissue of a dog; the incision healed at once, without doing him any harm. He injected the vomit into the jugular vein of a dog, which caused death in ten minutes. He then injected water into the jugular vein of a dog; the animal died in ten minutes. And now, to show how depraved a doctor can become when he devotes himself to science, he inoculated himself more than twenty times with black-vomit; no ill effects followed. He put the same into his eye, and no harm ensued. Not content with this, he evaporated black-vomit over a fire, inhaled the steam, made the residue into pills, and ate them. He then drank a half-ounce of black-vomit, immediately after his patient had ejected it, and repeated these trials a number of times without any ill effects. He was fully convinced by these experiments that yellow fever was not contagious.

Dr. Jackson and a majority of the medical profession in Philadelphia believed that the epidemic of 1820 had its origin not from abroad, but in the city. The outbreak of 1853 in Philadelphia was ascribed by many physicians to a local origin; and as late as 1879, Dr. Burroughs * sharply reviews the opinion of the commission which pronounced

---

* *Virginia Medical Journal.*

yellow fever an exotic disease. Although for two centuries there had been frequent communication with and importation of merchandise from localities infected with this disease, it did not appear in Brazil * until 1849. In 1859 there died in Rio Janeiro four thousand one hundred and sixty persons of yellow fever. In 1855 and 1856, and for six years ending 1868, there were no deaths in that city from the disease. Since then the city has never been free from yellow fever.

The yellow fever in New York in 1794 was pronounced by the Rev. Dr. Dickenson † not a scourge from the hand of the Lord, nor an imported evil, but "the offspring of their own domestic filth."

Dr. Cartwright attributed the fever in Mississippi to a mass of putrid bacon. Dr. Caldwell said that it seemed fairly referable to some putrid fish, and he had seen it caused by putrid oysters and hides.

It was said that the yellow fever at Pensacola in 1822 was caused by the importation of putrid fish; but Dr. Merrill showed that one hundred of the United States troops were exposed in the most uncomfortable quarters, in the midst of a whole cargo of this fish, for four weeks, and not a case occurred among them until three weeks after it was removed. At the same time, however, a battalion of infantry, stationed a mile distant from the putrid fish, in a dry, airy situation, was sorely afflicted with violent bilious fever. The sailors, too, who had been living on board for a number of months, close to the fish, arrived in good health, and suffered only in common with the other inhabitants at Pensacola, three or four weeks later. In Newbern the fever was ascribed to stable-manure and street-sweepings that were thrown into the water to extend a dock.

To prove the filth and fecal origin of yellow fever, the sanitarians call our attention to the exploits of General

---

* Reference Hand-book Medical Science.   † La Roche.

Butler in New Orleans, who, they loudly declare, "stamped out" yellow fever in that city in 1862 by removing the filth. To those who listen merely to the assertions of the sanitarians and of the general himself, who has told exactly how he did it, it appears that, either by revelation or by intuition, he actually possessed some sanitary prevision not vouchsafed to the rest of mankind. Dr. Chaillé* says he will not contest the general's claims as a warrior, a Democrat, a Republican, and, again, a Democratic statesman; but he protests against the validity of his fame as a great sanitarian. Dr. Chaillé says that in 1861, when General Lovell was in command in New Orleans, many of the civil and military inhabitants were unacclimated; yet no death occurred from yellow fever, and, as far as is known, there was no case of the disease in the city during that year. Dr. Chaillé cannot understand why General Butler should be honored by men of science as an eminent sanitarian, while the better sanitary results of General Lovell are ignored altogether.

The fact is that in both years there was no opportunity for the disease to arise from imported cases. But, as is shown in the following table, there have been many other years when New Orleans has been nearly as exempt as in 1861–62:

Year.	Number of Deaths.	Year.	Number of Deaths.
1831	2	1833	1000
1836	5	1837	1300
1840	3	1841	1325
1845	2	1847	2804
1851	17	1853	7849
1860	15	1854	2425
1865	1	1855	2670
1877	1	1867	3107
1882	2	1878	4056

* *Sanitarian*, vol. x.

Is it logical to assume that the fecal matter in New Orleans caused only one death from yellow fever in 1877, and suddenly became so active that the next year it caused four thousand and fifty-six deaths; or that in 1851 it caused seventeen deaths, and two years later became so operative as to destroy seven thousand eight hundred and forty-nine people?

Dr. Dowler,* of New Orleans, declared that yellow fever had never happened in the parish prison in that city, and concludes, "There is, if we may reason from what is known, but one certain method of escaping yellow fever in New Orleans,—incarceration." The disgustingly filthy condition of this prison had been often admitted to be a reproach to the city; and that, too, by its own people. And now comes the most brilliant of all sanitary paradoxes. We read in the Louisiana Board of Health Report (1880-1883) that the exemption of the parish prison from yellow fever at New Orleans has been due to isolation, and to "the exhalations of ammonia from the excrement of the bats which inhabit the garret of the building in vast numbers; and also from the urine of the prisoners!" Between the years 1855 and 1870 the admissions to this prison averaged from five thousand to seven thousand annually. Why the excrementa of bats and men should prevent yellow fever in the parish prison of New Orleans, and be competent to cause an epidemic of the same disease just outside the walls, will probably always remain one of the mysteries of Sanitary Science.

But, wherever it broke out, the bad sanitation of the place was held up to us. The keen-eyed sanitarian could find somewhere a privy, a cesspool, a stable, some kitchen refuse, or some decayed wood; or else he would learn that somebody had been stirring the filth in the saturated soil; and

---

* La Roche.

he called the discovery of any of these things an investigation.

In 1870 the disease visited New York harbor. Some ships arrived that year from infected ports, and were moored at the Brooklyn docks. The man unversed in Sanitary Science, whose mind has not yet quite grasped its inflexible laws,—that two and two make five,—will say that the disease appeared, of course, in the slips of the rotting wharves, and among the filthy purlieus, streets, and lanes, which border on the water. Not at all. Those places were untouched by it. It broke out on the rigidly-policed and cleanly-kept military post of Governor's Island.

The shrewd sanitarian looks for the cause, finds it, and records it on page 337 of the New York City Board of Health Report for 1870. We are solemnly told that during ebb-tide there was a current setting from the docks where the infected vessels lay towards Governor's Island; and things might have been thrown overboard from them and lodged on the beach. In this way we have the yellow fever of 1870 accounted for at that island, which General McDowell said was exceptionally clean; and which, Assistant Surgeon Guild said, had a drainage "so perfect that pools of stagnant water are never known on the island." But why the people who were continually going from the infected ships to the shore, and landing cargoes, and coming in contact with the inhabitants, did not communicate the disease to the crowded and filthy localities in Brooklyn; or how the things which were *perhaps* (for no proof was offered) thrown overboard, and which *perhaps* touched Governor's Island, could gather venom in their transit through East River, was never explained.

Though this idea of the filth-origin of the disease was common among those who made no investigation, it was forcibly contested by many able observers, who sustained their arguments by incontrovertible facts. Indeed, the author has found no instance where an epidemic of yellow fever has been investigated with care in which the proof was not conclusive that it had no relation to filth or fecal deposits. Dr. Bancroft,* in 1810, denied that filth had anything to do with the production of yellow fever. Dr. Revere † showed conclusively that putridity had nothing to do with causing it in Baltimore. La Roche quotes Deveze, who says it has always been observed that curriers, tanners, soap-boilers, candle-makers, and, in general, all who breathe an unwholesome atmosphere, are not liable to the disease. Dr. Rush noticed that it rarely attacked butchers, and that scavengers escaped it entirely. La Roche concludes, from the facts that yellow fever is often generated where there is no animal decomposition, and that dissecting-rooms, slaughter-houses, knackers' yards, cemeteries, and other places where putridity is at its height are recognized as innocuous, that this fever does not result from animal decomposition. Savaresy ‡ (*fièvre jaune*) ascribed the cause of the fever to the air vitiated by a noxious and unknown influence.

The subject received more serious scientific investigation after the epidemic of 1878. It appeared at Grenada, Mississippi, which stands on an elevated plateau, and is so situated that every rain washes the streets and gutters clean. The house where it first appeared "was in good sanitary condition." Yet only five white persons who remained in the town escaped the disease. Colonel T. S. Hardee, United States Engineers, said, "This recent epidemic of yellow fever (in New Orleans, 1878) was more virulent and prevalent, and to a greater extent, in parts of the city that were entirely paved and well drained, than in those parts where the streets were filthy, and where the drainage was imperfect." Major Walthall,§ whose experience as a nurse was greater than that

---

* Yellow Fever. † La Roche. ‡ 1809.
§ Proceedings American Public Health Association, 1879.

of any other man in the country, did not believe the sanitary condition of a city had anything to do with the spread of yellow fever. He had "seen it most virulent in the cleanest and purest sections, while the dirty and filthy portions were comparatively exempt." Judge Clapp said that Holly Springs was "proverbial for its cleanliness and beauty; located on a high, sandy ridge; yet the town was decimated by the scourge."

Dr. Hargis, of Pensacola, said that the idea that the fever was caused by filth on land was a delusion. Dr. Cockrane,* in a special report on yellow fever in Grenada, Mississippi, in 1878, says that the town was in good sanitary condition. He lays down the proposition that it is caused by a specific poison; that this poison is not in any way the product of ordinary filth, and has no necessary association with filth as such. In Baton Rouge it was a noteworthy fact that "the portions of the town where the worst sanitary condition was observed were the least infected, and suffered least." Dr. Campbell declared that if filth were piled to the second-story windows, we would not have yellow fever, unless it was introduced.

The delegation from the Mississippi Board of Health † says, "The abundant experience we have just had leads us to believe that, while disinfectants, cleanliness, good drainage, etc., have their merits in the prevention of some other diseases, they are utterly powerless to prevent yellow fewer. Localities that were in the best sanitary condition were visited by the disease, which spread rapidly, and was as fatal as in other localities."

Dr. Herrick, of New Orleans, asks, "If its causes are in sanitary conditions, why does not the fever show a decided preference for foul localities?" Major Walthall said that in

---

* *Sanitarian*, 1879.
† Proceedings American Public Health Association, vol. i.

1853 the disease broke out in New Orleans in June; therefore the people of Mobile had ample warning. Cleansing and fumigation were thoroughly carried on; but it arrived there on the 12th of August, and swept the city. He added, "Where yellow fever has prevailed the worst has been in the cleanest places;" and, comparing the condition of a place in the year in which yellow fever raged with its condition in other years, "It has been found to be much cleaner at the time yellow fever became epidemic than at any other time. The unhealthy districts of a city, as regards ordinary years and circumstances, are the most exempt from yellow fever. . . . The city of Pensacola is remarkably clean. It is built on a bed of clean sand, and has no filth; yet yellow fever has prevailed there in its severest types." Dr. Selden, of Norfolk, said that imperfect sewerage, upturned earth, etc., had never been the cause of yellow fever in Norfolk.

The Committee of the Public Health Association, consisting of Dr. Sternberg, Dr. Choppin, and others, who examined the evidence collected respecting the epidemic in the towns on the Mississippi, say, after a hasty examination, "The committee cannot find there are any uniform local conditions. Yellow fever has prevailed in cleanly places and in filthy places, in high places and in low places, among the rich and the poor." Dr. Sims, of Chattanooga, said that filth was not the cause in that city. Dr. Harrison had visited many localities in the South during the epidemic of 1878. "I found it was not in the most filthy parts of a city that yellow fever originated. It was not in the most filthy parts where it was most malignant."

Dr. Atkinson, of the Tennessee Board of Health, said, "If the utterances of members of this association shall go out to the world, assuming definite shape, as they have been advanced and expressed here, that the filthiness of a community has no influence in the spread of yellow fever, without challenge, it will paralyze all sanitary effort." Dr.

Atkinson's language was energetic, and his rhetoric superb. He yielded the floor to Dr. Bell, who remarked on the overflow of lands, and added with some pathos that he should be very sorry "that anybody could believe that with filth they might be indifferent." He said, "Let it never go abroad that filth is to be indulged in because in some one place it did not happen to be heated enough to develop disease." Dr. Atkinson and Dr. Bell omitted to tender any facts to controvert the reports of the two committees, or the unimpeachable testimony of the individuals that filth or fecal matter had no relation to the rise and spread of yellow fever.

Dr. Thornton reported that in 1879* the yellow fever broke out anew in Memphis; there were fifteen hundred and thirty-two cases and four hundred and eighty-five deaths that year. He says there was not a street in the city in an unsanitary condition. "The city was as cleanly as any city in the country of the same size; perhaps more so."

Dr. Heineman,† long a resident of Vera Cruz, says that for twelve years Vera Cruz has had a good water-supply; the baths are well attended, the streets are the best-policed in Mexico; yet in those twelve years it has had three frightful epidemics of yellow fever. He adds, "It therefore evidently follows that filth, *per se*, has nothing to do with yellow fever."

Surgeon Clements, U.S.A., on the epidemic in New Orleans in 1867, says that scrupulous precautions were taken with fecal discharges. He learned "no fact which lent support to the suggestion that, like cholera, it might be considered a fecal disease."

Dr. Nott, who, on account of long experience with yellow fever at the South, was appointed to investigate the epidemic

---

* Proceedings American Public Health Association, 1879.
† Report National Board of Health, 1880.

in New York harbor in 1870, said in his report that this disease may traverse a half-dozen States in an epidemic, "showing the same virulence on the arid, clean hill-top as in the filthy alleys of a city."

Dr. Joseph Jones,* of New Orleans, says, "Yellow fever may prevail in one or more cities, and at the same time be entirely absent from other cities in the same zone and subjected to very nearly the same hygienic conditions. The mere absence of yellow fever from a particular city proves nothing as to the mere sanitary condition and measures of the city enjoying the immunity."

The National Board of Health † says that New Orleans has a soil supersaturated with filth; "yet since 1858 there has been a very marked diminution in frequency of yellow-fever epidemics, though the filth has increased annually. The board adds, "If yellow fever be a fecal disease, it is difficult to understand why an increase of fecal saturation and deposits should be accompanied with a decrease of the disease."

The epidemic in Florida in 1888 showed no preference for filthy localities. Jacksonville, where it raged the worst, was a clean city. It was far superior, in respect of cleanliness, to most other places of the same size in the country. The fever prevailed in Fernandina; but Dr. Starbuck says ‡ that the sanitary condition of Fernandina, before the epidemic appeared in Florida, was excellent. " Still, when the fever broke out at Jacksonville, every lot and street in Fernandina was thoroughly cleaned and disinfected with lime, sulphur, iron, and pyroligneous acid." The city was never cleaner or more healthy; yet by the 20th of September, five weeks after it attacked Jacksonville, it was raging in Fernandina." The disease was severe at Gainesville. Assistant Surgeon

---

* Quain, Dict.   † 1880.
‡ Report Supervising Surgeon-General Marine Hospitals, 1889.

Martin * reports "the sanitary condition of the city excellent, and the general health exceptionally good." Mr. Joseph Voyle, C. E.,† says that in Gainesville "the infection began in the cleanest places in the city, in the ordinary meaning of cleanliness." The deduction which he makes from the epidemic is that the conditions which favor the rise and progress of yellow fever are beyond human control; "precautions may mitigate, they cannot obviate" them.

CHAPTER XIV.

Cholera.

OF all the so-called zymotic diseases, none has been so positively imputed to filth and fecal deposits as cholera. After this theory was adopted, numberless diverse and confusing hypotheses were brought forward to account for the action of filth in producing the disease. Not one of these hypotheses had any value, for none was founded on experiment or scientific observation. It was not very difficult for ingenious speculatists to say that it was caused by specific organic matter, in which was hidden the active principle of the disease which was called *cholerine;* or to say that emanations arose from fecal discharges which, under "favoring conditions," gave rise to "the choleriginous principle;" or to tell us that it arose from want of ozone; or for a great many others to talk of fecal deposits spreading the disease by undergoing "certain decompositions" under "certain circumstances." After we had read all these theories and

---

* Report Supervising Surgeon-General Marine Hospitals, 1889.
† Proceedings Quarantine Conference, 1889.

many more like them, we were no wiser about the cause of cholera than we were before they were written.

Though this disease was described two hundred and fifty years ago, it did not appear in Europe until the third decade of the present century. For two hundred years it had ravaged Burmah, Siam, Cochin-China, Malacca, Sumatra, and Java; and though intercourse was common and frequent between those countries and the Western nations during this period, with every opportunity for the disease to be transported by people and merchandise, it did not transfer itself from those localities until nearly 1820, when, through some unknown cause, it slowly commenced its travels westward.

In 1819 it had reached Bombay; it did not touch the Mediterranean shores until 1823; it then halted on the confines of Europe for a few years. In 1830 it was on the Caspian Sea, and reached Moscow; in June, 1831, it arrived at St. Petersburg; it was at Vienna in August, and in September of the same year it reached Berlin. In October it appeared in England; in January, 1832, it was in Edinburgh; at London in February, and at Paris in March. It then crossed the Atlantic. Early in June it broke out in Quebec; it was at Montreal on the 14th; in July it had reached Albany and New York. From thence it swept through the South and West, to the Gulf of Mexico and Yucatan.

It was declared that it came to Quebec by an emigrant-vessel, but there were physicians in that city who asserted that they had cholera-patients before the arrival of the ship. The first case in New York City was on the corner of Gold and Franklin Streets, on the 25th of July. Two days after, an aged woman, who had been in Bellevue Hospital for three years, and who had had no communication with persons outside, took the disease. Other cases followed in different parts of the city, until the epidemic became general; but there is no record that the earliest cases were connected with each other or with any newly-arrived emigrant-vessel.

This epidemic did not entirely cease in the South and West of the United States until 1836, and it lingered much longer about the villages and small towns than in the more crowded cities. Dr. Whitney says, "In no section of the States have greater numbers, compared with the whole population, fallen victim to it than in the fertile and sparsely-settled prairies of the South and West."

The cause of cholera, as it was manifested in this widespread epidemic, was the subject of an exhaustive investigation in France, undertaken by a commission nominated by the French government. It was published by order of that government in 1834. The commission * say of cholera, "It comes without any known cause; it disappears without any real reason." Previous to its arrival, the Board of Health in Paris inspected all insalubrious places and houses; closed up four hundred and two of the latter in a single district, on account of bad wells and privies; supervised all markets, theatres, coffee-houses, and cemeteries; all filth was removed, and chlorine was poured into the ditches and streets. The people were congratulated on the good sanitary condition of the city, compared with previous times. Eighteen days after the first appearance of the disease there had been between twelve thousand and thirteen thousand cases, and more than seven thousand deaths. The streets that suffered most were generally where the population was most dense, and composed of the poorer classes.

The commission says, "Investigation has confirmed the fact that everywhere the cholera sported with human provisions and gave the lie to opinions the most generally received, and rendered questionable what seemed to be most firmly accredited. It was often in the most salubrious and

---

* Rapport sur la Marche et les Effets du Choléra, Paris, 1832.

least exposed villages that its ravage was most severe, while it left scarcely a trace of its passage in localities which were always considered as sources of infection and disease. . . . It is difficult to have an exact idea of the filthiness of Gentilly. . . . Shut up in a narrow defile, it is traversed by the Bièvre, whose waters move sluggishly, mixed with the impurities of a multitude of wash-houses, wool-cleaning establishments, dye-houses, chemical laboratories, and other factories; . . . smelting-houses of grease taken from bones; but, above all, cloth-dressing establishments, so many in number that the spectator might be led to believe that all the followers of that business had fixed their habitations in the same village. . . . A great many of the wells of Grand Gentilly are so saturated by infiltration from the sewers of Bicêtre that the water cannot be used, even for the commonest household purposes." At Clichy insalubrity is at its highest point; the ponds and ditches in the fields surrounded by houses are full of soap and lye. "The streets themselves are common sewers, and each step one takes is amid stagnant waters." Yet the mortality in these two villages was eleven for Clichy and twelve for Gentilly per thousand; "while many communes in which the sanitary committee found nothing to blame lost thirty-five, thirty-seven, forty, fifty, and fifty-five per thousand." . . . " Can the mortality presented by Clichy be attributed to its being sheltered against the northeast wind, *and, above all, to the ammoniacal exhalations ?*"

The loss of the civil population in Paris was 21.8 per thousand. At St. Denis was reported an unhealthy establishment, where skin-dressing was carried on, and where twenty cows were kept. Urine and water had no outlet; they ran into a cesspool and were pumped out into the highway, corrupting the air for a mile with the stench. "That house had but one cholera-patient, and nobody in adjacent buildings was affected, though one was a school which had eighty boarders and forty day-scholars." The disease did not rage among butchers, though cow-houses, sheep-folds, and slaughter-houses were plenty. Bercy exhales noxious and repulsive effluvia; yet there was "but one case of cholera within its limits." The epidemic spared the village known as Brèche aux Loups and the Rue de la Lancette, which were constantly overflowed by the filth of laundries, so that the houses were surrounded by pools and the highways rendered impracticable. "Nothing could surpass the stench of that sink when first visited."

In the village of Colombe was a glue-factory, where the basins presented a large surface of fetid exhalations. "Its immediate vicinity proved less unhealthy than the rest of the village." The same remarks apply to the communes of La Villette and Grenelle. The villages Pantin, St. Germain-des-Près, and Belleville, which received the fetid emanations from the *voirie* of Montfauçon, are placed in the category of those that suffered least,—seventeen, eighteen, and nineteen per thousand.

Out of one hundred and fifty-four workmen who were employed in making poudrette from the choleraic fæces brought from Paris, only one died of cholera. During the epidemic the inhabitants residing nearest places where the animal manures were kept were not even attacked. The occupation of these people might warrant the belief that habit protected them; but those who lived and worked all around the *voirie* were nearly as exempt as those employed in it. The military posts in and about the city which were most carefully policed and cleanly maintained suffered a greater mortality than Paris at large,—namely, more than twenty-five per thousand.

The disease remained quiet in its home in the East until 1848, when from some unknown cause, and without any change in the condition of the people of Europe or of America to invite it, it again started westward by about the same route as before, and raged on the Continent and in Great Britain with greater severity than in 1832.

Dr. Sutherland, of the Board of Health, made a report to Parliament on the cholera of 1849 in Great Britain. The faith in the dirt-origin of all infectious diseases was so firmly grounded in England at that time that he would have been a bold man indeed who had run counter to it. Wherever the disease appeared, Dr. Sutherland was able to find a privy, or cesspool, or garbage, or a pile of manure, or drunkenness with filth, or a pump that somebody had complained of, although maybe the people had been drinking its water for a century without experiencing the slightest harm. A graveyard caused it in Bristol, a mud-flat in Cardiff. It raged in Inverness; here the town was clean, and now the subsoil water was accused. But why the subsoil water should be so particularly deadly during this hot and dry summer was not explained. Overcrowded cemeteries caused the disease in London.

This report has no scientific value, in that it omits all account of those places in Great Britain which had no cholera, or only a few cases, but which were just as obnoxious to it, by reason of exposure to precisely those influences which were said to be its cause where it did prevail. Birmingham was one of the filthiest towns in the kingdom; but this city was remarkably free from cholera, both in 1832 and in 1849. Though it raged at Billston, ten miles distant, and though there was constant communication between the two places, only thirty-one cases were reported at Birmingham in 1832. This city also escaped the epidemic of 1865. The Royal College of Physicians were not willing to accept Dr. Sutherland's report as final, and they appointed a committee to investigate the epidemic of 1849. The committee's report contrasts with that of Dr. Sutherland, but corresponds with that of the French commission on the cholera of 1832. They say (through Dr. Baly) that the epidemic fell severely on comparatively few localities. Densely-populated regions on rivers and sea-coast suffered most. Three large seaports, however, almost entirely escaped. Many large inland towns in a defective sanitary condition had but few deaths. "The sanitary condition of many of these towns would afford no reason for their escaping." "Of the multitude of smaller towns that escaped, very many were at least as insalubrious as places of the same size which were severely attacked." While cholera is generally associated with poverty, overcrowding, deficient ventilation, imperfect drainage, want of cleanliness, etc., "it is equally certain that in some public institutions, and even in some private houses, except where *perhaps* there is want of ventilation, the conditions of insalubrity did not exist, the disease caused a large mortality in proportion to the number exposed," while "some of the localities which presented the worst features of insalubrity escaped altogether or suffered only in the slightest degree," and "there must be (p. 24) some other condition essential for its action besides the known conditions of insalubrity; and this unknown condition must have been absent in entire districts, in others only in limited spots," and "it would be an error to suppose that the spots it first manifests itself are always the worst in a sanitary point of view." A part of Manchester is known as little Ireland. "It is occupied by the most squalid and indigent Irish emigrants; it has numerous pigstyes, undrained houses and cellars, and a population crammed to suffocation, and is in the worst sanitary condition of any part of Manchester," yet only five deaths occurred here from cholera. Reading has a population of twenty thousand. "It is built on a bog." Apparently the most favorable conditions for cholera exist here, there being no sewers, much filth on the surface, and pigstyes and slaughter-houses abound, yet only seventeen died in Reading from cholera. On the other hand, at Margate Infirmary there were nine deaths from cholera out of two hundred and twenty people. Here it is "remarkably healthy on a chalk cliff, close to the sea." At Southampton the disease was most violent in a high situation detached from the town. It broke out at Torton Barracks, where the "supply of water is excellent, and there are means of ventilation and cleanliness." It appeared in Boston workhouse, which is well ventilated, clean, and has a good dietary. It broke out in Seaham harbor, where "sewerage and drainage are excellent." At Stafford "the most favorable condition for the attack of cholera existed;" there are no sewers, and slaughter-houses and pigstyes abound. In Stafford there were but two cases of cholera, and no death. "There was nothing in the locality to account for the outbreak at Hertford jail. Prison generally healthy." At Bridgenorth there is "no drainage; filthiness is extreme, and there is no ventilation." There was but one case of cholera in Bridgenorth.

Dr. Graves* boldly expresses his doubt of the accuracy of Dr. Sutherland's report, that cholera affected undrained and filthy localities only, for the course of the epidemic in Ireland in 1849 contradicts it entirely. Some of the healthiest localities in that country were the worst affected. During the epidemic in Dublin, it did not appear in the poorer and less healthy districts any more than among the richer and more healthy. The best ventilated and drained streets were not more exempted than those in an opposite condition. In Tralee the best part of the town was the part chiefly devastated. The well-clad, the rich, and the temperate were carried off, while the poor and intemperate, who lived in the lower quarters, escaped. In the lunatic asylum at Limerick the cholera claimed many victims, yet the institution is a model of cleanliness, free from bad odors, and not uncomfortably crowded. The diet of the inmates is good and regular, and their minds, on account

---

* *Dublin Medical Journal*, vol. x.

of their mental condition, not harassed by fear of disease. Everything here was in contrast to Limerick jail, which abounded in nuisances and was greatly overcrowded, yet the patients in the asylum suffered severely, "while the prisoners in the jail altogether escaped."

Some districts in Dublin, which suffered in 1832, escaped almost entirely in 1849. Church Street, which is notorious for bad sewerage and bad situation, is an example, while the village of Castleknock, three miles from Dublin, an elevated site, well drained, which was untouched in 1832, lost half its inhabitants in 1849. No town is better drained than Parsonstown, but the inhabitants suffered in far greater proportion than other places in Ireland. The village of Bray is on the site of a granite mountain, is celebrated for its health, and is a resort for invalids. It has good drainage and no filthy population. "Yet this village was awfully scourged by the cholera in 1849," while three adjacent villages, in lower, confined, and wet positions, escaped nearly altogether. Carlow and Bagnalstown are extremely well situated, on dry soil; they were decimated by the cholera, while many towns and villages notoriously impoverished and unhealthy escaped during the existence of the epidemic. In Dublin, in Patrick Street, and the adjoining parts, although those districts are the most densely inhabited, the worst drained, and the most filthy to be found in the whole city, there was very little cholera. Dulick has sixteen hundred people of the lowest class, "who live in a shocking state of filth and wretchedness," yet only two cases of cholera occurred there during the epidemic. Two villages near Dublin, in most healthy localities, built on a very dry soil, with good drainage, lost a fourth of their population.

Dr. Graves says, "If we take a world-wide view of the progress of cholera, we shall find that its prevalence is unconnected with any physical peculiarity, either of climate, soil, temperature, water, air, or food."

In 1854 * the epidemic in England had no relation to sanitary condition. While cholera caused 53,000 deaths in 1849, it only caused a little over 20,000 in 1854. In both years it preferred the crowded towns in the low grounds near the coast, afflicting very nearly the same places. Oxford, Brackly, Towcester, Potters, Norwich, Milton, Margate, Ramsgate, and parts of Essex and Cambridge suffered in 1854 more than in 1849. There were one hundred and forty-five districts, with a population of more than two and a half millions, where no case of cholera occurred. Some towns which suffered heavily in 1849 almost entirely escaped in 1854. Cholera caused 5275 deaths in London in 1832, 14,137 in 1849, and 10,738 in 1854, showing plainly that sanitary condition had nothing to do with the mortality in that city. The cholera death-rate of 1849 was at the rate of 41 per 10,000; that of 1854 was 22 per 10,000. Nine of the districts around London suffered more heavily in 1854 than in 1849.

Liverpool had 1523 deaths from cholera in 1832, with a population of 165,000, while in 1849, with a population of 255,000, it had 4173 deaths, yet what is called its sanitary condition was much better in 1849 than it was in 1832.

Nine towns, including London, with a population of 1,984,802, had 8845 deaths from cholera in 1832. The same towns, with a population of 2,920,118, in 1849 had 25,894 deaths, yet much improvement had been made in their sanitary condition between 1832 and 1849.

Dr. John Snow † says that at Lambeth are many of the causes which are supposed to promote infectious disease, yet the deaths from cholera in that district during the epidemic of 1854 were 29 per 10,000. In the sub-district of Kensington, less densely populated, the deaths were 126 per

---

* Seventeenth Report of the Registrar-General.
† *Lancet*, vol. ii., 1856.

10,000, and in Clapham, a genteel, thinly-inhabited district, the deaths were 103 per 10,000. Again, the sub-district of Saffron Hill, with the slaughter-houses, catgut-factories, and knacker's yards of Sharp's Alley, and the Fleet ditch flowing through uncovered, the mortality from cholera in 1854 was only 5 per 10,000, while the Belgrave sub-district had 60 deaths per 10,000.

Dr. Parkin * investigated the cholera epidemic in Jamaica. He says that at Green Island it was prophesied that the people would die of cholera like rotten sheep on account of the foul condition of the marsh. It was so unbearable that he was obliged to close his windows, with the thermometer at eighty or ninety degrees. There were only thirty-five cases of cholera here and five deaths. He says this "pestiferous spot" gave the most favorable results of any on the island as regards cholera. But on the dry calcareous hills, where decomposition of organic matter was least, fifty, sixty, seventy per cent. of the people were destroyed.

On the 9th of November, 1848, a ship left Havre for New York with three hundred and eighty-five passengers. There was no cholera at Havre when the vessel sailed, and all on board were in good health. Sixteen days out the disease appeared. On the arrival at New York the vessel was quarantined, but "it was said" that one person escaped from the station, came to the city, and was seized with cholera; he was returned to quarantine, where he died in a few hours. From this time on, during the winter, there were ninety-two cases in New York with forty-two deaths when the disease ceased. It appeared at quarantine again in April and became epidemic in the city soon after.

It spread over the Eastern and Western States, made fearful havoc with the overland emigrants to California, and did not entirely disappear from the country until 1855. The

---

* Report on Cholera Epidemic in Jamaica, 1851.

report of the sanitary committee of the Board of Health, on the cholera of 1849 in New York City, says that it broke out at "Five Points" about the middle of May. Cases followed in other parts of the city, which could have no connection with those at Five Points, and by the last of the month it was epidemic. During the earlier part of the season it raged worst in the lower wards, where many of the people were newly-arrived immigrants, destitute of the comforts of life. It also raged with great violence on the elevated and sparsely-built regions between Twenty-fourth and Fortieth Streets and from Eighth Avenue to North River, while it was declining in the lower wards. The Board of Health did not ascribe this epidemic especially to filth.

In 1865, without any known cause, and with no more change in the habits of the people to invite it than in its previous tours, the cholera made another circuit of the globe. In April, 1866, a ship arrived in New York with cholera, and was quarantined. On the 1st of May a case broke out on the corner of Ninety-third Street and Third Avenue; on the 3d of May one occurred in Mulberry Street, five miles distant from the first.

Dr. Harris, who reports on this epidemic, says, "How it chanced that an industrious and temperate housewife, residing on the hill-top that overlooks the southern end of Ward's Island, became the first victim, may never be so explained as to remove just doubts of its exotic origin." This case was accounted for, because "it was said" that a few days before the woman had put some privy contents on a garden. The house was cleared of tenants, but in three days they returned; no other case followed in this dwelling. Dr. Harris says that the relation of the second case "to any exotic cause, if there was one, was undiscovered." Three other cases followed, wide apart from each other, and without any known connection with any newly-arrived immigrant, or with each other, except that the fourth case washed the linen of the third. It is distinctly stated that no local cause was assigned for any of these cases, except the privy on the corner of Ninety-third Street. No claim was made that it caused the cases from two to five miles away. The seventh case was in a house and block which were "the best models of neatness and comfort in the city." "No local cause for such disease existed there." None of the cases which at first occurred in Brooklyn "were traced to a sick immigrant, or to anything pertaining to the recent immigrants in the city." Dr. Harris says it is needless to speculate on the introduction of cholera into New York and Brooklyn that year. He only states the facts.

There were no more cases in New York until June 4, when "an elderly gentleman, somewhat feeble," was taken; "no special cause was discovered." Up to the middle of this month cases appeared in spots widely separated from each other and differing much in their salubrity, though there was a tendency for the disease to locate itself in the lower, crowded, and filthier portions of the city. Though the filth-theory of zymotic disease had been rapidly gaining ground in our country, the faith that filth is the cause of cholera, or that it propagates the disease, was badly shaken by this epidemic. That it did not rage among well-to-do and presumably cleanly people and in sparsely-settled localities as it did in poor, crowded, and filthy districts was true; it was no less true that everybody in the well-built districts who could possibly leave the city had departed. In this epidemic Dr. Wendt * says, "There was but little diarrhœal mortality in the best and cleanest parts of the city, partly because this population with their children always go to the country in hot weather."

Dr. Harris says, "The Fifteenth, Sixteenth, and Seventeenth Wards in Brooklyn, and the entire regions south

---

* Cholera, 1885.

from Greenpoint to Newtown Creek were menaced, but untouched." This part is notorious all over the country for its filthiness. Dr. Harris* says, "Whoever is conversant with the history and habits of cholera would safely predict that if its presence and ravages depended mainly or primarily upon general atmospheric conditions, or if, in our cities, it depended exclusively on local filth, a miasmatic soil, or unhygienic conditions of domestic life, then in the regions of Brooklyn here pointed out the pestilence would have committed worse havoc than in any other section of that city. Yet in these three most filthy and miasmatic wards here mentioned, with a population exceeding forty-seven thousand, rivalling the foreign population of the pestilence-stricken Twelfth Ward, which was only three miles distant, there were but eleven cases of cholera. . . . In that particular section south from and adjacent to Newtown Creek and its junction" (which was overflowing with filth), "less than twenty-five cholera cases occurred in the whole season."

From Bowling Green to King's Bridge cholera selected its fields and fearfully menaced all foul places. "But it did not reach all such places, and from several it did reach it speedily disappeared to return no more. . . . The cleanly, well-drained, and well-built districts escaped, with but three or four exceptions." But the people from these districts who were able had for the most part fled the city, especially the women and children.

It broke out, early in July, on Governor's Island, where the drainage was perfect and the sanitary arrangements complete.

At Blackwell's Island it raged, although "throughout its entire extent it is at all times kept in a good state of sanitary police." Three hundred and sixty deaths occurred

---

* Page 186.

here. It raged at the Emigrants' Refuge and Hospital, where the sanitary condition "is generally unexceptionable." At Randall's Island the grounds and edifices "are all models of sanitary care and strict police," yet the cholera ravaged the island.

This report of the cholera in New York in 1866 needed only to have been a little more thorough in order to correspond exactly with that of the French Commission on the epidemic of 1832.

One of the most important facts of the epidemic of 1866 was the sudden outbreak in the Kings County Penitentiary. The building is on a "summit of a hill, is noted for its cleanliness and excellent ventilation, and every precaution that intelligence could devise had been taken to secure its healthfulness." Fourteen per cent. of the inmates were taken with cholera in a single night. The profound sanitarian who reports on this outbreak discovers the cause, and records it in this report. At the foot of the hill is a stable, where are some cows and a pigsty. On the night of the outbreak, he says, there was not breeze enough to lift the vapors from these over the hill, and "in these vapors were the germs of the pestilence!" How the vapors from the cow-stable could have been lifted as high as the prison, if there was no breeze, is another of the impenetrable sanitary mysteries.

The committee * which reported to the Suffolk District Medical Society, on the cholera of 1866 in Boston and vicinity, says that the educated and uneducated, the wealthy and the poor, the old and the young, the prudent and the imprudent, the resident and the non-resident, furnished victims to the disease. Though the majority of cases were in the most unhealthy localities, locality does not afford complete protection from this pestilence.

---

* *Boston Medical and Surgical Journal*, February 28, 1867.

Dr. Ingraham,* of Chicago, says that in 1866, in New York City, he studied what classes or trades were least subject to cholera. He found that men who worked in horse-stables (and also the families of these men) suffered less from the disease than any other class of people; and he ascribes their immunity to the exhalations from the fæces and urine of the animals. Dr. Hamilton ascribes the cholera of 1850 at Suspension Bridge to upturned soil. That filth was not the only cause was proved, he said, by the fact that in Buffalo those living in dirty shanties by the lake escaped. During the epidemic of 1884 † it was shown that the tanners in Italy were free from cholera. The streets where tanneries were located were altogether spared, and the plague was less virulent in those towns where tanneries were plenty. Yet there is no occupation where putridity is continually at a higher point than in these establishments.

In all of the cholera epidemics which have visited our country the military posts have had no immunity, although they are supposed to be in the best sanitary condition, so far as cleanliness and discipline are concerned. In 1832–33, all through the North and West, from Buffalo to Fort Leavenworth, the cholera prevailed among the United States troops. In 1849 it raged among them in Texas, along the Rio Grande, at Governor's Island, and at Carlisle Barracks. It broke out even at West Point that year; it persisted at Jefferson Barracks and at Fort Leavenworth in an epidemic form for seven years, from 1849 to 1855. It was epidemic on Governor's Island in 1849 and in 1852–54. It raged again at Governor's Island and at Hart's Island in 1866. It broke out on the steamer "San Salvador," which sailed from New York in July, 1866, for Savannah, with four hundred and seventy-six troops. They were put into camp at once, presumably a clean one; but there were two hundred and two

---

* *Medical Record*, vol. xxvi.   † London *Lancet*, vol. ii.

cases and one hundred and sixteen deaths. In Jackson Barracks that year in New Orleans there were one hundred and seventy-three deaths, and the disease was carried to the forts below the city. During the epidemic of 1866, out of a total of twelve thousand seven hundred and eighty men in the United States army, there were two thousand seven hundred and eight cases of cholera, and twelve hundred and seven deaths. Again in 1873 it did not spare the United States forces. It raged in San Antonio, in Kansas, Kentucky, Indiana, and Virginia. All this occurred where the best sanitary conditions are supposed to prevail, and among troops directed by a superior order of medical talent.

The epidemic of 1873 first broke out at New Orleans, and was confined to the Mississippi Valley and the Southern, Middle, and Western States. The first twenty-five cases in New Orleans were subjected to close examination, and it was proved that none of them were imported. All were residents of the city. The history of the outbreak of 1873 shows that it afflicted towns without any regard to their sanitary condition. Some most seriously visited were reported free from any local cause which is supposed to excite cholera; while others which were notoriously filthy, like New Orleans, yielded only a few deaths. During the eight months that the epidemic prevailed, New Orleans, with two hundred thousand inhabitants, lost only about two hundred and fifty from cholera.

One peculiarity of this epidemic * was that it was most malignant in small country towns and villages. It prevailed greatly on the thinly-settled Kansas prairies. There were no sewers or drains to lay it to in Kansas, neither were the people specially filthy there. But the sanitarians were ready with a reason; they said the cholera persisted in Kansas because animals that died on the desolate prairie

---

* Wendt, " Cholera."

were left unburied, and the effluvia from them caused the disease.

During the epidemic of 1884–85 in the south of Europe, the cholera, as in France in 1832, " sported with human provisions." It chose a southerly route to enter Europe this year; it was located mainly along the Mediterranean; Russia and Germany escaped entirely. Paris alone of the northern cities suffered. If the outbreak in Europe in 1884 was a part of the Eastern epidemic, then the disease made a sudden bound from Egypt to Toulon. The "fissure" by which it entered Toulon has never yet been discovered.

Nobody disputed that Toulon, where it first broke out in 1884, was filthy; but it soon reached Marseilles. Marseilles had suffered terribly from cholera in 1865;* but during the nineteen years that followed it had been renewed in pavements and sewerage. Her water-supply was equal to any on the Continent, except that of Rome; her markets were carefully inspected, and in sanitary condition and regulations she was unsurpassed in excellence by any European or American city. The health of the city was good. At the time of the outbreak at Toulon the deaths at Marseilles were less than two-thirds the normal death-rate for that season. The outbreak here was sudden and simultaneous in all parts of the city, and several of the earlier deaths occurred in the cleanest and handsomest quarters. Long before the first rumor of trouble extra precautions were taken to have the town cleansed and made ready. But the disease came and swept the city. The sewers which flowed into the tideless Mediterranean were declared to be the cause. On the other hand, the records showed that the old port of Marseilles, which was the main cesspool for its sewage in 1865, was almost exempt from cholera that year. Only one death occurred on all the shipping there assembled. The consul

---

* United States Consular Report, 1885.

says that to explain "this paradox" "it is said that the miasm of that particular part acted upon the homœopathic principle, as an antidote to the venom of the epidemic." Why the miasm should prevent the disease in 1865 and cause it in 1884 was never explained. The consul says, "All that energy and liberality could perform, all that sanitary science could suggest, has been done; but the pestilence is here and defies restraint." Toulon, says the report, is extremely filthy; but "filth alone does not create cholera. If it did, Toulon would have a yearly epidemic." Lyons escaped,[*] as it did in 1832, 1849, and 1853. Only one death from cholera occurred there during the epidemic of 1832, and there were only twenty-seven cases in 1884. Her escape at any time was not due to cleanliness. In 1849[†] cholera broke out at the military hospital at Lyons. "Contrary to expectation, it is confined to the garrison." "Who would have believed before 1832 that Lyons, a damp, uncleanly, unhealthy city, full of workmen the majority of whom are in a wretched state of want and filth, would escape the scourge?" And, the writer adds, how strange, now that it has appeared, that it should be confined to that part of the people "who are decidedly in far better condition than the bulk of the population." Nobody has yet given a satisfactory reason why Lyons, Versailles, and Birmingham have never suffered from epidemic cholera.

The disease at first skipped Genoa, and the people expected to escape. The city was clean, and the consular report says that disinfectants were used everywhere. The habitations of the lower classes were inspected daily. No city "could take more precautions" than did Genoa; "great pains were taken day and night to keep the city clean." Genoa was "in as healthy a state as it is possible for human

---

[*] Bulletin de l'Académie de Médecine, 1884.
[†] London *Lancet*, January, 1850.

agency to make it." But in a few weeks its inhabitants were flying with terror in all directions. Consul Fletcher writes that cholera at Genoa seeks and finds its victims in every part of the city. The hitherto healthiest localities and the widest and most airy thoroughfares have it.

It passed by filthy Leghorn and other filthy cities in Italy in disdain, but struck no less filthy Naples, which was ravaged by the disease. A correspondent of the London *Lancet** writes that the mode in which cholera attacked the people of Naples has been very perplexing to the sanitarians. The narrow, ill-ventilated streets, where human beings teem like rabbits, have been precisely those places in which cholera has numbered the fewest victims; while the localities which are in marked contrast have filled the hospitals and cemeteries. "The ill-fed, badly-clothed, unwashed inmates of the rookeries have come off very lightly in comparison with the well-dieted, comfortably-clothed, not uncleanly bourgeoisie." It is naively remarked, " All this, however, should not shake our faith in the efficacy of pure water and cleanly dwellings."

Turin and Milan, Venice with her canals loaded with filth, Florence, and Rome escaped, or were only lightly visited. Let those who know by observation what those cities are in respect of cleanliness decide whether *that* cleanliness protected them. It avoided Palermo that year; but the consul writes, "The sanitary condition of the city had little or nothing to do with its escape." It was the filthiest of all the Italian cities. Active measures were taken to put it in good sanitary order. The drains were disinfected, the houses of the poorer classes whitewashed, the slaughter-houses and markets were inspected, the sale of unripe fruit and suspected food was forbidden. Nevertheless, the next year the pestilence came and threatened to sweep its inhab-

---

* *Medical Record*, vol. xxvi.

itants from the earth. The consul writes that, notwithstanding the bad sanitary condition of Palermo, the general mortality there is comparatively small; "the deaths between the years 1872 and 1881 being only 26.9 per thousand, and for the last few years even fewer; while the mortality of Rome, Turin, Venice, and Milan was 33.9 per thousand. Cholera raged at Gibraltar, although this place was pronounced in excellent sanitary condition.

Hyères is a town of twelve thousand people. It is distant a few miles from Toulon, and has constant communication with that city, which was the centre of the pestilence. Its sanitary condition* is truly deplorable. There is a ditch here which is called a sewer by compliment; it is three feet deep, and covered with flag-stones. The hot sun acts on the stagnant sewage below, which rapidly ferments. The fall is insufficient, so that "in one place the sewer had an inclination upward." It takes the overflow of numerous cesspools, and, as none of the conduits are trapped, the foul air rises into the houses. Cesspools abound on all sides, and "are, for the most part, built immediately under the houses." Untrapped pipes draw the foul air into the dwellings. The soil-pipes of the privies, being within the house, have a higher temperature than the outside air, and are active ventilators of the privies into the houses. There are no means of excluding effluvia from the dwellings, which are impregnated with the stench from the cesspools, brought up by the soil-pipes, and also by the sewer-gas drawn up through the waste water-pipes. "The prospect may well awaken the keenest alarm." The writer says that if one or two cases of cholera are imported, then there will be no safety for the inhabitants.

Such was the sanitary condition of Hyères, July 19, 1884. So like Newport in sanitary condition, it also resembled it in

---

* London *Lancet*, 1884, vol. ii.

salubrity, for it had long been renowned for its healthfulness. It was especially favored by the English, who had found out by wide experience that a residence in Hyères recuperated their health. In the same volume of the *Lancet*, just two months and one day later, is an article on the "Immunity from Cholera at Hyères." We are now told that it has "climatic advantages," though it is only eleven miles from Toulon, and may be considered a suburb of that city. The population of Hyères has been suddenly augmented more than twelve per cent. by fifteen hundred fugitives from Toulon, some of whom fell ill of cholera after their arrival. "None of these cases determined a local outbreak." It is still in the same unsanitary condition. Most houses have cesspools; the pipes are untrapped, and such sewers as exist are, for the greater part, defective. The streets are washed three times a day and some disinfectants are thrown about in them, which, however, will not quite explain why Hyères has no cholera. "Something more than this is requisite to account for the remarkable immunity of Hyères." We must search deeper.

We concentrate the rays of Sanitary Science on this subject in their greatest effulgence and find a solution of the mystery. The editor of the *Lancet* wrote, with apparent gravity, that "the probable explanation of this phenomenon (the immunity of Hyères) lies in the *granitic nature of the soil*." Given a "granitic nature of the soil," and open, reeking sewers, overflowing cesspools, soil-pipes and waste-water pipes ventilating privies and cesspools into the houses, and impregnating these with their repulsive effluvia, are entirely harmless. But why this "granitic nature of the soil" was not discovered and disclosed in July, so that the minds of the people at Hyères could be composed, instead of tortured about the cholera, was never explained.

At a sitting of the Paris Academy of Medicine, September

1, 1884,* the epidemic in one hundred and four places was reported on. In seventy localities it was ascribed to importation and contagion. Twice it was caused by soiled clothing sent from infected localities; and in seven instances it was ascribed to a stream of water flowing from such localities. It appeared in twenty-five places in which the physicians could find no imported case, nor any manner in which contagion bore any part in its transmission. Once appearing, its spread is accounted for by soiled clothing, by contact with patients, and by residence in houses which contained cholera. It prevailed in fourteen places in which the customs and habits of the people, in respect to filth and fecal matter, conformed to all of the prescriptions of hygiene. In ninety towns it could be said there was uncleanliness, and fecal matters were not properly cared for; but the report takes no notice of the hundreds of towns in France, like Hyères, in which this condition of filth and fecal matter was present, which were untouched by the disease, though, like Hyères, they afforded a refuge for fugitives from infected parts; and, unlike Hyères, they were not favored by a "granitic nature of the soil." Dr. Shakespeare † says, "It was not for the lack of grossly unhygienic circumstances that the cholera did not ravage the greater part of France during 1883–85."

At the sitting of the Paris Academy of Medicine, October 7, fifty other places were reported on where cholera had prevailed. In thirty-nine of these it was assigned to importation or contagion. The physicians of the remaining eleven towns were firm in their belief that it was generated on the spot. They could find no case of importation, nor any trace that it had been communicated by contagion. In many of the places where no contagion could be discovered, persons of various professions and in different walks in life, living

---

* Bulletin de l'Académie de Médecine.
† Cholera in Europe and India.

remote from one another and without any communication, were simultaneously seized. A few members of the Academy were positive that the disease could and did arise spontaneously; that in many instances neither importation nor contagion would account for it. M. Ricord expressed his disbelief in either, basing his opinion on observations of the epidemic of 1832 and those which had followed it. He thought that it might arise spontaneously, under some epidemic influence not yet understood; and that a quarantine was useless and vexatious, except such a one as would quarantine healthy people out of an infected district. That, he said, would be truly hygienic.

The alleged causes for the disease in France this year are bad water, want of ventilation, topographical situation, want of personal care, excesses, fear, imprudence, extreme heat, exposure to cold, filth, and particularly fecal matter. The last was specially insisted on as a prime cause, and no great opposition was made to it by the members of the Paris Academy. But at a sitting on the 26th of August, M. Bouchardat said that it had been the rule, in latter times, to assign great importance to breathing air charged with emanations from putrefying matter, as a cause of *cholera nostras*. While he admitted the reality of this cause, he thought it was much less dangerous than was supposed. He then made the remarkable statement that, in his long career in Paris, comprising a service of twenty-two years at Hôtel Dieu and his inquests of the Conseil de Salubrité, he had never known a death from *cholera nostras* among the great number of men employed in the depots for dead animals, nor in the depots for fecal matters, nor among the cleaners of privies, nor among workmen in sewers. He thought we ought not to exaggerate fæces and putrid matters as the cause of *cholera nostras*. He added that he had never known a case of this disease which arose from bad water.

At a sitting of the Academy, the 11th of August, M. Le

Roy de Mericourt declared that he could not see the relation of fecal matter to cholera. For ages fecal matter had been accumulating at Marseilles, yet no cholera appeared there until 1832. All of the towns on the Mediterranean shore were, in respect to fecal matter, like that city, but they were not' afflicted with cholera. The crews of ships are often decimated by the disease, but there is no accumulation of fæces on shipboard. At a later sitting M. Mericourt related an epidemic which occurred in 1854 among the French fleet in the Black Sea, and which destroyed in eight days eight hundred men out of a force of three thousand. M. Guérin replied that a certain change must take place in the fæces (*il faut donc que des conditions particulières viennent altérer les matières fécales*) in order to produce an epidemic. M. Mericourt asked his colleagues of the biological and chemical section whether accumulated fecal matter could acquire by time such pernicious qualities. M. Bechamp replied for this section, saying that all experiments proved that noxious germs escaping from the body lost their infective power (*ne tardent pas à perdre leur morbidité*) so soon as putrefactive changes commenced. Cholera germs, he said, could not exist in putrefying matters.

Whenever the cholera has afflicted Milan, which has happened on three different occasions, the people employed on the irrigated fields which receive the city's sewage and fæces have always escaped. When the disease raged at Edinburgh and Leith in 1865,—as we have seen on a previous page (68),—not a case occurred in the vicinity of the meadows which received the fecal excreta of the cholera patients in those cities. In 1884, during the cholera epidemic at Paris, Gennevilliers received one-half of the city's sewage; not a case of cholera occurred among the three thousand people in that commune, although they were daily exposed to cholera discharges.

In the winter of 1891–92 cholera assumed an epidemic

form in the East, and began to creep slowly but stealthily westward, taking a northern route, as it did in 1832 and in 1849. It ravaged Persia and Russia during the spring and early part of the next summer. The middle of July it had reached St. Petersburg. Without another note of warning, at one bound, it struck Hamburg, the first death from that disease occurring there about the 18th of August. By the 20th of the month there had been eighty-five cases and thirty-six deaths. The 21st of August there were eighty-three cases and twenty-two deaths. The next day there were two hundred cases. There was much running to and fro of bacteriologists, and it was the 22d of August before a microscopical examination determined that the disease was true Asiatic cholera. Before bacteriology would permit that it should be called by its right name there had been three hundred and sixty-eight cases. The first was hardly proved when the epidemic was wide-spread. There were six hundred and seventy-one cases on the 26th of August, and two days later eleven hundred and two, which was the highest number seized on any one day. Later, with but slight variation, the number of cases declined daily, so that on September 3 there were seven hundred and eighty. On the 10th the number had fallen to four hundred and thirty-nine, and on the 17th of September there were three hundred and thirty-eight cases, on the 24th there were one hundred and twenty-six, and forty-three cases on the 1st of October.* The number after this on any one day was not higher than forty-three, and with one or two exceptions it continued to decline until October 19, when there was one case and no death. From the 17th of August to the 20th of October there had been 17,988 cases and 7608 deaths.

It is not known how cholera effected an entrance into

---

* *Münch. Med. Woch.*, October 29, 1892.

Hamburg. The sanitarians said, with one accord and with a great uproar, that it was caused by the filth of the city, which they described as a place "where dirt reigns king, to whom all pay homage." Cholera was a filth-disease, therefore Hamburg must be filthy, or it could not entertain cholera. When the disease attacked the low part of the town, it was the filth in the soil that attracted the microbe. When it raged among the overcrowded, the poor, the wretched, the drunken, which it did more than it did among the rich, it was the filth in which the former lived that the microbe sought for an abode and nutriment. When the disease seized the rich, the well-to-do, and the temperate, which it did often enough, those of them who had not fled the city, the reformers ascribed it to the filthy water. Those who knew the least of cholera, and who perhaps had never seen a case of the disease, talked the loudest about it. On both sides of the Atlantic the buccal flux was something tremendous. Out of it the sanitarians distilled the maxim: cholera, being a filth-disease, is carried by filthy people to filthy places. Nothing could be more manifest to the sanitarian eye than the cause of the epidemic in the great German port. The problem explained itself. The filthy Russian Jew had transported the filth-disease, cholera, to filthy Hamburg. Out of this huge volume of words they condensed this other precept, which, if strictly followed, would be an unfailing protection against cholera,—boil the water and keep clean. And the reason why Hamburg in the course of nine weeks had eighteen thousand cases of cholera and nearly eight thousand deaths, was because it had not defecated with precaution and delicacy, and had forgot to boil its water. Probably none of these sanitarians had ever seen Hamburg. What were the facts? It was a healthy city. Its death-rate for 1891 was 23.5. It was one of the cleanest towns on the continent. Pettenkofer said it was the type of a well-drained city. The London *Lancet* of October 1 said, "Here was a city with fairly good drainage, not overcrowded, no famine, but little of the direst sort of poverty, a creditable condition of house and personal cleanliness." Dr. Seibert * says that he has known Hamburg for more than twenty years; has visited the finest streets and the smallest alleys; the houses of the well-to-do people as well as those of the poor. It is one of the cleanest of cities, and he intimates that the superior neatness of its streets, houses, and people is to be contrasted and not compared with that of New York. Besides, the houses in which the disease first broke out were new and in a far better hygienic condition than those of the older parts of the city. There was nothing, in fact, about Hamburg's sanitary condition to favor the appearance of cholera. The Jew, who, since the tenth century, had had nearly every epidemic in Europe laid to his door, proved his innocence through an *alibi*. He had not been to Hamburg at all for more than a month previous to the outbreak, but had been embarked from another point. When this depot was carefully searched for cholera it was proved that there had been no case of that disease among the Jewish emigrants. From Hamburg the disease was carried to other German towns, to Belgium, where there were over eleven hundred cases and nearly six hundred deaths, to Holland, England, and the United States.

It is probable that medical men were never quite so puzzled regarding the origin and spread of cholera in Europe as in 1892. It broke out in the suburbs of Paris, at the prison of Nanterre, on the 2d of April.† Soon after it appeared in twenty-six different communes. In those first attacked it is certain there was no imported case. Although there was constant communication between the city of Paris

---

* *New York Medical Journal*, December 10, 1892.
† *Revue d'Hygiène*, July, 1892.

and the outside communes, cholera was present in the latter for four months, and had caused four hundred deaths before a case broke out in the city. The mortality was appalling; in some communes it was as high as ninety per cent. of those seized. In July, one month before any case occurred at Hamburg, cholera appeared in Havre and became epidemic. Here were twelve hundred and sixty cases and four hundred and ninety deaths. The mortality was higher than at Paris; of the first fifty cases, forty-eight died. The origin of the epidemic at Havre is unknown.* It was ascribed to the arrival of the "Rugia" from Hamburg, but there had been seventy-two deaths from cholera at Havre before this vessel appeared in the port. It broke out in the insane asylum at Bonneval;† here were forty cases and twenty-eight deaths. No trace could be found of its entrance into this institution. It appeared at Rouen; the sanitarians again raised the cry of filth. Dr. Bataille ‡ says no cause could be assigned for its presence at Rouen. There was no imported case. During this epidemic in Europe there was a general disposition everywhere on the part of the sanitarians to traduce the water, though in not a single instance was there any proof that any of the inculpated fluid had received the cholera infection. It was said that Professor Koch had expressed his opinion (*gutachtung*) that cholera dejections had in a certain way (*gewissermassen*) got into the Elbe, and had been carried up-stream by the tide to a point above the intake of Hamburg's water.§ We have looked in vain for this "opinion" over Professor Koch's signature. His fame is not founded on his "opinions," but on his investigations. As late as October 27,

---

* *La Normandie Médicale*, September, 1892.
† Mouvement hygiène.
‡ *La Normandie Médicale*, August 15, 1892.
§ *Deutsche Med. Woch. u. Aertz. Cent. Auz.*, August and September, 1892.

1892,* no cholera bacilli had been found in the Elbe, or in the reservoirs or pipes which held and carried this water for the people of Hamburg. The irregular distribution of the cases, and the course of the epidemic, proved that it could not be explained on the water-theory. If Hamburg's cholera was caused by the Elbe water, no more positive proof can be given of the utter worthlessness of bacteriological examinations to determine the character of a water-supply. If, at any time during this epidemic of eight weeks or more, which caused nearly eight thousand deaths, specimens of this water had been given to bacteriologists to examine, they, not knowing its source, would not have pronounced it an unwholesome beverage. By municipal ordinance all water which was consumed in Hamburg was commanded to be boiled. This was begun almost immediately after it was shown that cholera was in Hamburg. Boiling the water must have been general by the 25th of August. A few days later the number of cases began to diminish, and doubtless this epidemic will go into sanitary history as having been arrested by boiling the water. However, we cannot close our eyes to the fact that nearly seventeen thousand cases of cholera out of a total of nearly eighteen thousand broke out in Hamburg after the water was boiled. But the germs had now spread through the city. What shall we say, then, of the disinfection that was practised? Dr. Seibert † tells us that it was performed with scrupulous exactness by a corps of trained disinfectors; there was rubbing and scrubbing, and steaming and burning and fuming, not under the direction of the tattling sanitarians, but supervised by scientific men like Koch and Rahts, genuine hygienists and conscientious investigators, who have the respect, if not the entire confidence, of medical men every-

---

* *Deutsche Med. Woch.*, October.
† *New York Medical Journal*, December 10, 1892.

where. Previous to 1892, Hamburg had had sixteen visits from cholera; it had not caused more than sixteen hundred and seventy-four deaths at any one outbreak. In 1873 it destroyed one thousand and one people. Since that time the hygienic condition of the city had been improved, in so far as its people had better and more varied food and raiment, more commodious shelter, less work and more wages, a general advancement in material comforts, and a higher moral and intellectual progress. Yet in spite of these advantages, when it was smitten by cholera in 1892 it lost eight or ten times as many of its people as it did twenty years before when no attention was paid to scientific disinfection. Regarding the cause and prevention of cholera, the epidemic in Hamburg in 1892 has only deepened the mystery.

The condemnation of the river-water induced the citizens to bore wells, and although these had long before been pronounced not only unfit, but dangerous to use, and although they were only twenty or twenty-four feet deep, and took the surface-drainage through the old soil of the city, the well-water was now shown, " by chemical and bacteriological experiment," to be good.*

In the suburbs of Paris, as late as July, M. Proust declared that the tie which connected all cases was the water of the Seine, though the Seine water inside the city did not produce the disease.† Towards the last of July it broke out in Paris, and its progress there had no relation to the Seine water. M. Ach. Lavache ‡ shows that the two zones which were most gravely affected by cholera in Paris in September received their water, one from the Seine, the other from the Marne and the Dhuys. The zone supplied by the Seine, with a population of 513,576, gave one case of cholera to

---

* *Aertz. Cent. Aus.*, October, 1892.
† *Revue d'Hygiène*, July, 1892. ‡ Ibid., October.

3292 people during the first half of the month, and during the last half one case to 5035 people. The other zone, supplied with water from the Marne and the Dhuys, having a population of 531,532, gave one case of cholera to 2242 inhabitants during the first half of September, and one case to 5537 people during the last half of the month. In either case this poisoned water of the Seine, which more than a half-million of people were drinking, affected only one person in 3292 during the first half of September, and not one in 5000 during the second half of the month. Dr. Gibert,* in reviewing the causes of cholera in France, denies the influence of the Seine water. He has never seen polluted water cause epidemic cholera in Europe; filthy water may cause colic and diarrhœa, but the cholera that travels from place to place, *never*. If Seine water can cause cholera, then, says Dr. Gibert, pull down sanitary barriers,—they are useless and expensive,—and away with the luxury of grand international conferences, where we so copiously discuss preventive methods of cholera.

The tables of the ravages of cholera in India would seem to show that it has no claim to be considered a filth-disease. The deaths in Bombay † for seventeen years, 1866 to 1883, were as follows:

1866	23,027	1872	15,642	1878	46,743
1867	5,143	1873	283	1879	6,937
1868	6,348	1874	37	1880	684
1869	52,330	1875	47,555	1881	16,694
1870	2,666	1876	32,117	1882	7,904
1871	5,281	1877	57,252	1883	37,594

It cannot be that these figures represent the filth of Bombay; that it was quite dirty in 1866, and that in 1867 it was five times cleaner, to grow more than ten times as dirty

---

* *La Normandie Médicale*, September, 1892.
† Shakespeare, "Cholera in India and Europe."

in 1869, when there were 52,330 deaths; that the next year the country was very clean, when only 2666 deaths were reported, again to rise to more than 15,000 deaths two years later. It is not complimentary to our common sense to tell us that in 1874 the filth was well-nigh all removed, when there were only 37 deaths, and that the next four years it was horribly dirty, when more than 180,000 died of cholera in Bombay; or that in 1880 it suddenly grew clean again, when only 684 died; and that in 1883 was a new accession of filth that caused 37,954 deaths. It would disgrace the anility of a country neighborhood to account in this way for the variation of cholera deaths in Bombay. The average annual death-rate from cholera per 10,000 in India varies, in the twenty-seven districts, from 6.05 to 49.51. Are we to understand that these figures represent the comparative cleanliness of the districts? For example, is Noakhally more than eight times as nasty as Dinagepoor?

Dr. J. M. Cunningham * says the causes of cholera are not to be settled by theoretical discussions, however clever or learned, but by a study of the facts. Though an earnest advocate of abundant and agreeable water, proper drainage, and every other measure which can possibly add to human comfort, and thereby to human health, he declares that "the cause of cholera, what governs its distribution and its relative incidence in different places, is still as inscrutable as when the disease first appeared," and "it must be remembered that the distribution of cholera, as may be proved beyond all doubt, is not regulated by conditions of filth or cleanliness;" and that quarantine, isolation, and disinfection have utterly failed to prevent or arrest outbreaks among European troops, even when carried out under the most careful superintendence. And the Indian Sanitary Commission † says no

---

* Epidemic Cholera in India, 1872-75-79.
† London *Lancet*, 1891.

satisfactory explanation is given of its almost simultaneous appearance in many centres of a district, nor of the strange exemption of certain areas environed by cholera.

But the final blow to the filth-origin or propagation of cholera was given by Koch himself, at the very time when he announced his discovery of the bacillus, called the comma bacillus. He premised * that we should suppose that, in their struggle against pestilence, people would start from a scientific basis; but this had not been done, especially with cholera. He showed that its bacillus flourished best at a temperature between 86° and 104°; below 63° the growth is slight, and it ceases below 61°. He was inclined to believe that if comma bacilli were brought into a putrefied liquid containing putrefactive bacteria, they would not come to development; and that if they were put into a sink or cesspool they would die, and "there would be no necessity of disinfecting." If there was a trace of acid in the fluid in which the bacilli were placed, they were stunted; if the fluid was acid in a marked degree, they died at once. He showed that the sulphate of iron, which had been so lauded as a disinfectant, actually exerted the opposite effect; and he says, "The process of putrefaction that goes on of itself in the cesspool is sufficient to kill the comma bacilli." Sulphate of iron only stops the putrefactive process and preserves them. Subsequent experiments have shown that wherever filth abounds and putrefaction is going on, with the presence of putrid bacteria, the comma bacilli are quickly destroyed. This helps to solve the question why those who, in times past, have been exposed to filth in its most repulsive forms have so often escaped the cholera; and why it has seemed to endure longest about those places where disinfectants have been freely used, and where putrefactive changes are least observed. In speaking of the

---

* *British Medical Journal*, 1884.

benefits that were to arise from the discovery of the cholera bacillus, Koch said, "But, above all, we can deduce this advantage: that an end will at length be put to the fearful squandering of disinfectants; and that millions will not again, as in the last epidemic, be poured into gutters and cesspools, without the slightest advantage."

Three hours after these bacilli were dried they died, unless they were in compact masses; and then they did not live longer than one day. The common water-bacteria always suffice to destroy the comma bacilli. "Everywhere," Koch says, "where I was able to come across a liquid containing bacteria, I examined it in search of comma bacilli, but never found them in it." He could not find them in the sewers of Calcutta, nor in the extremely polluted water of the Hooghly, nor in a number of the tanks which contained very dirty water. He did not believe that they could multiply in well- or river-water,—certainly not if either had a temperature as low as 61°,—because neither contained the concentrated nutriment for their increase. He could only imagine that they could develop at some point in perfectly stagnant water, where there happened to be vegetable or animal matter suited to their growth. But when the water has a rapid motion, or is subject to frequent change, the conditions of growth, even if the nutriment is there, are less easy, or do not occur at all. In but one instance did he find the comma bacillus in water, and this was in a tank where the soiled linen of a cholera-patient had been washed.

Surgeon-Major D. D. Cunningham [*] records his experiments with the cholera bacillus in earth mixed with human fecal matter. They extended over a period of time from the 16th of December to the 21st of March. He says, "In the above experiments we find that very large quantities of comma bacilli introduced into fecally-contaminated soil, and

---

[*] Scientific Memoirs, Medical Officers in India, 1887.

exposed to conditions similar to those to which the bacilli entering the soil at Calcutta are normally liable during the period of year dealt with, *failed to multiply, and, on the contrary, rapidly and completely disappeared*" (italics ours). He further continued these experiments, and adds, "The above experiments having clearly shown that comma bacilli rapidly and permanently disappear from portions of soil, whether pure or fecally contaminated, exposed to the ordinary conditions prevalent in Calcutta, an attempt was made to ascertain whether any other materials to which they are likely to gain access were more favorable to their continued existence." Cow-dung was now taken. In its normal state (that is, unboiled) the development (of the comma bacilli) is either entirely repressed or very much enfeebled, showing "that in this medium the commas have no capacity for the assumption of a resting condition." In media, such as fresh cow-dung, which appear to afford everything necessary to their continued vitality and multiplication, the presence of other organisms normally present in the medium is sufficient to repress or absolutely to suppress further development, and in other media, such as human fecal matter, even the aid of artificial measures calculated to place them at an advantage in the struggle are insufficient to be of any avail in securing their continued existence." And again, in 1889, Surgeon Cunningham * records five experiments made with the comma bacilli in water, and six experiments in earth. In fairly clean, unboiled water they disappeared in four or five days at the temperature of the air in Calcutta. In foul, unboiled water they disappeared in four days; *in foul, boiled, sterilized water they lived for twenty-five days.* Mixed with garden-earth they disappeared in from nine to twenty-six days. Mixed with garden-earth and fæces they were gone in six days; *while if the garden-earth and fæces*

---

* Scientific Memoirs, Medical Officers in India, 1889.

were boiled, sterilized, they were "*still present after forty-seven days.*" Surgeon Cunningham says, "Taking the case of the present experiments on their behavior (the comma bacilla) in soil and water of various degrees of pollution, the evidence was entirely in favor of rapid extinction and against the development of any specially resistant forms. And under normal circumstances they are singularly incapable of holding their own in the struggle for existence," and that "they very rapidly succumb" if they are not placed in a medium that has "not been specially sterilized for them."

Dr. Watson Cheyne[*] says the comma bacillus is never found in putrefying materials. Nicate and Rietsch[†] could find no comma bacilli in the discharges of cholera patients which had resumed the fecal odor and color. When the material in which they put the bacilli began to take on putrefaction, they rapidly disappeared. "The rapid disappearance of the microbe under the influence of putrefaction is a real fact." These repeated trials demolish entirely the filth-pathology of cholera in so far as the germ of that disease is concerned.

Dr. Macnamara[‡] says that the cholera bacillus quickly disappears in the drainage of cesspools.

Dr. Sternberg[§] says that the cholera bacillus grows in bouillon, especially in the incubating oven, and in "sterilized milk." An acid reaction of the culture medium prevents its development as a rule. "It also multiplies itself to some extent in sterilized river- and well-water." It preserves its vitality in sterilized sea-water and multiplies there, "but in non-sterilized sea-water it dies out within two or three days, the rapidity with which it disappears de-

---

[*] Cholera in Europe and India, Shakespeare.
[†] *Revue d'Hygiène*, 1885.
[‡] History of Cholera, 1892.
[§] *Medical Record*, October 1, 1892.

pending upon the number of saprophytes in the water." "It dies out in a few days in milk or in river-water which contains numerous saprophytic bacteria." "In competition with the ordinary putrefactive bacteria the cholera spirillum soon disappears," "so it would multiply more rapidly * in water not containing a large amount of organic material than it would in sewage."

These experiments with the cholera bacillus in water destroy what remaining faith there is in such legends as the Broad Street pump. They quiet apprehensions that an epidemic can be excited through water which is at a temperature below sixty-two degrees, or by a flowing stream, or by water which is well supplied with the bacteria which are common to it.

We are forced to one of two beliefs. If we cleave to the filth and fecal origin of cholera, we must forsake the germ theory. If we hold to the one, we must despise the other; and however great we may conceive our merit to be in hurrying to bow down to it, the new idol will not cease to repulse our oblations until we have cast down the old.

CHAPTER XV.

Diphtheria.

THE filth-origin of this disease was even better established in the minds of the sanitarians than was that of cholera or yellow fever or typhoid fever. Like the latter, it owes its description and its name to a Frenchman.

In 1818 † the military legion of *La Vendée* imported it to Tours, a town of twenty thousand inhabitants, where it

---

* *Sanitarian*, October, 1892.   † Bretonneau, 1826.

raged for two years. Bretonneau says that it was new to most of the physicians of Tours; he himself had seen only two cases of croup. He traces the disease to remote times, but says that since the end of the sixteenth century it had shown itself in every region of the old and new continents. He studied it with great care, made sixty autopsies, and believed that it was identical with croup and with scorbutic disease of the gums.

Except to rather lightly touch on its contagiousness, he says nothing of its cause. He asks, "May it not be developed spontaneously and afterwards transmitted to others?"

Guersant, Bouchert, Trousseau, and Empis followed Bretonneau with their observations. The first two mention no cause for the disease. Trousseau * says that when he first saw it at Tours, he thought wet and cold might be at least exciting causes; but he was persuaded they were only accessories, and was certain it did not depend on seasons or localities or conditions. He saw it in towns which were most remarkable for their salubrity, while villages that were situated in the midst of marshes were exempt. The habits of the people, too, offered no solution for its cause, and meteorological influences gave no explanation.

Empis † says that its etiology is one of its most obscure points, and that the different authors who have written about it are unable to discover what conditions are most favorable for it, and to what influences its visitations are to be attributed.

As with typhoid fever and the other so-called filth-diseases, the sanitary reformers made no new discoveries in the etiology of this disease. They did not even make any investigations, but their loquacity was boundless in ascribing it to filth. At first they said it arose from lack of sewers. When the sewers were put in, the disease raged

---

* 1835. † 1850.

with greater violence than before. They then said the sewer-gas caused the disease. The remarkable circumstance about diphtheria is, that in extent and fatality it has more than kept pace with the "gigantic strides" of Sanitary Science. It was mentioned as a cause of death in connection with scarlatina in the registrar-general's report in 1857, but did not appear in the list of diseases until 1859. In our country, previous to 1860, the mortality from it was very light. In 1866 it caused three hundred and thirty-four deaths in New York City. As sanitary laws were passed and sanitary regulations were enforced, the mortality kept steadily rising, until for the five years ending 1888, the average number of deaths yearly from croup and diphtheria in New York City was two thousand two hundred and eighty-three. This progressive mortality from diphtheria in New York is the type of what has been observed respecting it all over the country since the rise of Sanitary Science. In spite of these facts, however, none of the genuine sanitarians have renounced their faith in its filth-origin, neither have they ceased their malicious prosecutions on the ground that it is purely a filth-disease. In 1874, Dr. Farr* says, "It is a remarkable fact that of diphtheria out of the same number born more die in the healthy districts of England than in Liverpool; the proportions are one thousand and twenty-nine in the healthy districts and four hundred and forty-two in Liverpool out of one hundred thousand born."

In 1876 a severe epidemic of diphtheria visited Salem. The Massachusetts Board of Health investigated it. Here, not only the inhabitants of uncleanly sections, but families whose surroundings were unexceptionable, were attacked. Those who inquired into this epidemic report,† "We do not

---

* Registrar-General's Report.
† Massachusetts Board of Health Report, 1877.

find the source of diphtheria in filth and insufficient sewerage." It occurred in Lynn, the same year, and not infrequently under the most favorable hygienic surroundings. Its favorite haunts in Lowell were in low and filthy localities, though many of these were exempt, and it happened also in high and healthy places. The Massachusetts Board says that the opinions of physicians agree that the causes of diphtheria are, in part, "telluric or meteorological, and at present unknown." It adds that it may be fairly questioned whether "the great attention now given to filth-diseases and drainage, important as they are, does not often mislead people into overlooking other and potent sources of ill-health."

The report says that there are few better-drained places than Denis; yet diphtheria prevailed there for ten or twelve months. The physician at Easton writes that he is satisfied it is not a filth-disease. In Fitchburg it appeared under all conditions, "where there are sewers and where there are none, where it is high and where it is low, where it is wet and where it is dry. No portion has been exempt." In Haverhill and Hanover nothing could be found to account for it, with the exception of one case. In Holyoke it was most fatal in what, from its topography, seemed to be a healthy location. It prevailed in Northfield, where "the hygienic surroundings of nearly every house were as good as could be asked for." In Waltham, "the dirtiest parts of our town have had the fewest cases." There was a scathing epidemic in Gloucester in 1876. The report said that it had "been most prevalent and fatal in the lowest, worst-drained, and most filthy parts of the city, where the most improvident and poorest classes are obliged to live;" but some "of the worst and most overcrowded localities have often escaped almost entirely, while many houses have been invaded where the sanitary conditions were the most favorable in the city." At Littleton, "all cases occurred in dry, sunny, good houses and with well-to-do families;" and at Malden it prevailed on a hill where was "a fine natural drainage." At Natick it "was as often fatal in a high, airy, sunny locality as in a low hovel." In Clinton it prevailed equally in good and bad surroundings.

The Massachusetts Board, in summing up the evidence, says that "the special connection between this disease and filth is not so clearly made out;" and that there is an unknown atmospheric condition which, if not necessary, is at least important, to produce diphtheria. In 1878 * the disease prevailed in Ayer, in the best class of houses as well as in the poorest; on high, low, sandy, pine, and oak land; and was as fatal in one locality as in another. In Haverhill it raged the whole year. "The highest and driest neighborhoods have been visited, while low and filthy localities have escaped. In Holyoke, this year, the diphtheria was most prevalent in the highest, driest, and best-drained localities." The physician who reports from Rockport says, "I do not believe that filth or unsanitary surroundings have any other influence than by lowering the vitality." In Southbridge none of the cases could be clearly traced to local causes. It has not been possible to discover any cause for the disease at Ware. The physician in North Adams "cannot find the causes that develop diphtheria." One fact is established,— "that high elevations are more subject to it and with greater fatality. In Adams the mansion is no more exempt than the hovel." In Berkeley, "the hygienic surroundings where it has occurred are as favorable to health as those where the disease has not prevailed." In Manchester it was among the better class of people, and rarely among those who were uncleanly.

The Massachusetts Report for 1879 relates that diphtheria occurred at Winchester in seventy-two different places.

---

* Massachusetts Board of Health Report.

Sixty of these were carefully examined, and no unsanitary condition was found. In two cases the ground-water was complained of; in two others slops were thrown on the surface. In Orleans it was ascribed to picking over rotten potatoes. In Haverhill, this year, it prevailed in the most airy and healthy localities, while the densely-populated portions are exempt. In 1878–79 diphtheria was unusually destructive in Boston. The commission that reported on it say,* "The greatest number of deaths was in East Boston, a section not noted for unwholesome local conditions; much of the territory is high land, well drained, and occupied by a thrifty class of people." District Nineteen is rural in character, and presents apparently salubrious features. It has a death-rate from diphtheria greater than the city at large. In District Four, which "has for many years been synonymous with bad material and moral conditions," the inhabitants had a comparative immunity from diphtheria. The next section, District Five, a location formerly a mill-pond and now poorly sewered, is settled by poor people crowded in tenements. "In this insalubrious territory, presenting in its filth and in its compact population just the conditions for the spread of a miasmatic, infectious epidemic, the death-rate was lower from diphtheria than in any other portion of the city,—an anomaly most difficult to explain." A careful examination of every dwelling was made by an inspector, "especially chosen for his fitness," who reported that forty-seven per cent. of the premises "presented nothing objectionable." In thirty-nine per cent. the drainage was defective; in three per cent. the cellars and yards were dirty; in the remaining eleven per cent. sunken lots, stagnant water, and filthy dumps were observed. The committee say, "It is then our duty, in view of the concurrent

---

* *Boston Medical and Surgical Journal*, May 22, 1879.

testimony, to reject the idea that filth fosters the origin and dissemination of diphtheria."

The Massachusetts Board * reports an epidemic at Taunton. "Every house in which the disease occurred was visited again and again, to discover if any local conditions could be the cause, but none were found." In West Falmouth was an epidemic: the "cause appeared to be a too free communication between the sick and the well." At Maynard it was undoubtedly due to filth, because "many wells must be in close proximity to vaults and cesspools," though there is no record that any examination discovered this condition.

In the *Medical Record*, vol. xii., is an account of an epidemic at Burlington, Vermont. Drs. Calderwood and Thayer, assisted by Dr. Bowditch, of Boston, spent a number of days investigating it, "but without avail, no assignable cause being found." Dr. Conley† writes that his village and surrounding country have just passed through the worst epidemic ever known there. After a careful study of the disease, he is convinced that the cause cannot be found in the effluvia of decaying animal and vegetable matter, conveyed either through air or water.

The New York State Board of Health reports diphtheria occurring, "aided by polluted wells and foulness of dwellings, and yet in numerous other cases invading the most salubrious homesteads." It is at Benson, a straggling hamlet; "the region is high and well drained." At Chautauqua it occurred on the most elevated ground. At Indian Lake and other places in the Adirondacks, Dr. Curtis said that no local cause could be found; "but it was said" that the surroundings at the school-house were unhealthy. Sandy Hill was invaded. It lies along the margin of an abrupt elevation; has wide, cleanly streets, houses well

---

* 1884. † *Medical Record*, vol. xix.

built, in neat and roomy yards. Few houses have cesspools, and the slops are thrown on the ground. Dr. Curtis ascribes some cases here to polluted water. There is no report that any of the water was tested; but he says, "It is well known that polluted water is often used with impunity for a long time, and then suddenly becomes operative in producing diphtheria. The causes for this are not altogether clear." Dr. Curtis * reports the disease at Walden. Here he finds in the yards a privy, well, and cesspool,— "established data to work upon. Given these," he says, "and we have defiled water, defiled air drawn into the houses, and diphtheria." Dr. Curtis † investigated the diphtheria at Moravia. Here was an uncleanly cheese-factory; and he remarks that the relation of bad sanitation to this disease is too well known to need comment. "Its germs perish in a clean place, and thrive in a filthy one." He found it at Nassau, "on an elevated site;" but there are privies and cesspools here. The sanitary conditions are bad, though no worse than they have always been, nor worse than are found in many "or perhaps most other villages." "Here are facts," says our philosophic sanitarian, "very instructive and interesting to the student of disease." Its continued existence in Nassau is due to "failure to properly destroy the disease-germs, and to accumulations of filth, where we know these germs grow most actively, and even perhaps may develop *de novo*."

The fifth report of the Michigan Board of Health says that diphtheria prevails at Lansing, on account of low water in the wells. Dr. Conner writes that he does not know how to trace the disease in Detroit. At Port Sanders it is not due to any want of sanitary precaution. At Lapeer, Manistee, and Kalamazoo it could not be traced to unsanitary

---

* New York State Board of Health, 1886.
† Eighth Report New York State Board of Health.

conditions. At Monroe it is attributed to "lake influence." Dr. Caulkins, of Thornville, writes that it is more difficult to find the cause of diphtheria than that of any other disease. At Grand Rapids, Dr. Griswold could gain nothing reliable to show that bad hygiene was the cause. At Hastings "the most cases have occurred in families possessing the best sanitary condition in the city." At Albion the sanitary condition of all the houses where diphtheria occurred was good. It broke out at Sturgis, but the sanitary condition was excellent. At Ann Arbor it prevailed under good hygienic conditions. In Kent County the cause could not be traced.

The seventh report of the Michigan Board of Health names ten places where diphtheria has occurred, and declares seven of these towns to be in good sanitary condition.

The ninth report of the Michigan Board of Health relates that diphtheria occurred in the deaf and dumb asylum. There were one hundred and thirteen cases. It broke out in different rooms; "nor could any existing cause be detected in or about the building."

The eleventh report says that the disease occurred in the asylum for the insane, but there were no sanitary defects. The thirteenth report says diphtheria has prevailed in Michigan for twenty-five years. Seventy reporters this year say that the mode of introduction of the disease is unknown. One hundred and forty-nine attempt to trace it to a previous case, or to assign a local cause. The board says that some of the opinions expressed are conjectural. Then follows a list of causes; among them are dampness, impure water, filth, decaying vegetables, decaying log-house, washing sheep, an open well, threshing musty wheat, overflowed land, etc.

The New Hampshire Board of Health * say that diph-

---

* Third Report.

theria "is one of the worst types of filth-disease known to civilization." Twenty-nine doctors who report on it that year mention no cause. Twenty-six say they could not trace the cause. Three are particular to say the sanitary condition was good where diphtheria occurred. A good many mention contagion as the only cause. Four ascribe it to polluted water. Twenty-one say "bad sanitary condition." Others give a variety of causes,—as cold, a swamp, a bad drain, etc. Dr. Fowler, of Bristol, thinks, "Possibly it may be of divine origin, as a reminder of our shortcomings." Dr. Blaisdel could not trace a single case to bad sanitary conditions. Dr. Hill reports forty-one cases; he had "not been able to trace a single case to bad sanitary conditions." Dr. Davis had "never been able to trace this disease to any cause." Dr. Chase said his cases could not possibly be from bad sanitary conditions. Dr. Pierce's cases were all where the best sanitary conditions prevailed.

The fourth report of the New Hampshire Board presents a similar account of diphtheria. Dr. Chase believes it exists independent of filth. Dr. Kimball says, "It is too much to ask what is the most common cause of the disease."

In the first report of the Connecticut Board of Health diphtheria is reported at Winsted; "the cases were on high ground and clean surroundings." It occurred in Norwich in a healthy locality. Dr. Brownson[*] reports a severe and extensive epidemic in New Canaan. "In the first cases no cause for the outbreak could be found; the disease broke out in one of the most salubrious localities in the town." It was exceedingly fatal "in some families where the hygienic surroundings were all that could be desired." In the Connecticut report for 1888 are the replies of physicians to a question concerning the cause of diphtheria. Seventy who allude to the disease make no reply to the question as to

---

[*] Connecticut Board of Health Report, 1886.

the cause. Twenty-five said that the sanitary condition where it occurred was neither good nor bad; three said it was defective; fourteen said it was good; thirteen said good in some cases and bad in others; five said it was fair. Dr. Campbell writes that there was no clew to its propagation. Dr. Frost said that in Waterbury it had no reference to sanitary condition. Dr. Paddock said that in Norwich the sanitary condition was, in many cases, perfect. The secretary of the Connecticut Board of Health [*] investigated an epidemic of diphtheria at the Fairfield County Orphans' Home. "A very careful inspection of the whole premises was made, both of house and surroundings; the most scrupulous cleanliness was found to exist in every part. No explanation of the cause of the disease could be discovered." The cases were isolated, the premises were carefully disinfected and fumigated, yet in a few days after the disease broke out anew in the same building.

In the Connecticut report for 1890 thirty-nine physicians report the sanitary condition to be good where diphtheria occurred. Twenty-six say the sanitary condition was bad. One says it was "perfect" in some cases and bad in others. Some said it happened under all conditions. Bad plumbing is mentioned as a cause in three cases, a bad well in one case, and a bin where were rotten potatoes in another.

The same board's report for 1891 gives the result of an inquiry among physicians for the cause of diphtheria. Sixty-eight made no reply to the question if any cause could be found. Twenty-six said plainly there was no cause. The sanitary condition was good—in some cases noted perfect—in twenty-one instances; it was bad in twenty, and sometimes good and sometimes bad in seventeen cases. Among the unsanitary cases were mentioned a filthy hen-roost, a bad well, a cesspool, a broken pipe, etc.

---

[*] 1889.

The Minnesota Board of Health * sent a circular to physicians in that State inquiring the cause of diphtheria, and requesting a report of the sanitary condition of places where it occurred. There were two hundred and twenty replies. The condition was stated in only twenty-six instances. Eleven of these were reported good, ten were bad, and five good and bad. Diphtheria had prevailed extensively in Minnesota for ten years, and the State Board of Health had been wrestling to find the source. It caused forty-two deaths in that State in 1872, and the mortality rapidly increased until, in 1881, there were thirteen hundred and ninety-seven deaths reported from diphtheria. The Minnesota Board of Health seems in despair regarding its origin, for at a meeting of the American Public Health Association, Dr. Hand,† the president of the board, offered the following resolution, which was duly passed: "That as we know little or nothing of the origin of diphtheria, we request the National Board of Health to continue their investigations as to the causes of the disease." In the Minnesota report for 1883-84, it is related that diphtheria occurred in thirty-two places; only one reports a bad sanitary condition as the cause.

In Ohio, Dr. Hutt‡ reports diphtheria at Waverly, where it prevailed among all classes. "Those who prided themselves on keeping their premises scrupulously clean were no more fortunate than those who did not. The home-comforts of the majority were all that could be desired." The Ohio Board § received two hundred and eleven replies to the query for the origin of diphtheria. One hundred and nineteen say the origin is unknown, or that there was no known exposure. Eighty-eight report the sanitary sur-

---

* Report, 1879-80.
† Minnesota Board of Health Report, 1879-80.
‡ Board of Health Report, 1886.   § 1887-88.

roundings good, sixty-two bad or not good, and thirty-nine fair.

The California Board of Health * inquired of the physicians in the State whether diphtheria could be traced to local causes, bad sewerage, etc. Seventeen replied that they could not trace it, in their practice, to local causes, filth, or sanitary surroundings. One traced it to impure water, two to imperfect sewerage, one to a slaughter-house. One physician wrote that it had no respect for condition in life. Dr. Smith said that in Cloverdale it raged most virulently where everything seemed most conducive to health. Another doctor writes that it has shown no preference for unhealthy localities. Another says, "Strict inquiry failed to develop any unsanitary condition." Another says, "It attacked the cleanly as well as the filthy; it had no respect for person or place." The California Board expresses official surprise because so many declare that no cause can be found in unsanitary surroundings, "in view of the strong ground taken on this subject by the leading sanitarians of the world." The secretary of the board reinforces "the leading sanitarians of the world" by saying that, while he has been writing the report, some diphtheria has been brought to his notice, where, within eighty yards of the house, there is a drainage canal and an enclosure for ducks and chickens which is quite filthy!

Dr. Earle, of Chicago,† investigated this disease through correspondents in the extreme Northwest. He found that it prevailed in the mountains and prairies, where were the purest air, the purest water, and the purest soil, with the same malignancy that it did in the cities. One physician reports to him that the worst diphtheria he ever saw was in a habitation upon sandy soil, in the hills, fifteen miles from any point of infection.

---

* 1878–79.   † 1888.

An official report of the examination of premises in Boston,* where one thousand and twenty-one cases of diphtheria occurred, showed that five hundred and thirty-four of the houses were in bad sanitary condition and three hundred and ninety-three in good sanitary condition. A column of ninety-four is given under the head of "not examined." If we assume that no examination of these was made because the board knew they were in good condition, and add them to those reported in good sanitary condition, we have four hundred and eighty-seven houses in Boston in good sanitary condition where diphtheria occurred, against five hundred and thirty-four in bad sanitary condition. It is not unfair to presume that many of those living in houses reported in bad sanitary condition were poor and destitute and overcrowded, and that contagion was an important factor. In 1888, of twelve hundred and thirty-two houses examined in Boston, where diphtheria occurred, five hundred and ninety-six were in bad sanitary condition and five hundred and twenty-one in good sanitary condition. No examination was made in one hundred and fifteen cases. At Boston, in 1889, of fifteen hundred and fifty-two houses which were examined, where diphtheria happened, seven hundred and seventy were found in good sanitary condition and seven hundred and eighty-two in bad sanitary condition. In 1890, in the same city, the premises where diphtheria occurred were found to be in good sanitary condition in six hundred and twenty two instances and in defective sanitary condition in six hundred and nineteen cases.

That medical men in France are no more certain of the causes of diphtheria than they were when Bretonneau and Trousseau investigated the disease, is shown by the fact that at a sitting of the Société de Médecine Publique, May 23, 1888,† M. M. Brouardel and Dr. Du Mesnil asked that

---

* Report of Boston Board of Health, 1887. † *Revue d'Hygiène.*

a commission be appointed to inquire into the causes of diphtheria. The reason they give for making this request is that "there is no epidemic disease of which we have less knowledge as regards its propagation."

The English sanitarians, who for thirty years have been trusting to the "slightest inspection" and to the "moment's thought," to designate the cause of this disease, as they did for the fever at Wolverhampton and other places, have ascribed it at one time to want of sewers, and then to the presence of sewers, to filth in general, to milk, and, when these failed, to cats and to poultry. They now find it breaking out in 1891 * at Salford, where the sanitary condition is perfect, the milk pure, the cats and poultry in sound health,—for it is distinctly stated that no sickness prevails among the domestic animals,—and these EMINENT SANITARIANS concur with the medical officer at Salford in the belief that "the necessity for an exhaustive inquiry as to the origin and maintenance of diphtheria in this country is growing daily and has already become a matter of daily concern." It bursts out in St. Bartholomew's Hospital.† The drainage is accused; but the medical and surgical staff assert that this is in good condition, and that the cause for the outbreak is unknown. Dr. Alfred Carpenter ‡ stated that fifty years ago it was unknown, but year by year it had steadily increased, and that the carrying out of sanitary measures "had failed in the case of diphtheria." Dr. R. Thorne, in the Morley Lectures on Diphtheria, notices its increase in London and the large cities, the number of cases having nearly doubled in ten years, and he remarks that this calls "for a searching inquiry into the natural history of the disease, especially as regards its causation and prevention." At the International Congress of Hygiene in

---

* London *Lancet*.   † Ibid., April, 1891.
‡ *British Medical Journal*, September, 1891.

London (1891), as the result of a discussion of the causes of diphtheria, the sanitarians in their despair offered a resolution, which was passed, that "European governments should make a comprehensive and systematic inquiry into the causes of diphtheria."

The causes of epidemic and infectious diseases have long engrossed the anxious thought of medical men in general; and in seeking for the causes of such diseases every realm in nature has been thoroughly explored by the most penetrating minds in the profession; and since they all openly confess or tacitly admit that the causes are unknown, and since they are acknowledged to be men of keen perception and sound judgment, we need not apologize for our lack of confidence in those other men who have organized themselves into a sanitary hierarchy, and who profess to have discovered the why and wherefore of infectious and epidemic diseases, and to know the causes thereof as exactly as we know the causes of railway accidents.

To be sure, the sanitarians are, as a body, far more conspicuous than the physicians, and vastly more arrogant and dictatorial. They perambulate, at public expense, from one sanitary convention to another,—each man saluting his fellow as "the Eminent Sanitarian." They concoct and ripen panics about the public health; they usurp the functions of the attending physician; they dictate offensive legislation; they call for heavy fines and even imprisonment upon those who transgress their sanitary codes; and, impatient of the slow course of the law, they recommend, as we have seen on page 27, mobbings, hangings, and burnings. But not one of them ever made a discovery of any consequence, or even any investigation that merits the epithet *scientific ;* and in attempting to solve profound and intricate problems affecting the welfare of the whole human race, they have put forth the most puerile theories and proclaimed the most ridiculous conclusions.

We respectfully submit to medical men the question, whether it is not inconsistent with the dignity of their profession, and with the duty which they owe to their fellow-men, that they shall longer tolerate by their silence such charlatanry.

CHAPTER XVI.

Epidemics.

THE history of epidemic disease is coeval with the history of mankind. The obscurity which surrounds the causes of its appearance is as dense as it was six thousand years ago. Too vain to avow their ignorance, the ancient wise men pretended to discern its origin in the movements of the heavenly bodies. Spiritual and devout men, who felt oppressed with the burden of their own sins or the sins of others, recognized in such diseases the righteous power of an offended Deity.

Diemerbrock * says the particular causes of plague are very intricate. It comes by divine appointment as a just punishment for our national sins; but, secondarily, it proceeds from some secret, malignant, and virulent seeds, consisting of very subtle and volatile particles which disperse their contagion through the atmosphere. Dr. Nathan Hodges † says the plague is "an aura that is poisonous, very subtle, deadly, and contagious, chiefly arising from a consumption of the nitrous spirit of the air." Von Helmont says that the remote, crude, and first occasional matter of the pestilence is an air putrefied through continuance; or, rather, a hoary, putrefied gas, which putrefaction of the air hath not the eight-two-hundredths part of its seminal body.

---

* Stanton's Translation. † London, 1665.

Dr. Mead said that the plague was due to the putrefaction of animal substances in the East, but that no kind of putrefaction can produce it in Europe.

If we compare these definitions of the causes of epidemics in early with those of later times, we find that they do not differ at all in substance, and not very much in language, from those given by the modern sanitarian. Although the latter affects to have emancipated himself from the superstitions of the past, and to despise the belief in divine mediation in the production of pestilence, he has no better theory to offer for its appearance. He talks learnedly of the "septic ferment," the "morbific principle," and the "mephitic gases" as the active agents in causing disease; but there is no more proof of their existence than Von Helmont had of his "hoary, pestilential gas," or Diemerbrock of his "secret, malignant, and virulent seeds." Admitting all that is claimed at the present day for the germ-theory of disease; that in reality it is proved beyond a doubt; and, as if this were not enough, admitting the more delicate hypothesis of the influence of the ptomain which, we are told, is generated by the germ, and we still have no explanation of epidemics. It is still hidden why the germ should be so rampant or the ptomain so energetic in one year as to destroy millions of people, and then suffer a condition of atony and quiescence for a century, and perhaps disappear altogether.

Most, if not all, of the ruinous epidemics which appeared before the eighteenth century were said to have been preceded, or followed by, or accompanied with, great celestial, meteorological, or telluric disturbances,—meteors, comets, eclipses, tornadoes, extreme cold, stifling heat, drought, floods, earthquakes, and volcanic eruptions. The great pestilence which swept over Europe in the fourteenth century was heralded, Boccaccio says, by an unfortunate conjunction of planets, with Saturn in the ascendant; and the astrologers therefore predicted great calamities to mankind.

Boccaccio was not certain whether the plague at Florence was caused by this unhappy meeting of the heavenly bodies, or whether it arose from our iniquities, for which a just God was about to punish us. Ample warning was given of its approach. Wise sanitarians held much counsel, and published instructions for preserving health. Many religious processions were made, and prayers were offered to placate an angry God. All filth was removed, and all suspected people were denied access to the city. Those who were sick were closely confined to their houses. No human endeavors, however, could stay its progress. This particular epidemic, of a disease which had been known for nearly thirty centuries,* took its rise in Cathay in 1346, swept over India and Turkey, penetrated Egypt, and appeared in Sicily in 1347.

In 1348 it arrived in Florence, where it spread like fire among tinder, and carried off one hundred thousand of the inhabitants. No medical aid vanquished its symptoms. To talk with one afflicted, or to touch his body or clothing, would communicate the distemper. It seized the inferior animals also. Boccaccio saw two swine rooting among the clothing of a patient which had been thrown in the street. The animals were attacked and died in a few hours.

To avoid the pestilence, some people went on shipboard and sailed out to sea; some shut themselves up in their houses and lived in the most abstemious manner; some moved about from tavern to tavern, eating and drinking freely, singing, joking, and laughing. Others took a middle course, living temperately, and following their usual occupations; but the pestilence smote alike the fugitives, the over-cautious, the reckless, and the temperate. Human affections and human laws were set aside; citizens and friends distrusted each other; husbands and wives deserted

---

* Papon, "De la Peste."

their homes and separated; sisters and brothers no more recognized the family relation; children forsook their parents, and, says Boccaccio, worse than all, parents abandoned their children.*

In 1348 † there appeared in Paris, towards the west, says an old chronicler, a great and clear star, which seemed nearer the earth than other stars; it separated into many divergent rays and disappeared. It was thought to be the presage of the terrible scourge that was approaching to desolate France the next year. During the winter of 1347–48 the pestilence progressed slowly, but in the spring it took on a frightful energy. Three-fourths of the people of Avignon succumbed to it; Narbonne was depeopled; in some places in Languedoc and Provence only a tenth of the inhabitants were spared. Petrarch's Laura was one of the victims. It advanced from town to town, sparing neither age nor sex.

In 1349–50 it overran Germany and England. It is estimated that in four years one-half of the inhabitants of Europe were swept away. In London ‡ fifty thousand were buried in one graveyard; one hundred thousand died in Venice; Lubec lost ninety thousand of its people. It extinguished the colony of Danes in Greenland. It behaved sometimes in a most capricious manner. Here and there portions of the country, like Milan and Brabant, escaped entirely.

In London, in order to control the pestilence, dwellings were searched, slaughter-houses were suppressed, stink-pots were burned in the squares, cannons were fired, church-bells were tolled, and the streets were filled with the fumes of tar, brimstone, and vinegar. It is interesting to notice that,

---

* Il Decameron, Introduzione.
† Histoire de France, H. Martin, vol. v.
‡ Brooke's History of Pestilence in London, 1772.

except tolling of church-bells, these same prophylactic measures were adopted by the sanitarians in Jacksonville during the epidemic of yellow fever in 1888.

There were sanitary experts in France in the fourteenth century, and some of them discovered the cause of the pestilence in the atrocious designs of the Jews, who, rumor said, had poisoned the wells. They were arrested in great numbers and put to the question. In their spasms of torture they confessed the crime and implicated others. Wholesale convictions were the result, and the robbing, hanging, and burning of thousands followed. But the obstinacy of the race was never more conspicuous. The Jewish women had witnessed, in mute despair or silent rage, the torture and burning of husbands and brothers; but when the Christians laid hands on their children to baptize them, the Hebrew mothers tossed them into the flames and leaped after them in triumph.

When the pestilence had passed, men and women married *à l'envi;* women conceived beyond measure; none were sterile; everywhere were *femmes enceintes*, and many gave birth to two and even three living children. The world was, in a way, renewed, and there became a new age. "But, alas!" says the historian, "this renovation brought no permanent betterment to mankind; for men only became more avaricious, and peace was established neither in State nor Church." *

Fifteen years later, in 1362, the plague repeated its visits; and now it seemed to avoid the cities, and raged in mountainous districts where the air was admitted to be pure. For three hundred years this particular plague made the tour of Europe at irregular intervals. In the seventeenth century there were more than fifty epidemics of the disease in the different states in Europe. Between 1500 and 1530

---

* Histoire de France, H. Martin, vol. v.

it ravaged Switzerland and decimated Geneva. M. Mallet* tells us that he found some curious records of the plague in that city. The sanitarians of that day had their sanitary codes; and, notwithstanding that the reformers of our times boast of the "gigantic strides" of Sanitary Science in the care of epidemics, the precautions now taken are almost precisely the same as those used three hundred years ago. They consisted then in a pest-hospital, and in the pretended disinfection of the houses containing the sick or the dead. In those days the trust in fumigations was as deeply grounded as it is at the present time, and the process was probably as efficacious then as now. The Board of Health convened every day. It forbade convalescents to go abroad, unless they were attended by a guide; at home it compelled them to close their windows, except at night. Assemblies of people were forbidden; no one was allowed to bring fresh fruits to market. The schools were closed. Heads of families could not leave the city, and none but children and pregnant women could stroll in the fields. Mallet says that it is difficult to conceive the benefit of this regulation respecting the heads of families, for it only concentrated the people and increased the misery. Privies and stables were supervised, and all nuisances were removed; pigs and geese were banished from the town. In 1568 the Board of Health at Geneva issued an order to kill all the cats and dogs in the city. Aromatic woods were burned in the streets, and preservatives of all kinds were in the drug-shops, at the pleasure of the public. In 1530 it was pretended that a plot was discovered which showed that the hospital authorities were engaged in spreading the pestilence. The supposed author of the plot, Caddoz, at first denied all knowledge of it; but, when put to the torture, he confessed his own guilt and accused others. All were hanged, drawn,

---

* Notice sur les Anciennes Pestes de Genève.

and quartered. The sanitarians had a convincing proof that this "aggressive sanitation" was efficacious; for the historian says, " Immediately after, the plague ceased."

"Aggressive sanitation" worked so well in 1530 that, when the pestilence reappeared in Geneva in 1545, it was tried again. A new plot to spread the disease was discovered. The supposed authors at first denied all complicity, but under the torture confessed; and seven men and twenty-four women were burned alive. The surgeon and the *enterreur* of the hospital had a worse fate. In 1568 they were still burning, in Geneva, the spreaders of the pestilence. Mallet, in commenting in 1835 on these transactions, says that in Paris, in 1832, during the cholera, an insensate mob tore to pieces some people who were suspected of being enemies of the public health.*

In 1576 the plague raged through Northern Italy, attacking some cities and leaving others untouched. It puzzled the doctors in those days to find out why Padua and Verona should be swept by it, and Vicenza, which lay directly on the road between them, should escape. In 1603 it carried off two thousand people daily in Paris. The disease here was ascribed to the filth. Mr. Webster, sagacious man that he was, confesses his inability to comprehend "why the filth in Paris did not produce the plague in other seasons." Whether it was this or some other epidemic which desolated the American shores before the landing of the Pilgrims, it is impossible to determine; but Captain Dennan, an English sailor who cruised along the northern coasts in 1619, found whole Indian villages destroyed, which the year before were populous. Nearly all of the people had died from some pestilence.

The plague subsided after 1665. It was not seen in England after that year, and soon ceased to afflict the Continent,

---

* See note at end of chapter.

until, in 1720, it again broke out at Marseilles. Its disappearance from England in 1665 was ascribed by the sanitarians of that day to the great fire in London, and the subsequent rebuilding of the city, when the streets were broadened and made more airy. Mr. Webster is a little incredulous on this point, for he cannot see how the great fire in London should cause the entire abatement of the plague in Germany, France, and Spain.

De Foe[*] says that his "friend, Dr. Heath," was of the opinion that the plague might be known by the smell of the breath. "It was the opinion of others that it might be distinguished by the party's breathing upon a piece of glass, where, the breath condensing, there might living creatures be seen by a microscope, of strange, monstrous, and frightful shape, such as dragons, snakes, serpents, and devils horrible to behold." De Foe himself, however, questions the truth of this, for he says that there were no microscopes at that time with which to make the experiment.

The plague visited Milan in 1630. In "I promessi Sposi," Manzoni gives, probably, as truthful an account of this epidemic as can anywhere be found. Just how it arose was the subject of much dispute; indeed, there is hardly an epidemic on record the rise of which has not been the subject of more or less disagreement, and often of wrathful altercation, between the contagionists and the non-contagionists. The usual sanitary measures which were in vogue were adopted in Milan. Houses were fumigated, the bedding and clothing of the dead and convalescent were burned, dwellings were closed, and the sick were removed to pest-hospitals. None of these means offered the least resistance to the scourge. At length the belief sprang up somehow, and rapidly gained ground, that certain diabolical arts were practised by bewitched persons to spread the disease with venomous oint-

---

[*] Plague of London.

ments. In the cathedral a board partition had been oiled. This partition and some seats were afterwards taken out and placed in the square. The cry was sounded that they had been covered with the poisonous grease. The president of the Health Council with three officials visited the spot. They saw nothing to sustain the foolish report; but instead of boldly proclaiming the nonsense, they ordered the partition and seats to be scraped and rubbed. This procedure, undertaken, no doubt, to quiet apprehension, and to prove the caution of the health authorities, made a certainty of what was before only a suspicion. The next day there went abroad a rumor that would brook no contradiction. It was to the effect that houses in all parts of the town had been smeared with some yellowish filth, which was surely the deadly ointment. It was never known whether this was a wild dream, or whether some stupid or malicious person had daubed the houses.

The sanitarians in authority visited them and pretended to examine the stains; they experimented with the stuff on dogs without any bad result to the animals, and treated the matter as of no consequence. But the story of the venomous ointment had gained so strong a hold on the public mind that the city was in an uproar. The public buildings, the chains, the door-knockers, were anointed. The story flew from mouth to mouth that the poison of the ointment was sudden, keen, penetrating; it was the venom of toads and serpents, and the *slaver* of patients who had perished from the pestilence. If any one smiled incredulously at these stories, he was blind to the danger; if any one denied them, he was an enemy of the public health and an accomplice of the *untores*. Foreigners were arrested and brought to judgment; if a suspicious-looking person halted on the street, he was one of the anointers, seeking a place to smear his ointment; if he walked along, he was disseminating it. The remains of San Carlo were now transported through the streets of Milan in grand procession, in the hope that they might exorcise the evil spirit of pestilence. For a few hours a fanatical security possessed the populace; but in a day or two the mortality doubled, and reached three thousand five hundred deaths in twenty-four hours. This sudden bound in the number of deaths was not ascribed to the crowding together of the people in procession, but to the *untores*, who had mingled with the multitude and applied the ointment to their clothing. Something must now be done. The sanitary authorities were equal to the emergency. However much they might doubt these crazy tales, they now issued a decree promising reward and immunity to those who would expose and denounce the anointers. In the early twilight of the morning a maid-servant saw, from a second-story window, a man passing along the street. He seemed to be touching the walls of the yards and houses. She gave the alarm that he was one of the *untores*, and the whole neighborhood was aroused. The people thought that they discerned the ointment on the walls, so they lighted wisps of straw to burn it off, and scrubbed and scraped and whitewashed the walls anew. The police were notified; but when the reputed anointer was arrested, he proved to be a sanitary inspector. When arraigned, he denied all knowledge of the unction. He was put to the question, that he might confess, and divulge the names of his accomplices. Even in the most acute paroxysms of pain, the unhappy wretch sustained his denials. Not until the *Tribunale della Sanità* had promised complete immunity did he yield. In a spasm of exquisite torture he spoke the name of a poor barber, who, he said, was to receive much money from certain persons if he would manufacture and spread this ointment. The barber and his family were thrown into prison, and their premises were searched by the sanitary authorities. Twelve cents in money were found. Barrels, bottles, and pots were peered into. At last the searchers found that one of the vessels contained some dirty suds, in which was a slimy sediment. The proof was complete! The barber was indicted, but denied the charge against him. Bruised in body by the torture, which was applied again and again, and broken in spirit by his confinement, he accepted the promised immunity, confessed the crime, and named his confederates. These in turn were arrested, imprisoned, and tortured. Notwithstanding the exemption which had been promised, the sanitary inspector and the barber were savagely executed with horrible torments, protesting in their dying moments their own innocence, and retracting the charges they had made against others. The barber's house was demolished by the authorities, and in its place was erected a column, *La Colonna Infame*, which stood for one hundred and fifty years a monument to the ignorance, folly, and wickedness of mankind.*

The plague appeared as an epidemic for the last time in Western Europe in 1720, at Marseilles. The same precautions were taken to prevent its approach, and the same sanitary measures were taken to subdue it after it arrived; and the same fruitless results followed their execution as had succeeded them in previous epidemics. It raged in that city for five or six months. The bishop ordered a religious procession, and appeared himself as the scapegoat, loaded with the sins of the people. Barefooted, with a halter about his neck and a cross in his arms, he celebrated a mass before the people. As this abasement of clerical dignity had no power to abate the sickness, he made an exorcism against the plague a few days later. Three days after, it had so increased in virulence that neither Church nor State sufficed to control it.

The habits and customs of the people of Europe in the fourteenth century did not vary, to any great extent, from

---

* Storia della Colonna Infame, Manzoni.

those of the eighteenth. Yet the modern sanitary orators of both sexes, lay and clerical, tell us that this wonderful dispensation, which held the world in terror for nearly four hundred years, was caused by accumulations of filth, which reached in Asia and Europe precisely the point of activity to develop the pestilence in its greatest virulence in 1346-50. Moreover, these collections of filth oscillated at irregular intervals, between periods of repose and frightful activity, for four centuries, and then suddenly ceased to have the least power to produce the plague. With flaming eye, dishevelled hair, and frantic gestures, the same orators tell us, in almost so many words, that this pestilence, this " Black Death of the Middle Ages," will surely reappear, unless we endow them with power and furnish them with money to avert the calamity.

The plague broke out again in the East in 1856, and until 1878 there was much apprehension that it would again devastate Europe. Three or four government commissions were sent to investigate it. During the twenty years which followed its outbreak in 1856 a number of limited epidemics occurred. Dr. Tholozan, physician to the Persian Shah, in his account of those which arose in 1856, '58, '59, '67, '74, '77, says that he has compared the epidemics of plague in those countries where sanitary systems were in operation with those where the barbarism, ignorance, and fanaticism of the populace are opposed to all hygienic measures. "I have shown thus, not without surprise, that the existence or not of a sanitary administration has made no difference in the intensity, duration, and occurrence of epidemics. We do not know, in reality, why these plagues disappear, any more than we know why they appear at certain epochs." He hopes to render a service to Sanitary Science by giving occasion for a revision of its methods, removing defects, and seeking for more humane, more practical, and more serious means of action than have hitherto been taken.

Dr. Hirsch, who was one of the German Commission to investigate the outbreak of plague in Astrakhan in 1878, says that the "suggestion is scarcely worth discussion," that it was due to unsanitary condition. While in this respect there was much to be desired, the sanitary condition was in many places much better where it appeared than it was in those places which escaped. It was no worse anywhere than it had been in previous years, yet there had not been a trace of plague in that country since 1807. "Indeed," he says, "there are many places in Germany and in other countries in Europe which are more unwholesome than those places which were attacked in Astrakhan." He adds that his opinion—namely, that it did not arise from local conditions—is held in common with that of other European commissioners who investigated this epidemic. Dr. Zuber,* the French commissioner who investigated the same plague, says that the disease first broke out at Vetlianka, not in the most uncleanly part of the town, which was in a detestable condition, but at the opposite end, where the best hygienic conditions prevailed. A dozen of the best houses in the town, in which the plague first started, were burned. He rejects the idea that it was caused by personal uncleanliness, and although the reigning belief has become a dogma, that filth generates and fosters it, and it is a heresy to deny it, yet if he is asked, "Has the epidemic struck uncleanly countries, villages, quarters, houses, or people, in preference to cleanly ones?" he is obliged to answer, "No." "The progress of the epidemic was independent of the ordinary conditions of age, race, profession, or even meteorological and hygienic conditions." Dr. Tholozan † says of the disease in Khorassan that it broke out in salubrious localities which offered none of the

---

* Hygiène Publique.
† Comptes-Rendus de l'Académie des Sciences, 1877-78.

conditions supposed to be necessary for the development of pestilence; the half-nomadic inhabitants living on an open plain, or on the sides of mountains, engaged in agriculture and grazing, were struck with the disease in its gravest forms. Dr. Tholozan studied the plague which prevailed for the twelve years ending 1878. He says it can arise and rage just as well in dry, elevated, and cold countries as in low, wet, and marshy lands; on the summits of mountains as well as at the delta of the Nile; that, like cholera, diphtheria, and typhoid fever, it takes on an epidemic form at certain times and places which actual science does not know the cause; neither can it arrest its progress nor hasten its decline. Dr. Tholozan * declares his disbelief in the efficacy of so-called sanitary measures; that experience has proved that no prophylactic means contribute to its disappearance or decline; that, like all epidemics, it has its periods of spontaneous rise, progress, and abatement, uninfluenced by sanitary cordons, quarantine, isolation, disinfection, or ventilation; that it frequently spreads when all these measures are carried out with military severity, and that it often disappears when none of them are adopted. He concludes that no prophylactic measures, however thorough they may have been, have ever contributed to the disappearance or diminution of plague, and that it, like all other epidemics, has its times of invasion and development, of spontaneous increase and decline, sometimes slow, sometimes rapid, with isolated or multiplied explosions, sometimes grave, sometimes slight; but alike uncontrollable, because its cause is as unknown as that of the cholera, diphtheria, or smallpox.

Throughout the eighteenth century the people in Europe and America suffered from circumscribed epidemics, which were quite as mysterious in their origin, arrival, and depart-

---

* Les Trois Épidémies de la Peste.

ure as was the great pestilence of the Middle Ages. Dysentery, fevers, diphtheria, small-pox, and measles made great havoc in town, city, and country. The new world suffered equally with the old. The early settlements in America were often threatened with destruction by pestilence. Philadelphia was but seventeen years old—a sparsely-settled village—when it was stricken with a scorching epidemic. The earlier historians of New England record, almost yearly, epidemical invasions of contagious disease.

Although the human race suffered local epidemics for the hundred years which followed the entire disappearance of the plague in Europe, it enjoyed a reprieve from those which desolated vast continents, until about the year 1820, when the cholera began to advance westward. The sanitarians tell us that it was the filth, and that alone, which invited it; we had not done our slaughtering and bone-boiling with sufficient care; we had neglected to sweep out the stables; the privies were extraordinarily full, and a subterraneous communication had been established between them and the wells, and the cesspools had been uncleansed. The cholera, therefore, selected 1832, 1849, 1865, and 1873, those being the precise years when the filth was in its highest activity—its perihelion—to make its destructive circuits of the globe.

Filth has not failed to account, not only for the worldwide epidemics, but for those which were limited to States, cities, towns, and small districts, and even for every sporadic case of so-called zymotic disease. The sanitarians never considered it incumbent on them to explain why the filth undulated between violence and repose, why it sometimes lay dormant for two centuries, as at Newport and Stamford, or why it was quickened into furious activity in a few weeks, months, or years, as in California and Colorado.

But for those wonderful displays of an unknown power, which in former centuries, and even as late as the middle of our own, threatened to sweep the human race from the face of the earth; those terrible scourges which, taking their origin in the direction of the rising sun, spread westward over the globe, now sparing in their caprice whole provinces and cities only to involve these in destruction on the next visit; for these awful exhibitions of an invisible energy, which made the apostate pause in his unbelief and the devout man to question whether the Great Disposer of events was animated by the spirit of Good or the spirit of Evil; these mysteries the sanitarians offered to solve with reasons which were silly enough to exasperate the mind of an idiot: they said it was purely a question of scavengering; we had not been quite as squeamish and delicate in performing the acts of urination and defecation as we ought to have been, and it served us right to be visited, therefore, by these pestilences.

The modern sanitarian seems incapable of viewing anything but filth as the cause of infectious disease, epidemic or sporadic. Heredity, race, temperament, sex, age, climate, food, raiment, shelter, work, rest, play, wealth, comfort, poverty, civil condition,—married or single,—trade, profession, habit—all of those circumstances and conditions, alterable and unalterable, which the medical practitioner considers of such hygienic importance to the well, and which he tries to estimate at their full value when called to prescribe medicine or regimen for the sick—pass for naught with our modern sanitarian. It is enough for him to know whether the drains have the proper declivity, and whether they are trapped, to prevent the backward pressure of the "hoary pestilential gas" or the "septic ferment."

When the pestilence arrives, he does in Jacksonville in 1888 as his prototype did in Florence in 1348, and in London in 1665. He musters his forces, invades the houses of the citizens, gets out his stink-pots, fires his cannon, fills the air with the fumes of pitch, tar, sulphur, and vinegar; and after the epidemic is spent, he tells us that he has stamped out the disease, and he congratulates the people that Sanitary Science has made " gigantic strides."

For nearly a thousand years we have records, more or less obscure, of epidemics of influenza, for the most part affecting mankind, but occasionally seizing the inferior animals. Although the organs affected both in man and animals are the same, and the symptoms almost identical, there seems no reason to believe that the disease is mutually transmissible. Though the cause of these epidemics in man and beast has been the subject of much speculation, it has never been explained any more satisfactorily than the cause of other infectious diseases. Between the years 1580 and 1610 six extensive epidemics of influenza are recorded. An interval of repose followed, lasting for nearly forty years, until 1647, when the disease again appeared. For the next century and a half the longest intermission between any two epidemics was five years.

After an immunity of nearly forty-five years, the disease suddenly arose in the Russian dominions in September or October of 1889, and moved rapidly westward. M. Leon Collin* declared that it travelled on lines independent of human transportation, and he compared the velocity of its movement to that of light or electricity. It broke out simultaneously in two distinct quarters of St. Petersburg,— one rich, the other poor and crowded. During its presence in that city it is estimated that from one-third to one-half of the people were affected by it. By the middle of December every European country had been invaded by it. People were attacked by it in Boston, December 10,—as early as they were in London. It seemed to radiate from Boston to all parts of Massachusetts. It spread rapidly all over the United

---

* Bulletin de l'Académie de Médecine, Paris, 1889.

States. Dr. Lee* says that cases of the disease were seen in Pennsylvania as early as November. By January, 1890, it had reached New Mexico and Central America, and later in that month it appeared in the Argentine Republic and Chili. In the latter country seventy per cent. of the population were affected by it. Neither season, nor climate, nor weather had any influence on it; prevailing winds offered no impediment to its progress. The next year it reappeared, sometimes raging with greater severity in the same localities which had suffered the previous season, sometimes striking with a double force those which had hitherto escaped.

In this epidemic the professional sanitarians, from whom we were led to expect so much, should pestilence invade us, and who had so long boasted of the resources of Sanitary Science and the "gigantic strides" it had made, were as helpless as they had always shown themselves in former epidemics. They were even less noisy than usual, and less sanguinary; for they began no prosecutions against the enemies of the public health. Except in one instance,—in a city of nearly a million of inhabitants,—where a Board of Health made a silly order which had no other effect than to derange and impede transportation, they offered no suggestion that could have any bearing on the disease. By common consent, but without making any investigations, they proclaimed that it was not a filth-disease; yet there was as much or more evidence to show that it depended on filth and defective drainage as there was that these caused typhoid fever. It prevailed extensively in filthy cities, as at New Orleans, and in filthy parts of clean cities. Chicago, with her slaughter-houses, was severely visited by the disease. It would have been easy enough for the reformers to have shown that at one time or another it prevailed near some pond, or stable, or slaughter-house, or in the vicinity of

---

* Journal American Medical Association.

some well or stream that contained nitrites. It would have been easy to show that the customers of some particular milk-dealer were affected by the influenza to a greater extent than others; but, on this occasion, the sanitarians seemed to have lost their inquisitorial character, and, except for some incoherent sanitary drivel as to whether it was a micrococcus, a diplococcus, a pneumococcus, or a ptomain that caused the disease, they were silent. They agreed that it was an infectious disease,—a germ-disease,—but they made no efforts to suffocate the germ with sulphuric acid, or to smoke it out with other fumigations; they established no sanitary cordons, they invaded no private houses to isolate or quarantine the inhabitants,—they did not even recommend disinfection; but allowed the micrococcus, diplococcus, pneumococcus, or ptomain to travel how and where it pleased, destroying in our own country tens of thousands of lives. The people may be congratulated that the reformers in this instance were willing to hide their ignorance by slinking into privacy. In the study of epidemics, ancient or modern, one is forced to conclude that the meddlesome, teasing, persecuting regulations of the sanitarians of every age have been utterly worthless to avert or control any epidemic of any infectious disease, for the reason that they have failed entirely to determine the cause of such disease; and that epidemics are no more under human control than are biting frosts, scorching droughts, famines, cyclones, volcanoes, or earthquakes.

NOTE.—Fifty years later, in Vermont, it needed only a little more excitement than was actually raised, to have caused similar scenes. An epidemic of diphtheria * broke out in a small town in that State. The children of a school had drunk water from a brook. The rumor was noised abroad that it had been poisoned by a farmer who had thrown dead animals into it. After a pretended investigation, the secretary of the National Board of Health writes, "The farmer who has caused this wholesale slaughter of children is in danger of

---

* Report of the National Board of Health.

personal violence, or even of a trial of manslaughter." The secretary continues, " As one child after another dies in agony, becomes a repulsive, bloated corpse, and is hastily buried before it can fall to pieces, the wails of some of the mothers are heart-rending; and the feelings among the men, as the news flies around the neighborhood, is very dangerous for the cause of the disaster. . . . The indignation is spreading" into neighboring towns. The secretary writes that a farmer declares " there is lightning in the air."

So far there is nothing but noise, fright, panic. If we had our reformer from Michigan here, whom we have quoted on page 27, we should see a vindication of Sanitary Science by mobbing, hanging, and burning. After a while the fury which the sanitarians raised and kept up quieted down, and a calm inquiry followed. It was found out that diphtheria was prevailing in the town four weeks before the school assembled; that the disease was in adjoining towns in Vermont and in New Hampshire. No trace could be found of the dead animals with which the farmer poisoned the brook, and which formed the cause of the heart-rending wails of the mothers, for which he was to suffer personal violence or trial for manslaughter. All that could be discovered was that, a year before, a dead horse was buried seventy-five feet from the edge of the brook, and two rods above where the children drank. The report does not say whether the horse was buried ten feet above or ten feet below the water-level of the brook. The State chemist examined the water, and found that there was nothing in it detrimental to health; and, finally, somebody showed that the cause of diphtheria in that town was the want of ozone in the atmosphere!

CHAPTER XVII.

Boards of Health.

NOTWITHSTANDING the divinity that hedges the sanitarians; notwithstanding their exalted character when organized as Boards of Health; notwithstanding that behind the resonant voice of these philosophers presumably reposes a gravid and comprehensive intellect, unceasingly occupied in solving problems which have to do with saving our race from desolating epidemics, we hold it to be neither impious nor irreverent to lay down the proposition,—which seems undeniable,—that in so far as they have directed their efforts and consumed their energies on subjects which have no influence on individual or public health, and in so far as they have diverted the attention of the people thereto, just so far have they retarded and obstructed true progress in that branch of medical science which is devoted to hygiene, and just so far they have been a positive detriment to the public health.

In the present chapter we purpose to review briefly some of their transactions, limiting our comments to the more important boards which have been established in this country. The Metropolitan Board of Health was set up in New York City in 1866. It had jurisdiction, not only over the city, but over Brooklyn and some other towns. In its first report it proclaims its belief that filth and its concomitants cause the great destruction of life in New York City. The board opened its campaign by issuing that year thirty-one thousand orders against filthy alleys, privies, cesspools, urinals, stables, slaughter-houses, and bone-boiling establishments. It looked sharply after the markets, and explored the cellars in which people lived. The board says * that "man's inhumanity to man" is nowhere more observable in the sacrifice of human life than in the disposition of proprietors to rent basements and cellars for people to live in. Although some of the owners "are persons of the highest character," they make no effort to bring about any compromise with the applicants for a shelter by offering them a home in their own residences, but actually rent them the cellars! As an excuse for the bad condition of these, "the proprietors urged the filthy habits of the tenants, losing sight of the fact" that for the tolerance of these habits the owners "themselves were alone responsible." The board says that the only way to reform this condition is to hold the owner liable. He should make "weekly, or, if necessary, more frequent inspections." This would soon improve

---

* Appendix, p. 9.

the habits of the tenants; but the board does not say who would protect the owner from being ejected for his impertinence.

The cholera appeared that year. The board claims to have made a discovery, which, if true, is of vast importance to mankind. It says the fact that cholera can be "stamped out" by the liberal use of disinfectants "is positively confirmed and established. . . . The sulphate of iron destroys the germ of cholera." The expenses of the board that year amounted to $178,633.91.

The second report of the board claims a gain in life and health. By enforcing sanitary regulations, "the fatality and amount of diseases have decreased." The board issued this year fifteen thousand one hundred and eighty-six orders against filthy yards, alleys, privies, cesspools, bad wells, etc. It granted permits to keep cows, goats, and geese; to sell fruit, fish, meat, and vegetables. During the year a little event happened that would have disquieted a less firmly-rooted faith than that of the filth-origin of disease. Typhus fever broke out at the Reformatory for Young Girls. "There was no overcrowding, the water-closets and drains were in excellent order, and cleanliness universally prevailed." The board says that the fever was brought to the institution by some visitors; but no typhus is reported elsewhere in the city, and why visitors should bring the disease into the clean reformatory instead of spreading it among the filthy wards of the town is a little confusing to a man unversed in Sanitary Science.

The people in the tenement houses are still troublesome; but they are thankful to the health-inspectors, and the owners are more tractable under the discipline of the board. Some of them—the board does not say whether they are "persons of the highest character"—are hiding behind "middlemen," to whom they lease their houses, either because the owners do not relish the persecutions of the board, or do not want to undertake the education of their tenants in good manners. These middlemen are even worse than the owners, and we have a new illustration of "man's inhumanity to man," for in subletting they try to make all the money they can out of the tenants. The total disbursements of the board this year were $243,395.13.

The next report does not felicitate the people on a gain in life and health, but says, "The fact that the mortality among children is so great, leads the board to inquire if there is not something radically wrong in their management." It thinks that instead of removing the filth from the tenements it would be better to move those who live in them into the country. But the difficulty in the way is that they refuse to go; they prefer to live where they are. They are hard to deal with; they defeat the benevolent schemes of the board for their good. The very windows which it ordered put in, to provide fresh air, they have sealed up with paper, pasted across them. They clog the sewers with ashes and garbage. Filth abounds everywhere, and the only way out of the trouble is to hold the landlord responsible, "making him pay the penalty of his tenants' misdeeds." The board has crushed a good deal of opposition to sanitary reform; but it encounters an unexpected obstacle in the fact that some people deny that nuisances and offensive trades cause disease; and this doctrine "meets with countenance from persons of intelligence and, in many respects, of considerable knowledge in hygenic matters."

This year the Texas cattle-disease gave an opportunity to excite a panic regarding diseased meat. As has been seen on page 203, it only needed a hint from the board—for no investigation showed that the beef was the cause of the increase in bowel disorders—to deepen and widen the panic, which became quite general. The board is still pursuing the owners of stables and slaughter-houses, and the melters of fat. It says that the filth only requires "heat and moisture to convert it into poisonous gases, which are now known to be the efficient agents in producing the severest forms of diarrhœal diseases in cities." The wharves are in a condition perilous to health. They cover immense beds of putrescent filth, which is stirred up by every tide. Powerful disinfectants are poured into the gutters, and later in the summer—for the temperature has now fallen—the deaths "from that class of diseases produced by putrescent organic matter" begin to decline.

The death-rate this year was 25.45 per 1000, and the percentage of zymotic to total mortality was 29.81. In New York and Brooklyn there were two hundred and sixty-eight deaths from diphtheria this year. The registrar takes comfort that the death-rate in later periods of life seems to be declining, and he ascribes this to the spread of sanitary knowledge. The expenses this year were $164,401.89.

The next report says that the mortality of the previous year was less than that of any for the last four years, except 1867, and that the benefits of sanitary improvements "are amply illustrated." Besides vigilantly looking after the privies and garbage, the board has extended its warfare: it has reached out for the breweries and distilleries and gas-works, and has begun to pay attention to kerosene oil and smoky chimneys. Milk and water are analyzed; and cosmetics are chemically examined, among others, "Hagan's Magnolia Balm," "Eugénie's Favorite," and the "Superior Lily White." With few exceptions, all of the articles for the improvement of personal beauty and the concealment of personal defects "are highly dangerous to the health." The only things which the board has inquired into and found "dangerous to the health," are the cosmetics; but, strange to relate, it takes no measures to suppress their manufacture or sale. It continues, however, to fight the slaughter-houses and other establishments, which have never been proved to have caused sickness; indeed, they have not been investigated.

The death-rate this year was 25.13 per 1000, against 25.45 the year before, estimating the population at one million. Sanitary improvements are " amply illustrated" by the rise in the percentage of zymotic in proportion to total mortality, it being this year 30.50 against 29.96 per cent. of the previous year. The board says, "What the death-rate might have been without the intervention of the Board of Health it is impossible to compute."

In 1870 a new board was formed for New York City alone. In its first report we find that the death-rate of the city has risen to 28.79 per 1000 of the population, and the percentage of zymotic to total mortality to 30.60. Nothing is said now about sanitary improvements being "amply illustrated," but the increase in mortality is laid to the heat. The mean temperature of the five summer quarters ending 1870, was 74°, 70°, 75°, 72°, 74°, omitting fractions. A rise of two or three degrees in the thermometer amply illustrates the futility of sanitary improvements.

The board is in despair about the tenement-house people, who continue to give a great deal of trouble. They are poor, and filthy, and ignorant; they pay no attention to the instructions of the board; they allow their children to run at large, and to be out late at nights; and when they are sick they are not properly cared for. The pesky landlords show no disposition to go among the tenants to educate them, but seem rather to avoid them, by still renting the property to the "middlemen," who are reckless and grasping. This year two thousand five hundred and seventy-six prosecutions have been commenced against the enemies of the public health.

The board granted permission to keep cows, geese, goats, and swine; to sell poultry, vegetables, pork, fish, and beef; to blast rocks, to cart bones, to salt hides, etc., although what special diseases were caused by these occupations, or prevented by the permits, is not mentioned.

The percentage of zymotic to total mortality for the last four years has been 28.41, 29.96, 30.50, 30.60; and, in spite of all the permits, regulations, and prosecutions, the mortality from typhoid fever is "somewhat excessive," being nearly double what it was the year before. The expenses of the board this year were $169,478.27.

During the summer yellow fever broke out at Governor's Island; its cause is related on a previous page (page 286). There were eleven cases in New York City, nine of which proved fatal. Although some of these cases occurred before the board had any knowledge of the fever, and, as Dr. Nott, who reported on the disease to the board, says, though all "were quartered in crowded and filthy parts of the city," they did not communicate it to others. The board moved with energy on Governor's Island—seeming ambitious to pass into history, along with General Butler, for having "stamped out" yellow fever. It made a great alarm throughout the country about the disease, advising the evacuation of the island, the burning of certain buildings thereon, and other radical measures. Here we record the only instance in the history of sanitary reform in which common sense sustained itself against sanitary terror. If the police emissaries of the sanitary Robespierres could only have reached General McDowell, who was in command, the matter would probably have had a different ending, for surely *sans culotte* never felt keener triumph in gazing on the headless trunk of an aristocrat than did the sanitary terrorists in proclaiming as a foe to the public health a wealthy and influential citizen. But the general, calmly intrenched on Governor's Island, faced the terrorists with composure. He wrote to them that he had quarantined the island, at great inconvenience to the service, and that the only infraction of the quarantine had been made by the board themselves, who had persisted in visiting the island and then returning to New York. For good reasons, he refused to evacuate the place. As to burning the buildings, which the board recommended, while he did not deny that a conflagration of that kind was a good disinfectant and well adapted to a barbarous people, he had heard that there were other methods. He reminded them that the people in New Orleans and Mobile, after they had suffered from yellow fever, did not burn down the towns. In conclusion, he positively declined to comply with the recommendations of the board. It was now October. On the 10th of this month a frost came. The fever was ended, and the opportunity to "stamp out" the disease was gone.

The next year, although a great improvement in general cleanliness is announced, the report gives the death-rate as 28.6 per thousand, and the percentage of zymotic to total mortality has mounted to 31.01. The board is still pursuing the slaughter-houses, fat-rendering establishments, and stables, though no epidemic, and not even a single case of disease, is yet assigned to thĕm. Keepers of stables are arrested, but the grand jury refuses to indict them, although the board regards their business as "so notoriously injurious to health." This year three thousand and sixty prosecutions were begun against the foes of the public health. The expenses were $194,976.54.

The tenement-house class is still averse to leaving the filthy, crowded parts of the city. It is pleasanter for them to live there, the board says, than in airy and remote spots, away from the society of their friends and companions. Besides, the land-owners in the upper part of the city are not much better than the owners of the houses down-town; for the board says they are holding the land at a high price, and if the tenants were moved there, they could not afford to remain; and, altogether, there is a prospect that the selfishness of mankind will be a permanent check to sanitary reform. There was some fright this year about the cholera, which broke out in the little country-town of Woodridge, New Jersey. Consequently, extra precautions were taken, and cleanliness was enforced everywhere.

The fright was hardly past, when cerebro-spinal meningitis appeared in New York. The physician who reports on this outbreak says it is caused by "the excessive quantities of filth which had accumulated throughout the city." This is a little startling, because in a previous part of the same report it is clearly intimated that the removal of this filth was what saved the city from cholera that year. We are now told that the spaces under stores and dwellings are "surcharged with an enormous quantity of sewer-gas," and the condition of local drainage is such as "would necessarily produce the disease." But the same report says that the meningitis first broke out in the best and cleanest parts of the town, and "subsequently" developed in the lower wards, which were overcrowded, badly ventilated, and exceedingly filthy. The doctor says that the professional mind has become convinced that in sewage "lurks a gaseous poison, whose direct effect on the human system is typhoidal."

The third report also tells of the unrelenting war on the slaughter-houses, stables, and tenements. "In the interest of the public health" the board orders the removal of cushions from the horse-cars, because vermin find there a hiding-place. The board has also found out that there are a great many irregular practitioners in town, and, judging from the amount of business they do, it thinks the death-rate is increased by their incompetence. This year the general death-rate is 32.6 per thousand, and the percentage of zymotic to total mortality has risen to 36.19!

The board thinks it is time that saleratus, and baking-powders, and cream-of-tartar were examined. It keeps its official eye on the water, extends its vision to the confectionery and the tobacco and cigars. But now a new light dawns upon its view. It is suspected by somebody that some cases of syphilis have been caused by the rite of circumcision. There is no financial report this year.

The reports for 1874–75 tell us that the board is still fighting the slaughter-houses and stables. For the twenty months ending December, 1875, it had begun five thousand two hundred and eighteen prosecutions of the conspirators against the public health. A large area—nineteen square miles, to be exact—has now been annexed to the city. The general death-rate in the whole area of the city for 1874–75 was 27.57 and 29.47 per thousand. The percentage of zymotic to total mortality for the two years was 34.12 and 35.52. In 1871-72-73-74-75 there were 308, 224, 903, 1665, 2329 deaths from diphtheria in the city. The mortality from this disease was increasing enormously, while the prosecutions against the adversaries of the public health were going on, and the war against filth was being vigorously waged. Moreover, this disease, which the board said was peculiarly a filth-disease, created and sustained by it, was most steadily and fatally prevalent in the Nineteenth, Twentieth, and Twenty-second Wards. The registrar says that its persistence in that open section, on the highest ground; its prevalence in the neat as well as in the worst parts of the Seventh, Ninth, Eleventh, and Eighteenth Wards, and its rare presence in the most crowded and filthy quarters, indicates that "diphtheria does not as uniformly elect the most unhealthy wards."

The board says that there is great ignorance on the "vital question" of plumbing. "A siphoned trap, or the absence of a trap," has "undermined the health of adults and slain the little ones." Consequently, the most useful work of the bureau is now to look after the plumbing, and a plumbing-ordinance is called for. Defective joints have been found. A light was applied to one of these, and the sewer-gas was "of such a foul and poisonous nature as to extinguish the light and ignite the escaping gas, which burned with a blue flame." The sewer-gas is forced past the traps. Now the mystery increases, for in some cases there are defects found "caused by too much trapping." Some new creation is necessary to account for the rise in general mortality and in that from zymotic disease, and the sewer-gas comes in at just the right moment. The diphtheria having been caused, as the board said, by the bad plumbing, plumbing-laws were passed; but the more stringent they were made, the higher went the mortality from this disease. For the ten years ending 1885 the proportion of zymotic to total mortality reached nearly thirty-two per cent. In 1881 it touched the enormous figure of 34.93 per cent.,—a point much higher than it averaged before the board made any efforts to eradicate the causes of zymotic disease.

Let us now survey the result of the board's work for the first ten years of its existence. From the beginning its aim and purpose had been to reduce the amount of zymotic disease, which was regarded as preventable by all eminent sanitarians. The board having announced that it knew the causes of such disease, and only needed power and money to exterminate it, received unlimited power, and expended so much money that it apparently did not dare mention the amount for 1873–74–75. It made tens of thousands of prosecutions, and furnished plenty of offices to its friends, and the friends of its friends. Nevertheless, we find that the general death-rate was steadily increasing, and the percentage of zymotic to total mortality for the ten years ending 1875 was 32, 28, 29, 30, 30, 31, 36, 32, 34, 35, omitting fractions. In the mean time, the area of the city had been nearly doubled by the addition of the rural district of Westchester County.

The board ceased its reports after 1875. It no longer sought to excuse itself by words and accuse itself with figures, and for many years we had to be content with the vital statistics of the city. We may be pretty sure that there was no abatement of prosecutions and no decline of power, for in the New York *Sun* of June 10, 1891, we read that on the receipt of an anonymous communication, which stated that some children had been made sick by candy bought of Mrs. Katherine Vath, of 168 Avenue B, a sanitary inspector went to the store, "*destroyed a quantity of the colored candy, and took some to the Health Board for analysis*" (italics ours). "No children could be found who had become sick from eating the candy. It is thought that the complaint against Mrs. Vath's place was made by a rival."

In September of the same year the board destroyed tons of grapes, on suspicion that they had been sprayed with "Bordeaux mixture." No case of sickness had been charged to the consumption of grapes which had been treated with this mixture, although the crops of three previous years had been subjected to it. After the condemnation of the fruit the board proceeded, as it did after the destruction of Mrs. Vath's goods, to investigate, for the purpose of ascertaining if the grapes contained anything poisonous, or if the Bordeaux mixture itself was harmful. Chemical analysis sustained the experience which the people already had who had eaten the crops of the previous years. The grapes contained no poisonous substance. A great panic, however, had been raised, which came near causing a total loss of the entire harvest to the growers, and depriving the people of a delicious and healthful fruit. Sanitary Science, however, had made a "gigantic stride." How far this panic might have been carried we have no means of knowing. The Board of Health soon found out that this time it was not dealing with Mrs. Vath, but with a set of men, growers and dealers, who were determined to face its nonsense and violence even in the presence of a panic which it had raised about the public health.

After 1875 there was a steady increase in the number of offices, for in the report for 1890 are listed the names of one hundred and sixty-six officials, besides a sanitary police, consisting of forty-two patrolmen, two roundsmen, and one sergeant, and a little later an appropriation of nearly half a million dollars was granted for the expenses of the board.

During the last six years there has been a small apparent decrease in the amount of zymotic mortality in New York City. Croup, which before figured in the zymotic class, and which gave an annual average of seven hundred and sixty-three deaths for the last ten years, is now transferred to the column of respiratory diseases. Puerperal fever, which before was in the zymotic class, and which—with puerperal diseases, metritis, and peritonitis—caused for the last ten years an average of four hundred and two deaths, appears in 1889 under another head. Besides, the number of persons who are now removed to the hospitals, when they are seized with illness, is very much larger than formerly. There can be no doubt that many of these survive who would die if left in their own homes; and where the disease is of an infectious character the danger of contagion is largely diminished. The average annual number of deaths in the hospitals for the ten years ending 1880 was 4920; the average annual number for the ten years ending 1890 was 7369, the number in 1890 being 8316. By general vaccination, and by a cessation of the unknown epidemic influence of small-pox, the deaths from that disease, which averaged annually four hundred for the ten years ending 1880, were reduced to an average of ninety-six a year for the decade ending 1890.

The board will hardly claim that the apparent reduction of zymotic disorders by changing the nomenclature of diseases, and the real reduction by the voluntary entrance of sick people to the hospitals, by vaccination and the subsidence of epidemic influence, were due to the passage of plumbing-laws, to the prosecution of the foes of the public health, to the deluging of the gutters with disinfectants, or to the swearing-in of a squad of sanitary police.

Another way to estimate the work of the New York Board is by considering the rate of infant mortality. This, according to the sanitarians, is a finer test of the efficacy of so-called sanitary measures than the proportion of zymotic to total mortality. The total number of deaths in New York, for the ten years ending 1880, was 290,785; the total number of deaths under five years was 138,499, the percentage of infant mortality being 47.6. For the next ten years, ending 1890, the total number of deaths was 377,516; the number under five years was 163,364; the percentage of infant mortality for the latter period was 43.27. We should bear in mind that now a great many children who are taken with infectious diseases are received into hospitals designed especially for such cases. If, in addition, we consider the efforts of private benevolence,—the St. John's Guild, the King's Daughters, the Hebrew Sanitarium, the *Tribune's* and the *World's* Fresh-Air Funds, all of which must have a marked effect on infant mortality,—there is a beggarly residue for the New York Board of Health to claim, as a result of all the power which has been conferred upon it, and the money which has been lavished for its pretended sanitary improvements. Notwithstanding, the board could say, as it did in 1869, " What the death-rate might have been without the intervention of the Board of Health it is impossible to compute !"

There is another element in this calculation which it is not unimportant to consider,—namely, immigration. During the last ten years more than 300,000 people have come from Italy. The proportion of sexes in this number has been about eighty per cent. males and twenty per cent. females. The immigrants from Hungary have amounted to about 128,000, seventy-four per cent. being males and twenty-six per cent. being females. Previous to 1880 the immigration was almost wholly from Ireland, England, and Germany. The proportion of the two sexes from Ireland was about equal. That from Germany was about fifty-seven per cent. males and forty-three per cent. females. Of the 555,496 arrivals in 1891, 135,766 were registered as destined for the State of New York. This disproportion of sexes would not only affect the present birth-rate, compared with that previous to 1880, but also the death-rate of New York City.

For eight years ending 1882 the percentage of infant to total mortality was 46.63; for eight years ending 1890 it was 42.55,—a sharp decline beginning in 1883, when the immigration began to be plentiful from Italy and Hungary, being in the latter year 40.74, against 46.20 in 1882.

It is just at this period, too, that the decline commences in the proportion of zymotic to total mortality; and that occurs largely from diarrhœal diseases under five years, the number of such diseases being 3398 in 1883, and 4050 in 1882. For eight years ending 1882 there were 29,244 diarrhœal deaths under five years, while with a steadily-increasing population there are for the next eight years 28,242 deaths.

It is highly probable that, all over the country, there has been a decline in mortality from zymotic disease. This may be due to those unknown causes which, in by-gone ages, brought about the disappearance of certain epidemic diseases then prevailing; or it may be due to more intelligent separation of the sick from the well, and to improved methods of treating contagious diseases.

This diminution of zymotic disease, by other than sanitary measures, is illustrated by New Haven (page 92), a city of 100,000 inhabitants.

It is a noticeable fact that, in estimating the decline in the general death-rate,—which, after all, is not so great as was taking place before the rise of Sanitary Science,—and also the slight decrease of infant mortality, and of that from zymotic disease, the sanitarians wholly ignore the improvements which have been made in the methods of treatment. The medical profession, as a rule, is not given to much boasting about its own progress and achievements; neither is it over-sensitive to the praise of the public. Nevertheless, it is not content to rest quietly under the humiliating imputation that it has done nothing for mankind; and it is bold enough to assert that the general increase in longevity is due, in no small measure, to the introduction of new remedies and the more intelligent use of old ones. If we take into consideration all of these facts, there is absolutely nothing in the small gain to life and health which can be fairly claimed for the reformers.

In New York City, for instance, we have the spectacle of a Board of Health, clothed with despotic power, furnished with an enormous sum of money, supported by a battalion of officials, with a sanitary code of between two and three hundred sections. This board has its pathologists looking out for the microbes; its chemists seeking for the ptomaines, and searching the air, water, soil, ice, milk, meat, cosmetics, and saleratus for impurities; its meteorologist watching the weather; its inspectors of fish, flesh, poultry, vegetables; its inspectors of tea and coffee; plumbing-inspectors and general sanitary inspectors; disinfectors; inspectors of virus and vaccination; its inspectors of contagious diseases, with strict laws to compel the report of every case. Yet this city, possessing superior natural advantages for salubrity, and so protected by all the paraphernalia of Sanitary Science, not only shows no real decline in mortality on account of sanitary laws, but the highest death-rate of any city in the United States; that is to say, an average of 26.50 per thousand of its population for the last ten years.

Brooklyn and Boston, which come next to New York in sanitary regulations and sanitary prosecutions, show a death-rate of twenty-three and twenty-four per thousand. Philadelphia, which is comparatively poorly supplied with sanitary officials and sanitary laws, shows an average death-rate for ten years of twenty-one per thousand. In the Philadelphia Board of Health Report for 1889 we are told that "there is no systematic inspection of food" in that city; the milk inspection is a failure, and there are thirty-seven cemeteries wherein were placed nearly five thousand bodies in 1889; "vast multitudes of human remains lie crowded within these contracted places." Mr. George E. Waring, in a lecture* on the drainage question of Philadelphia, says that nearly all the sewers here, except the large main ones, are built as they were forty years ago; the lower half is laid without mortar, and the soil takes up the liquid sewage; the means for conducting house-drains is a sort of "go-as-you-please." The inlet basins of the sewers are clogged with filth "to putrefy before the eyes and under the noses of the people." This is shovelled into the street, and there left to drain until carried away by the contractor, or scattered by the wheels of passing carriages. The sewers are practically without any ventilation, and accumulation of gas is unavoidable. "If a judicious and proper connection of plumbing-work, with even tolerable sewers, exists anywhere within the city, it is the exception and not the rule," and probably not one soil-pipe in a hundred in Philadelphia has a good ventilation. Some of the houses are built over old cesspools which still receive the drainage of the new house, and the method of getting rid of the filth is through cesspools and privies.

Losing sight for a moment of the inflexible law in Sanitary Science that two and two make five, in order to ascertain how the inhabitants of Philadelphia are affected by her sanitary condition, we naturally turn to the vital statistics and compare them with those of New York, where, through sanitary law, there is protection against all of these multiplied causes of disease. Besides noting the general death-rate, which averages 21 per thousand against 26.50 for New York, we apply those other two unerring touchstones to prove the truths of Sanitary Science,—namely, proportion of zymotic deaths and infantile deaths to the total mortality.

---

* 1885.

In Philadelphia the percentage of zymotic to total mortality has averaged about 17 for the last ten years. It was 23 in 1881, when about 1400 people died of small-pox, and has been steadily declining since that year, reaching some years as low as 15 per cent., against 25 to 32 per cent. for New York. The percentage of infant to total mortality in 1890 was 37. Philadelphia expended for health purposes in 1889 less than one hundred thousand dollars. Chicago, which ranks with Philadelphia as regards sanitary law, has had an average death-rate for the last ten years of twenty per thousand; while New Haven, bereft of all sanitary measures, shows a death-rate of seventeen per thousand, and a more constant fall in zymotic mortality than any of the other cities.

It may be said that it is not proper to compare a city of 100,000 with one of 1,600,000 people. But this is what the sanitarians, over and over again, have invited us to do. Herein, they said, was the test of Sanitary Science; and they have repeatedly declared that a large, and even a crowded city, controlled by proper sanitary laws, is, and necessarily must be, healthier than a small one, or even a country village, which has no sanitary supervision.

We respectfully submit to scientific men, whose subsistence does not depend on the propagation of error, and whose reputation does not depend on the periodical excitation of a panic, whether, in view of these figures, it is not time to seek elsewhere for the causes of epidemic and infectious disease than in a siphoned trap, a disreputable privy, the casting of slops on an ash-heap, nitrites in water, or a bad smell from a slaughter-house.

In 1892 the approach of cholera was the signal for great sanitary activity by the New York Board. They said they were prepared to "stamp it out" if it arrived. Remembering that for twenty-five years numbers of epidemics had raged in New York without being modified in the least by the acts of the board; remembering, too, that distinguished medical men who had made a profound study of and had had a wide experience with cholera declared that its causes and prevention were unknown, and that neither quarantine, nor isolation, nor disinfection, even when carried out with military severity, had any effect to stay its progress, we confess our scepticism of the board's ability to restrain it; but the confident speech, and the jaunty air with which they delivered it, were calculated to inspire the people with the belief that the members of this board actually possessed some secret remedy or some sanitary prescience whereby they could halt and "stamp out" cholera. The first sanitary measure which the board took was to arrest and bring to judgment a number of milk-dealers who had been accused of selling watered milk; afterwards some spiteful man or woman complained of some respectable hotel-keepers for serving this milk, and these were arrested. The wretched occupants of some houses—said to be unsanitary—were turned into the street, with no shelter at all, or were forced to find lodgings which were perhaps more miserable than those from which they had been ejected. A good many men and women, penny peddlers of fruit,—fifty-six in one day,—were arrested on the charge that their fruit was either too ripe or not ripe enough. At the refuse-market, on Ninth Avenue, a small boy who had some stale muskmelons was sternly reprimanded by the medical gentleman who was acting as an escort to the board's garbage-cart. The young malefactor blubbered, and probably lied when he said he did not know the stock was bad. He was told that if he was bigger he would be arrested; though why a diminutive enemy of the public health should escape the vengeance of sanitary law was not explained. Another man was arrested for killing chickens. Nearly a hundred arrests of different people occurred in the course of three or four days. These sanitary measures called forth the applause of the people. Vessels, crowded with passengers, with clean bills of health, and which had cleared from perfectly healthy ports, were held at anchor in the lower bay. For many long days infected ships, loaded with anxious men and women, were detained, the well shut up with the sick. In some instances these passengers were drinking and using for toilet purposes water from the Elbe, which at that time was said to be causing hundreds of deaths daily in Hamburg. Had the cholera arrived at New York with the suddenness with which it reached Hamburg, these crowded ships, motionless on a motionless sea, the passengers oppressed with fear, forced to breathe a stagnant air, and deprived of exercise, would have offered the most favorable nuclei for the outbreak of cholera. It was not the fault of the health authorities that they were not decimated by pestilence, but because the conditions—the unknown conditions—for an epidemic were absent in New York in 1892.

When deaths occurred on the infected ships the corpses were cremated. It is practically of no importance whether the dead human body is thrown into the sea, tossed to wild beasts, incinerated in an oven, or put to moulder in the earth, the final result, a return to its original elements, is attained. There happens to be, however, a very tender and precious sentiment in the minds of some hundreds of millions of civilized people in favor of what is called Christian burial, and to go out of the way to treat this sentiment with disrespect is brutal. The burning of these bodies was something very impressive to the vulgar mind; it gave occasion for startling paragraphs in the papers; it intensified the fears of friends who were almost in sight of the crematory, and whose bodies were perhaps to be similarly treated. The testimony of physicians in 1832, in 1849, in 1852, and of the French commission in 1866, quoted on a previous page, would appear to determine the harmlessness of a body dead of cholera. At a sitting of the Academy of Medicine,* Paris, October 7, 1884, M. Ricord testified that during the progress of cholera in 1832 he continued his lessons in operative medicine to numerous pupils, many of whom were from abroad. He used bodies dead of various diseases, "*notamment des cadavres de cholériques*," and none of the assistants or spectators of these operations were seized with cholera. In 1854,† Mr. Radcliffe writes to the Board of Health of London that bodies dead of cholera do not receive the same funeral rites that other bodies do on account of fear of contagion from them, and he inquires if this fear is well founded. The board reply that while they recommend the speedy removal of corpses from dwellings, they do not consider such precautions as Mr. Radcliffe alludes to as necessary for the public health. At a sitting of the Paris Academy of Sciences,‡ January 7, 1867, M. Dumas stated that of eleven hundred persons engaged in burying the dead during the cholera epidemic in Paris, only one was stricken with the disease. He says disinfectants were used, but we now know that the disinfection practised in 1865 was inefficient and unreliable. We can find nothing recorded on this subject during the epidemic of 1884-85, except the official report of the statistics of cholera at Marseilles, which informs us that "the men who were employed to bury the dead, to disinfect the houses, and to clean the sewers enjoyed remarkable immunity." Would the physicians who ordered the cremation of these bodies in New York have hesitated to perform autopsies on them? Would they not have invited their friends and relatives, supposing them to be proper persons, to assist at and observe these ceremonies, and would they not have assured them that there was no danger from these bodies?

The authorities in New York behaved as if they thought

---

* Bulletin de l'Académie de Médecine, 1884.
† London *Lancet*, vol. ii.   ‡ *Comptes-Rendus*, 1867.

the public health would be protected in exact ratio to the annoyance, distress, and expense which they caused to individuals. It must be admitted that these "sanitary measures" had the sanction of five hundred years of trial. They were precisely those of the fourteenth and subsequent centuries. In all candor we submit the question to intelligent medical men if they were not either puerile or mischievous? The authorities at Hamburg had been blamed for a want of sanitary vigor, and Dr. Krause, the health-officer, by a kind of childish vengeance, was driven from office. Does any physician, or even any intelligent layman who has given the subject any reflection, believe that if, in Hamburg, all of the milk-dealers had been jailed, the hotel-keepers disgraced, the penny fruit-pedlers arrested, the small boys brow-beaten, destitute people thrust into the street, crowded vessels detained, or a crematory erected, that city would have been saved from the scourge of cholera? It should not be forgotten that men who possess the soundest intellects in the medical profession, and who have had twenty and thirty years of experience with cholera, declare that quarantine is useless to prevent it. But supposing that it be effectual. It was only a pretence in New York harbor in 1892. There were many fissures through which cholera could have crept in. The pilot of one of the worst-infected ships, who had been on board for twenty-four hours, went from the vessel straight to his home without disinfection. He was requested not to talk about it. Favored passengers managed to escape, with their baggage, from these ships in violation of law. Dr. Sanborn, the papers said, had orders to shoot any person who should attempt to board the ship on which he was placed. But a party—the names of eight were given—of medical men and others visited the infected ships and the sick on Hoffman and Swinburn Islands and came back to the city. There is no account that any precaution was taken with this party. The very persons who clamored loudest for strict quarantine were those who broke it. The fumigation of passengers and their baggage was a subject of ridicule.* The reason why New York did not have epidemic cholera in 1892 was not because of sanitary measures or quarantine, but because the cholera, which travels from place to place, did not arrive there in an epidemic form. When it did appear no one was more surprised than the health authorities themselves. There were cases enough, so that nine died between the 6th and 29th of September, all rather widely apart from each other except two. None of these had any connection with any case from abroad, "although untiring efforts have been made to obtain as complete a history as was possible in each case." †

The submission to the prevailing theory of cholera is so complete that no one is found intrepid enough to whisper the heresy that it may be caused otherwise than by the ingestion of the comma bacillus. No one, however, has volunteered to inform us how or when this bacillus arrived in New York City, or by whom the unwelcome guest was introduced, so that nine persons died of cholera, none of whom had had any communication with each other, with one exception, or had come in contact with any person or thing, so far as can be known, that was tainted with the cholera germ. So far as we can learn, there was no disinfection of bodies, clothes, or houses of any of these cases of cholera. Indeed, no one was presumptuous enough to call the disease by that name until permission was accorded by the bacteriologists. The medical practitioner seemed a most insignificant person, and, as the new order of things becomes established, it looks as though we may dispense with his services altogether. No one has told us how it happened that the cholera bacilli, which were generated in

---

* See note at end of chapter.
† Dr. Nagle, *Sanitarian*, November, 1892.

the bodies of these nine persons, lost their potency to spread the disease among those who surrounded and ministered to them when they were sick, and who buried them after they died, although no precaution was taken.

Notwithstanding our eagerness to proclaim allegiance to the prevailing theory, lest we be reproached for tardiness in not accepting it, as it becomes more complicated and paradoxical, there is danger, unless we adopt strong repressive measures, that before long common sense will lift its impudent head and begin to ask impertinent questions.

That this country is in grave danger from cholera cannot be overlooked. Whether it can or cannot be prevented by human efforts, would that the functions of administration might be in the hands of men who, if they are needy in purse, would scorn to replenish it by playing on the fears of their fellow-citizens, who feel secure enough in their reputation not to desire to build one up by inventing some clap-trap operations to dazzle the public, and who are independent enough to fold their hands and do nothing rather than engage in annoying their humble fellow-citizens— penny-pedlers of fruit, rag-pickers, and bone-collectors— by causing their wholesale arrest for pretended offences against the public health.

The Massachusetts Board of Health was established in 1869. In the organic act creating it, its duties were specified as follows: To make investigations into the causes of disease, and especially of epidemics; to ascertain sources of mortality, and the effects of localities and occupations on the health of the people, etc. The board began its work with great enthusiasm. It promised, in the first report, to scale the bounds of the law which formed it, for it says it "will adopt the maxim, Pagan though it be, that nothing which pertains to humanity, in its widest sense, will this board deem foreign to its aims." It opened hostilities, without delay, on the slaughter-houses at Brighton, as is seen on page 79. There is not a particle of evidence offered to show that these ever caused an epidemic, or even a single case of disease, but for ten long years the board waged an unrelenting war against them and against various offensive trades.

The author has searched in vain for evidence to show that slaughter-houses or their surroundings have any influence in causing disease. In the seventh report of the Michigan Board of Health, Dr. Hitchcock reports on slaughter-houses in that State. Circulars were sent to forty health-officers, inquiring for information respecting the condition of these places, and whether they had ever caused a case of sickness. With some few exceptions the drainage was reported as "none." The liquid filth was poured upon the ground under, or immediately around, the building, to saturate the ground and gradually to contaminate the *wells "from which in most instances the water supplied for the premises was drawn."* It seems a strange omission, but there is no record in Dr. Hitchcock's report that a single health-officer answered the only question which was of importance to the public health, namely, whether any disease could be fairly ascribed to these establishments. Not a single case or suspicion of a case of disease is mentioned in this report as being caused by slaughter-houses.

Dr. Johnson,* in a discussion on slaughter-houses, said that in Chicago "it is possible to bring thousands of persons to testify that these places are the most salubrious in the city."

During the last thirty years the author has known and treated many persons who have been employed in slaughter-houses, and many families that have lived near them. Though they have been repeatedly and justly complained of as great nuisances, yet he does not remember ever to have seen a

---

* Proceedings American Public Health Association, vol. iv.

case of sickness that was caused by them, or any disease in which the symptoms were in any way aggravated by proximity to these houses.

The second report of the Massachusetts Board deals with a variety of topics, such as poisoning by lead-pipe, trichinæ, homes for the poor, and employment of minors in the factories. In the third report are essays on arsenic in paints and in wall-papers, on the obstruction of mill-dams, on the use of intoxicating drinks, and on adulterations of food. We are told that the analyses have actually detected minute quantities of sulphuric acid in vinegar; and the report reveals the fact—which had been known for fifty years to every thrifty housewife—that burnt grain and beans and peas are mixed with ground coffee.

The board does not cease its hostilities against the slaughter-houses and bone-boiling establishments. In 1871 it secured the passage of a law giving it power to close at discretion all establishments of this kind in any town in Massachusetts of more than four thousand inhabitants. Why the towns of more than that number were thus favored is not explained; but the reason why these trades are likely to imperil the public health in towns of four thousand one hundred, and be harmless in those of three thousand nine hundred people, is, doubtless, one of the many secrets of Sanitary Science. The butchers resented the interference and were arraigned before the courts. They might have won but for a circumstance which turned out very advantageously for the board. A butcher died from blood-poisoning in consequence of a wound received while dressing an ox which was reported to have died of disease. It was said, too, that one-half of its carcass had been sent to the Boston market. Though a death from blood-poisoning might have happened from a wound received in a counting-house, or a cotton-factory, or an iron-foundry, the accident was so exploited that "wholesale alarm" was caused in Boston. People were said to be in danger of being poisoned by meat. It was declared that the city had no guarantee against diseased meat, and evidence was brought to show that such had been sold and eaten; but no proof was given that any disease had been caused by its consumption. This occurrence helped to gain a victory over the butchers, and hushed, for a time, the enemies of the public health; but they soon began to conspire against it, and contested every inch of ground. It must be kept in mind that no investigation had been made by the board to ascertain whether these offensive trades affected the public health. No charge was brought that they had ever caused a single case of disease in Massachusetts. This report discusses the causes of alopecia areata and pityriasis, though it is not stated how these diseases of the scalp could affect the public health.

The fourth report deals with beer-shops and prohibition laws, food-flavors, etc. The beer-shops are proved to be a nuisance, and prejudicial to the public health; but the board makes no effort to get a law passed which will allow it to close them at discretion, as it did the slaughter-houses. It does not even advise the suppression of the tippling-shops. In an article upon the food of the people of Massachusetts, the writer discusses the nutritive value of fried meat and baked beans. The latter are recommended as a diet, if well done. Our author says he has found out that the average time spent at meals in Massachusetts is from twelve to fifteen minutes. He protests against this custom; but neither he nor the board tells us how State medicine can suppress it. Rapid eating is certainly an injury to the public health, and, besides, it detracts from the dignity of the Massachusetts citizen; but it is so wide-spread that the entire able-bodied male population in the State could not furnish a sufficient number of detectives and inspectors to control it. Separate tribunals would have to be created to reach the offenders; indeed, the very inspectors and judges themselves might need watching. Besides, the Massachusetts Board, like the New York Board in relation to the cosmetics, is so busy with offensive trades and things which do not affect the public health, that it really has no time to devote to the suppression of rapid eating.

In the fifth and sixth reports the board tells of its wars against the slaughter-houses. In 1875 it proclaims that it has beaten the butchers at every turn. Decision after decision of the courts had gone against them; and this year the joyful announcement is made that the Brighton *abattoir* has taken the place of the filthy slaughter-houses in that town, which have for so many years imperilled the public health. The battle has been long and fierce, but Sanitary Science has finally prevailed. But right in the midst of the anthems over this conquest of State medicine comes the news that typhoid fever has broken out at Brighton! Brighton, which for so many years has been comparatively free from this disease! One physician has thirty-seven cases, another reports fourteen. We must search for the cause. It is not in the filthy slaughter-houses, for they have given place to the cleanly abattoir. It is not in the slaughter-house hog, which was wont to have the laurel-like wreath of entrails about his burly neck, as we have seen on page 80, and his body smeared with blood; there is entire reformation in his habits. The board says that if we must eat him, he looks more appetizing than he did.

We must go deeper for the cause. We must focus the rays of Sanitary Science, to ascertain why typhoid fever breaks out at Brighton, as we did to learn the immunity of Hyères from cholera. When we do so we find the cause to be—a faulty construction of the privies! We are at a loss to know whether this condition has been of long duration, and has suddenly become active to awaken the disease; or whether the privies of Brighton have been lately reproduced, and, in their *renaissance*, have suffered from some serious miscalculation of the architect. There seems to be a combination of causes to produce the Brighton epidemic, for a more thorough investigation has discovered some ash-heaps on which people have thrown slops!

As the years go on, the board considers the pollution of streams, and looks after the quality of the saleratus and the pepper. It enters into a costly and tedious investigation of oleomargarine. The parturient mountain labors with energy and the mouse is brought forth. The board shows, through this expensive inquiry, by a scientific process, what had been known to the frugal cooks of a half-dozen civilized countries for a score of years, namely, that oleomargarine is not an unhealthy article of diet. The board keeps prominently in mind the filth-origin of infectious disease, though this theory is dreadfully shaken every time it investigates the cause of an epidemic. In 1881 a typhoid-fever outbreak happens at Nahant, which has always been noted for its salubrity. The sanitary expert is sent for. He finds ample cause for the disease in the wells, which are too near the privies; but some of them had been there for a hundred years. The board is a little disconcerted, and, for once, comes very near distrusting its own expert. It says that Nahant has great natural advantages for health; it is in marked contrast to Lynn, which adjoins it, and which abounds in filth, but has no typhoid. The board says, "A comparison of these two neighboring towns shows what has been so often observed before, that if typhoid fever originates in filth, the filth is very often so concealed that only an expert can recognize it."

In 1884 the sanitary condition of Taunton is such that "it seems as if every condition was the most favorable to the causation of so great an amount of sickness as to arouse the population to a determined effort." The privy-vaults and cesspools are of the most primitive order, the deposits being open to sun and storm, and the ground filled with impure matters. Mill River is the trunk-sewer of the city, and the stench is strong in crossing it. The wastes of the mills bring their quota to the pollution; "but, so far as is observed, add nothing that is detrimental to health." There have been a few sporadic cases of typhoid fever here, but "no true epidemics of the disease." This is explained by the fact that the people spend a good deal of time in the open air, and that the factories are well ventilated. Taunton's death-rate is 15.04 per thousand. The report for 1891 informs us that the board was petitioned to abate a disgraceful nuisance of many years' standing caused by the overflow of the filth of various cotton-, woollen-, and paper-mills and tanneries on meadows adjacent to the Neponset River. Portions of eight different towns along the river were exposed to emanations from this nuisance. It is represented "that the stench arising therefrom is at times most unbearable, and is the cause of much sickness." Instead of prosecuting the offenders without delay, the board did an unheard-of thing under similar circumstances. It started an *inquiry*. It investigated the condition of the river and found that it had not been exaggerated; "it was pre-eminently a river which is polluted." Putrefaction here was at its highest grade. At one point the river contained two million four hundred and forty thousand eight hundred bacteria per cubic centimetre. The board then addressed a circular to the physicians who practised in the towns which were offended by this nuisance, asking if it "has an appreciable influence on the health of the people residing in the vicinity." Thirty-one replies were received. The shameful character of the nuisance was denied by none. One physician reported seven cases of disease in one year, and added, "Should say the condition of the river was a prime factor in their development." Another thought this condition "might provoke" malaria. Another said it "must be unfavorable," though neither of these specifies any case of disease.

Twenty-seven physicians declared that they knew of no disease which had been caused by this condition of the river, and many of them express the belief that it was innoxious. The opinion of one of these doctors was based on an experience of twenty-three years; that of another on thirty years' practice in the neighborhood. This investigation showed that the affair was entirely outside the domain of the public health. By prosecuting the owners of these factories, they, with their millions of capital and thousands of operatives, could be driven to another State. The Neponset River would then look nicer and smell sweeter. It was a question of economics and æsthetics, to be settled, not by sanitarians, but by intelligent citizens in town-meeting assembled.

As we are continually reminded by the sanitarians that the proof of the efficacy of State medicine is to be found in the vital statistics, we naturally turn to those of Massachusetts and compare the registration returns since the establishment of her Board of Health with those previous to its formation. For the twenty years ending 1869 the general death-rate in Massachusetts averaged 18.6 per thousand. For the twenty years ending 1889 the average general death-rate was 19.7 per thousand. For the ten years ending 1879 it was 19.9, and for the ten years ending 1889 it was 19.6 per thousand. Here, as elsewhere, there has been a decrease in mortality from certain infectious diseases. The average percentage of zymotic deaths to total mortality for forty-eight years ending 1889 was 25.22. The average percentage for the five years ending 1889 was 18.38. This seems to be about the average decline of zymotic mortality in our country without regard to sanitary measures.

Small-pox has almost entirely disappeared in Massachusetts in the last ten years. For the previous ten years, ending 1879, the deaths from that disease per ten thousand of the population had averaged 1.44.

Typhoid and scarlet fevers have also declined, while there has been a marked increase of diphtheria. There has been also a startling increase of mortality from cancer, heart-disease, pneumonia, apoplexy, and Bright's disease, and the comfort that we can derive from a study of the vital statistics of Massachusetts, in finding ourselves partially saved from small-pox, typhoid and scarlet fevers, is entirely swept away by the knowledge that we are all the more likely to be carried off by Bright's disease, pneumonia, heart-disease, cancer, and diphtheria.

The Connecticut Board of Health was established in 1878. Its organic constitution was like that of Massachusetts, as far as its duties were concerned. It at once issued an appeal to the clergy, inviting co-operation, which received from that body a most favorable response; and in 1880 it congratulated itself that its work had received "the indorsement by our State medical societies." The board announced in unmistakable terms its conviction that the prime cause of preventable disease was filth, to the removal of which it promptly devoted its energies. Slaughter-houses, offensive trades, privies, cesspools, and cemeteries received attention. The board superintended the removal of a dead horse near the town of Bristol, and the disposal of a heap of clam-shells which lay near Milford. For two years it sat in judgment on this pile of clam-shells, lest the public health of Connecticut should be imperilled thereby. In 1886 a physician at Unionville reported to the board that the water of a well in that town had already caused several cases of dysentery, two of them resulting in death. He had forbidden the use of the water, but said there was no law to reach the offenders. The chemist of the board analyzed this water and pronounced it of "good organic purity."

In 1888 the State Board, in solemn conclave, heard that the town-hall at Stamford needed scrubbing, and that the secretary of the board had flown to the rescue. The same year it discussed the building of a cesspool at the State Prison. This year the public health had a miraculous escape. It was in imminent danger on account of a foul urinal on one of the cars of the Shore Line Railroad. The secretary, who, fortunately, was on one of his excursions, smelled the closet and scented danger to the public health. He remonstrated with the conductor, who turned a deaf ear, and perhaps laughed in his sleeve. The secretary then wrote without delay to the president of the road. Probably the latter would have met a labor delegation with an unblanched face, and maybe would have turned his back square on a walking-delegate, but he quailed before the sanitary terrorist and wrote him "a courteous letter."

The board was deeply stirred over the sanitary condition of the summer resorts in Connecticut. It made frequent visits to them. With the artlessness of the callow youth who recounts his first travels to the interested parent at home, the board, in its reports, told the Legislature of Connecticut how it commended some resorts, and found others in a dangerous condition; all, however, were in equal health. It told the Legislature, also, how the inspectors of the board were fawned on by the proprietors of these resorts, who knew only too well the power of these visitors to empty their houses of guests, by intimating with a wry glance that bacteria might be prowling in the remains of yesterday's dinner.

Year after year seven Connecticut gentlemen, who have been engaged in the serious affairs of life until they have passed middle age, meet quarterly or oftener, at the call of their secretary, and discuss these and similar questions, without a sign of laughter.

The board seems oppressed with a sense of its responsibility. It says, in the report for 1888, that a well-organized Board of Health is as necessary for the safety of the lives of a community, as is a fire department for the safety of property. Just as a conflagration could be extinguished in its beginning, so could an epidemic, by equally prompt and well-known means.

But suppose that the fire-brigade were continually giving out false alarms; and suppose that when a genuine one was sounded the brigade should yoke its steeds, and, with shout and whip and spur, dash away like mad in the opposite direction, and suppose that, after the fire was spent for want of material, the aforesaid brigade were to make a report to the authorities, detailing their alacrity in extinguishing the flames, and asking for an increased appropriation for the fire department. This is what the sanitarians have done, and are everywhere doing, respecting disease.

From the vital statistics of Connecticut we learn that for the fourteen years preceding the formation of her Board of Health, the average annual death-rate was 16.3 per thousand. For the fourteen years ending 1890 which followed its organization, the average annual death-rate was 16.9 per thousand. Let us scan these returns a little closer. If we assume that it required five years for the board to get into working order, so that it had subdued the causes of preventable disease, it will be fair to compare the last nine with the first five years' work of the board. The average annual death-rate for the five years ending 1881 was 16.2 per thousand. The percentage of zymotic to total mortality was 23. The percentage of infant to total mortality for five years ending 1883 was 29.9. The average annual death-rate for the nine years ending 1890 was 17.50 per thousand. The percentage of zymotic to total mortality average, the last seven years, 20.82. The percentage of infant to total mortality for six years ending 1889 was 30. Yet all over the State panics had been fomented, health laws were in operation, prosecutions without number had been carried on, fines had been imposed, imprisonments threatened, privies and cesspools cleaned, bad wells closed, and sanitary improvements pushed to their fullest extent, except in New Haven, which was destitute of sanitary law, and soaked with filth, but which made a better showing, both in the general death-rate and decline of zymotic mortality, than any other part of the State.

The New Hampshire Board reported in 1891 that much sanitary improvement had been going on in that State; many causes of disease, and especially of zymotic disease, had been removed. Many bad wells had been suppressed; pure water had been introduced, and sewerage had been extended in the cities; nuisances had been carefully watched, and many had been removed. Sanitary codes had been formed, and sanitary laws properly enforced. We turn to the vital statistics to see the effect of these measures. In the seven years since registration commenced, the number of deaths from zymotic disease was for—

1884.	1885.	1886.	1887.	1888.	1889.	1890.
907	890	1095	1073	1097	1172	1155

The population of New Hampshire was 346,991 in 1880, and 376,530 in 1890. There has been an almost steady and continuous rise in the deaths from zymotic disease in that State as the so-called causes of such disease have been removed. The average general death-rate for three years—1884-85-86—was 17.33 per thousand. The board had been in existence five years in 1886, and had got well into operation, so far as sanitary codes and the adoption of sanitary measures were concerned. For the next four years—1887-88-89-90—the average general death-rate was 18.39 per thousand. The percentage of zymotic to total mortality for three years—1884-85-86—was 16; for the next four years it averaged 17.9 per cent.

For the fourteen years previous to the formation of the Rhode Island Board of Health, the death-rate in that State averaged 16.6 per thousand. For the next fourteen years the average death-rate was 18 per thousand. For the seven years ending 1883, Rhode Island's death-rate was 17.7. For the next seven years, after sanitary laws and measures were in full operation, her death-rate was 19 per thousand.

New York's State Board of Health was founded in 1881. Its duties, like those of the boards of Massachusetts and Connecticut, were to investigate the causes of disease, especially epidemic disease, etc. It took the same active measures for the removal and the suppression of filth. Whenever disease broke out in any town it could almost always find a privy or some slops or a pond or a bad well. If these failed, the resource which never failed was the "filthy condition of the soil." It was a little baffled in 1889 when an epidemic of typhoid fever broke out at Richfield Springs. Richfield has "a bill of health so clean as to put most of our village sanitary bureaus to the blush." Its system of water-supply and sewerage is "surprisingly satisfactory." "Provision is made for the rapid carrying off of surface-water and all forms of sewage. No unclean beasts are allowed within its borders during the warm season, and the suggestions of the State Board of Health are carried out to promote the salubrity of the township." High up among the hills, it has "a salubrious climate and healing waters."* Yet in this favored spot typhoid fever broke out. The State Board re-examined the town, thinking, perhaps, that previously it had overlooked something that now might account for the epidemic. The board reports that "the inspection showed an excellent sanitary condition." But its faith in the filth-doctrine of disease is still firm.

In the ninth report we are told that its methods for doing sanitary work are perfected, and it is "capable of meeting any emergency and putting down any epidemic disease almost before its presence is appreciably felt, and without the slightest disturbance to the business interests of the Commonwealth." The very next year the influenza appeared in New

---

* Dr. Nott, in *Harper's Magazine*, August, 1884.

York and spread over the country, causing directly or indirectly a greater number of deaths and a larger amount of distress than any other epidemic which had ever visited this continent since its discovery. An excess of five thousand deaths above the average number for the State of New York is reported for the month of January alone, which is ascribed to the effects of the influenza. There is no evidence that the board made any effort towards "putting down" this epidemic, or even made an inquiry into its cause. It remarks that the board is not in a position to deny or affirm to what it is due, and adds, "it is not a filth-disease." This last is a most impudent assertion, for, for anything the New York State Board of Health knows, it may be a filth-disease pure and simple. The board, however, disposes of this desolating epidemic with the profound reflection that "it will soon, doubtless, come to an end, but will not soon be forgotten." Although the same epidemic raged in New York for two subsequent years, the Board of Health did not make the slightest effort to ascertain its cause or arrest its progress.

The vital statistics show that New Jersey is not made any worse, at least, by sanitary law. The average death-rate in that State for the last thirteen years has been 19.13 per thousand. For five years ending 1883, it averaged 19.46, for the next five years 18.72, and for the three years following it averaged 19.23 per thousand.

The Pennsylvania Board of Health was established in 1885. In its first report it paid its addresses to woman, and especially desired that she should take an interest in the problems of sanitation. The emotional character of Sanitary Science, and the flattering insinuation that here she might find her true sphere, never failed to command woman's sympathies and active endeavors. In subsequent reports the Pennsylvania Board does not say what co-operation it received from the gentler and better sex. Like the other

State Boards, it refused all compromise with filth, and announced its decision to exterminate it.

The Michigan Board was one of the earliest that was founded. Its reports abound in essays from physicians, clergymen, and ladies, who strive for precedence in maintaining the sanitary excitement. Michigan's Board was no less uncompromising in its attitude towards filth, which it faithfully believed was the great generator of disease. In a former chapter (chapter iv.) we have given some copious extracts from the reports of that board.

The sanitary delirium swept over Minnesota and other Western States, and spread to Colorado and the Pacific. The health-officer * of one of the important cities of Minnesota, in relating to the State Board the difficulties which beset him, says, not, that much has been accomplished through investigation and experiment, but that "much has been accomplished through *terrorism*." Colorado reports a scathing epidemic of typhoid fever at Denver. The city is hardly ten years old; but there are cesspools there, and the cause of the disease is the saturation of the soil with filth. One incredulous and daring citizen sneers at this theory of the origin of the epidemic, and opens a communication between his house and cesspool, with an untrapped six-inch pipe. His family consisted of ten persons; they enjoyed the best of health during the epidemic.

In looking through the health reports of a dozen of the more important States, one is amazed at their barrenness in everything pertaining to the object for which boards of health were founded,—namely, care of the public health. All of them cling with an abject fondness to the tradition of the filth-origin of epidemic disease. They refuse to consider any other cause. None of them (if we except the Massachusetts Board) prosecutes any researches into its

---

* Minnesota Board of Health Report, 1883–84.

etiology that deserve to be called scientific. Two or three times, in investigating epidemics, the Massachusetts Board has come very near abjuring the old faith, and confessing its agnosticism respecting their causes; but to do so would be the first step towards winding up its affairs and going out of business. That this board has dealt with more fairness and intelligence, and with less apparent self-seeking,—especially in its earlier years, when it was animated by the Pagan maxim,—than any other, is plain to him who will take the trouble to compare its reports with those of other State boards. Indeed, if we except a few valuable essays in the Massachusetts reports, and two or three scholarly papers which are hid away in those of Connecticut, there is hardly anything in the pile of turgid volumes which make board of health reports in our country that possesses any scientific interest. They are full of empty boasting, loose assertion, idle conjecture, vain declamation, and abominable rhetoric. They are important only as they are bulky with facts which prove that the very basis of so-called Sanitary Science is a sham. The facts which its professors adduce place them in the light of a public prosecutor, who is pursuing an offender against the law, and supports the charge against him with the most villifying oratory, and at the same time proceeds to vindicate the defendant by introducing witness after witness to prove the falsity of the original indictment. They loudly proclaim the filth-origin of disease, and then heap evidence mountain high that shows the theory to be untrue.

There is no sign in any of these reports that a health board has ever foreseen an approaching pestilence that was not manifest to the ordinary citizen, or has ever added a particle of information to elucidate the causes of epidemics, or that the efforts and laws of any health board on the arrival of an epidemic ever modified its influence, abated its virulence, or prevented its return. We are constrained to repeat what Mr. Webster said a century ago, in his exhaustive study of epidemics, and which is rendered more certain by the investigations of the plague by Dr. Tholozan and by the history of all wide-spread epidemics of the last hundred years,—that there is "no sufficient evidence that health laws ever saved a country or city from pestilence in a single instance, but abundant positive proof of their utter inefficiency in a great number of cases."

> NOTE.—In a foreign port during the autumn of 1892, it was the author's privilege to witness the process of fumigation under sanitary order. A large number of passengers had sought this port to embark for the United States, because it was free from cholera. They were obliged to send their trunks to be fumigated twenty-four hours before sailing. This order caused them great annoyance and extra outlay, and, besides, it was an expensive operation for the steamship company, each fumigation costing, the agent said, four hundred dollars. Though these passengers had arrived from cholera-infected districts, they were allowed to take on board, without fumigation, all valises, hand-bags, rugs, and loose clothing. The trunks only were fumigated, and these were not opened. Nothing could be more ridiculous. While the fumigation proceeded, the company's agents laughed, the consul who made the certificate laughed, the doctor who superintended the operation laughed; the joke was of such magnitude that the very passengers who had suffered the inconvenience and extra expense joined in the merriment. After the doctor's jollity had subsided, he said that the ceremony was beneficial, in that it quieted the fears of the people, increased their respect for Sanitary Science, and heightened their reverence for its professors.

CHAPTER XVIII.

The Vital Statistics.

IN this chapter the author makes no attempt to throw any new light on the subject of vital statistics, his object being to show that these, which the sanitarians so boastingly display for their vindication, not only offer no support to Sanitary Science, but, when carefully examined, show incontestably that the so-called sanitary measures of the reformers have been utterly futile in the prevention of infectious diseases, and that these have spontaneously disappeared time and again when no such measures have been undertaken, while, on the other hand, they have arisen and increased in number and virulence in the face of the most rigid efforts to prevent them.

There are certain laws governing the mortality of mankind which are fixed and unalterable by human provisions. The law of heredity is well recognized by all physicians, but it is not sufficiently understood to permit any very positive conclusions. The law of sex is universal so far as we have any information. We know that death strikes the male child *in utero* oftener than it does the female, and that the advantage gained by the latter in conception is held for the first five years after birth to such an extent that it need not be despised as an element of prognosis in many acute diseases. This advantage of the female is no longer maintained after five years; the chances of life are now about equal for the two sexes, and remain so until the approach of puberty, when they are diminished materially for the female. After puberty the male continues to have a slight preference during the activity of the child-bearing period of the woman, when, as her climacteric arrives, the condition is again reversed, and she resumes the advantage of longevity, and retains it until the generation which we are considering will have passed away. On the whole, however, the death-rate, at all ages, of males will be nearly a steady one of more than two per thousand above that of females.

Legitimacy seems to be another of the unchangeable laws of mortality. The child conceived in the stealth of dishonor actually helps to conceal the shame; it is much more likely to be expelled from its mother's womb a mute witness, stillborn, than is the one conceived in wedlock. If born alive, the illegitimate child, on the whole, is about twice as likely to die under five years as the one honestly begotten. In

Plymouth, in 1891, the deaths of illegitimate children under one year were 330 per thousand, as compared with 180 per thousand of those born in wedlock. In Brighton the corresponding figures were 293 per thousand of illegitimate and 118 legitimate. In Blackpool there were 382 illegitimate deaths and 169 legitimate deaths per thousand under one year. There are certain districts in some countries where the difference in mortality between legitimate and illegitimate children is frightful, being three or four times greater with the latter than with the former. That illegitimate children are not reared with the same care that the others are is true, but this is not enough to explain the enormous difference in the relative mortality.

The vital statistics of our own country show that race not only has a positive influence in the development of certain diseases, but that the mean duration of life differs rather widely for different races. There are certain diseases which destroy the black race in much greater proportion than the white; while there are others which augment the mortality of the white race from which the black race is comparatively exempt. The Hebrews present, probably, the most remarkable example of the influence of race, both in the protection it seems to afford against certain diseases as well as in the promotion of general longevity. That they largely escaped the blasting epidemics of the Middle Ages was the groundwork for suspicion that they in some way caused their spread among the people. Within the last century it has been pretty nearly proved that the Jews everywhere not only enjoy special immunity from epidemics, but that the mean duration of life is longer among them than it is among any other race of civilized men. This difference is shown to be so great in our own country * that if it were not carefully substantiated it would be incredible. It is here proved that

---

* Census Bulletin No. 19, 1890.

the expectation at birth of life for the Jews is sixty-one years, while in Massachusetts the expectation of life at birth is not quite forty-three years; and that of one hundred thousand born for the Jews one-half would be living at the end of seventy years, while of one hundred thousand born in Massachusetts one-half would be dead in less than fifty years. Careful studies of this race in Germany, Italy, France, England, Africa,—indeed, in every part of the world,—show that it possesses about the same superior longevity as is shown by Census Bulletin No. 19.

The sanitary orator tells us that this superiority of health and life of the Jews is accounted for by the Mosaic sanitary code; that this inculcates cleanliness and the removal of filth; that the meat which is eaten by the Jews is inspected by a priest; that the walls of a house are cleansed after a case of leprosy, and that the diet of the Jew is sedulously guarded; that he washes his hands before eating; that Sanitary Science, in fact, carefully extended for forty centuries, has conferred on the Israelite this longevity and immunity from disease.

Is it a fact that the Jew is more cleanly than other people, that he is more frequent and regular in his ablutions than the Mohammedan and the Christian? Is it more efficacious to scrape the walls of a house after a case of contagious disease than to do as we do,—namely, fumigate the rooms with sulphur and carbolic acid? Does the resistance of the Jew to disease depend on his eating the pygarg and refusing the ossifrage, as the Mosaic code imposes; or is the Jew healthy because he regales himself with the grasshopper and the locust,—and the bald-locust,—but despises the humble oyster, the lobster, and the eel, and abhors every creeping thing that creepeth? If it be true that he is freer from disease and longer-lived than other men, is it not rather that it is because he is a born master; that he has been tried and toughened in the fires of persecution, and that, though trodden under foot of men for centuries, yet, through craft and cunning,—genuine statesmanship,—he has kept his race pure and himself in the ascendant in all that pertains to the intellectual man, and because he is more temperate, sober, and virtuous than other men? Are not his days lengthened because he honors his father and mother, and rests on the Sabbath-day? Drunkenness is a vice almost unknown among the Jews, and one of them is rarely convicted of crime. Does not this longevity, in great part, depend on the fact that the Jew is slow and careful in entering the married state, and that when he has taken on himself this condition he is prudent in begetting offspring?

The birth-rate of Jews in our country is given in the bulletin alluded to as less than twenty-one per thousand, while that for the whole country is about thirty-one per thousand, which indicates that the Jew, perhaps without being conscious of it, comprehends and practises the mild and humane philosophy of Malthus, which teaches that mere sexual irritability and sexual energy no more impart power and respectability to a nation than to an individual.

One of the most constant laws of mortality is density of population; everywhere people in rural life yield a death-rate of from three to six less per thousand than the inhabitants of cities, though these show a wide range of death-rate in different localities, illustrating another constant law of mortality,—that people surrounded by ease and comfort enjoy a superior longevity. Mortality is, again, regulated by civil condition. Marriage is much more favorable to long life than celibacy, although its influence varies widely for the two sexes. The registration returns of Scotland show that the comparative longevity of single women is much more assured than is that of single men, proving that man is much more dependent on his helpmeet for length of days than the woman is on man. From the fifteenth to the thirtieth year, the usual period of giving birth to the first infant, married women die in greater proportion than single women. From the thirtieth to the fortieth year the death-rate is slightly higher among unmarried than among married women. From forty to forty-five years they fare about alike as regards longevity, but from forty-five to seventy-five the difference is decidedly in favor of married women, there being 12.99 deaths of married women per thousand living between those ages, against 14.80 per thousand of unmarried women. With men the influence of celibacy is very striking. From twenty to twenty-five years, 6.80 per thousand living of married men die annually, while twelve per thousand living of those die of the unmarried. From twenty-five to thirty, 8.29 die per thousand of the married, and 14.32 of the unmarried. The corresponding figures for the married and unmarried men between thirty and thirty-five years are 9.60 of the first and 15.51 of the second, and from thirty-five to forty years 11.62 per thousand of the married die annually, and 16.85 of the unmarried. From fifty to seventy-five years, 44 per thousand living of the married die annually, and 55 of the unmarried. These observations, which were continued over a period of many years in Scotland, seem almost incredible, but they are entirely corroborated by the inquiries of Bertillon concerning the effect of celibacy in France, Belgium, and Holland.

It may be said that there is hardly a law of mortality, which is influenced at all by human agency, that does not admit of some important exceptions. Race, sex, age, celibacy, legitimacy, density of population, profession, and occupation offer probably the fewest exceptions. In surveying the mortality tables of nations, education seems to have a certain influence on human longevity. The death-rates of educated Prussia and Belgium are much lower than those of ignorant Austria and Italy. But uninstructed Ireland has a lower death-rate than enlightened Scotland and England. Civil and religious liberty seem to exercise a favorable influence on a nation's death-rate. This is, in general, lower in Protestant than in Catholic countries, and it might be argued from this fact that the insurrection of Martin Luther was a great sanitary measure and himself an Eminent Sanitarian. But Catholic Belgium has a lower death-rate than Protestant Prussia, and bitterly oppressed and intensely Catholic Ireland shows for the last thirty years a more favorable death-rate than free England, Protestant Sweden, educated Prussia, or any other civilized people on the earth.

In general, those countries which have a large proportion of illegitimate births, like Austria and Italy, show a high death-rate; but Denmark, Sweden, and Scotland furnish as great a proportion of illegitimate births as Italy, yet the number of deaths in those three countries is from twenty to thirty per cent. below that of Italy.

The extreme prudence, in forming conclusions, of those philosophers who have profoundly studied the subject of vital statistics, contrasts strangely with the reckless, haphazard, unscientific manner with which they are treated by the sanitarians. The same discomfiture attends these in their consideration of rates of mortality as in their elucidation of the principles of Sanitary Science. Indeed, the reformers tell us that the question of longevity is, like that of great pestilences, purely one of efficient scavengering, and that sanitary condition—that is, the prevention of zymosis—is the great law of mortality; and that by sanitary measures—that is, the removal of filth—the progressive decrease of the death-rate of mankind is to be maintained.

They stake the value of sanitary reform and so-called sanitary measures on the vital statistics, and parade them without reference to any of those almost innumerable conditions which affect the comfort, well-being, and, indirectly, the longevity of the race, and which the intelligent and reflecting physician or hygienist does not fail to consider in the study of mortality tables.

It is fortunate if, through ignorance or design, these are not grossly misrepresented by the reformers. A report of the board of health of a large city lies before the author, which professes to give the vital statistics and the conditions of life and health in that town. It calls attention to the fact that in one ward the death-rate is from twenty-two to twenty-four, while in another it is only twelve or thirteen per thousand. Besides, the infant death-rate, that sensitive test of sanitary law, is extremely low in the latter ward. Here, the report says, sanitary measures are perfected, cleanliness prevails, and these statistics are exhibited to show in particular the advantages of good plumbing, and the importance that this should be extended throughout the city by plumbing-laws, and that plumbing-inspectors should be appointed. Not a hint is given in this report that the board which issued it had the slightest knowledge that there are other distinctions in these wards; that the ward with the low death-rate contains proportionately three or four times as many people between the ages of fifteen and forty years as the other ward, or that a male-student population—instructors, undergraduates, graduates, and celibates—constitute at least two-fifths of the people in the former ward, or that the birth-rate in that ward is ten or twelve, while that of the other is thirty or forty per thousand.

Nowhere in this report does the board which issued it betray the least knowledge of the fact that in the ward with the high death-rate idleness, poverty, ignorance, debauchery, severe labor, imprudence, overcrowding, intemperance, and violence abound, while the other ward is noted for its small families, roomy houses and yards; for ease, comfort, intelligence, prudence, and sobriety. Yet documents like these are yearly emitted for the instruction of the people, and to inform scientific men of the progress of Sanitary Science, and from such reports the sanitary elocutionist draws the inspiration and material for his public discourses at sanitary convocations.

As evidence of the value of Sanitary Science its professors summon us to behold the striking difference in human longevity between present and past times; that the death-rate, which in the sixteenth and seventeenth centuries was fifty, sixty, seventy, and in epidemic years even eighty per thousand annually, is now, through the adoption of sanitary measures, reduced to twenty-four, twenty, and even fifteen per thousand. In listening to the sanitary orator, or in the perusal of sanitary literature, one would never guess that the reformers had any knowledge whatever that the general death-rate was steadily decreasing and the mean duration of life as steadily increasing for two hundred and fifty years before the great sanitary uprising. According to their story the air, water, and soil had been undergoing a saturation with filth as civilization advanced and population augmented, and it was only by a kind of miraculous circumstance that the evolution of the Eminent Sanitarian just at the right moment saved the world from destruction, and that decrease in the general death-rate corresponded to the rise and progress of Sanitary Science.

The fact is, that proof is at hand which clearly demonstrates that there was a greater proportional decline in death-rate for two hundred and fifty years before than there has been since the sanitary excitement, and that it was specially marked in infant mortality and in that from so-called zymotic disease. The vital statistics of Geneva reach back to 1549. Here was a mediterranean city whose population was little affected by emigration or immigration. In 1543 it had thirteen thousand people; this number was not doubled until 1828, nearly three hundred years later, when it had twenty-six thousand inhabitants. It therefore presented a field for the study of mortality rates, which, though not very extensive, was adequate to give correct information of progressive longevity, and M. Mallet says Geneva during those three hundred years may be considered as a fragment of the grand total of civilized mankind as regards vital statistics. He shows that the mean duration of life in the sixteenth century was eighteen years, in the seventeenth it was twenty-three years, in the eighteenth it was nearly thirty-three years, and in 1830 it was more than forty years. The great betterment in the duration of life during these three hundred years lay, Mallet says, in the diminution of infant mortality,—this being in 1830 only one-half what it was in the sixteenth century,—and in that from contagious disease, and these results are largely attained by the lessened birth-rate. Now,* he says, the marriage age is late; fecundity is at its minimum in Geneva; but longevity is greater; no city counts so few births and so few deaths, and her prosperity is a powerful aid to the argument of Muret, that, " La force de la vie dans chaque pays est en raison inverse de la fécondité," and of Ivernois, who declared that, " Si les hommes vivent plus longtemps il en naît un moins grand nombre."

Among the causes of this increased duration of life, Mallet enumerates, aside from the lower birth-rate, the cessation of epidemics like plague, a general improvement in the ease of the people and in medical science, better houses, more abundant and better food, vaccination, cleanliness, precaution against famine, a higher intelligence, and a higher morality. In France, Dr. Aubrien studied the vital statistics of the commune of Gault for two hundred years. The results are almost exactly like those of Mallet. From 1690 to 1700 the death-rate of Gault was thirty-nine per thousand. From 1860 to 1874 the deaths there were only sixteen per thousand. Gault, like Geneva, was favorable

---

* 1833.

for a study of this kind, because its population has not varied materially for two hundred years. Dr. Aubrien says that infant mortality is not more than one-third of what it was two hundred years ago; fecundity is less than one-half. But he says if less are born here in Gault they are better preserved.

Dr. J. L. Casper* showed that during the previous hundred years the mean duration of life in ten of the principal capitals of Europe had increased ten years, and that in London it had increased nearly twenty years. Dr. Casper set forth the proposition that the proportion of births to the population expresses almost exactly the mean duration of life; that in Prussian districts, where there is the largest number of births, this duration is 28.9 years; and in those districts where there is the least number of births, it is 32.6 years. In the most fertile parts of the Low Countries, where there is one birth to twenty-four, there is one death to 36.9 of the people; while in those provinces where the births are one to twenty-eight, the deaths are one to forty-nine inhabitants. He found a difference of six years in the duration of life between the most fertile and least fertile districts of France. The same condition prevailed in Belgium, Switzerland, and England; and in Russia, where the births are more numerous than in any other country (being between forty and fifty per thousand inhabitants), the mean duration of life was only twenty-one years. Dr. Casper says the law is that fertility is the regulator of mortality (Die Fruchtbarkeit der Regulator des Todes); that the duration of life increases or diminishes according as the fecundity is greater or less; showing that nature remedies an excessive fertility that may be detrimental to the next generation, and that man, through being placed in easy circumstances,—ease and vitality being synonymous terms,—by regulating the fertility of the race,

---

* 1835.

by the practice of virtue, by industry, sobriety, and economy, can control to a certain extent the duration of his own life, and become "der Herr seines Lebens," and that no doctrine is more pernicious (verderblicher) to the common welfare than the admonition, "Be fruitful and multiply." He showed that twice as many rich attain the age of seventy as poor, and that the mean duration of life for princes and counts was fifty, while that of beggars was only thirty-two years.

Dr. Casper called attention to the fact that locality had little to do with longevity; that the wet lands, foggy and moist atmosphere of many of the countries bordering on the northern seas differed widely from the dry, sandy plains of Brandenburg, but that the difference in the mean duration of life was almost nothing.

Quételet noticed the same increase in longevity and the same relation existing between births and deaths as did Mallet, Aubrien, and Casper. He says in Brabant and the two Flanders, "Il meurt et il naît le plus d'individus;" Great Britain produces less than our country (Belgium), yet her fruits are more durable; fewer people see the light there, but those who do are better preserved, and if fecundity there is less, the utility of the people produced is greater. We cannot too often repeat, says Quételet, that the prosperity of a state consists less in the multiplication than in the preservation of the individuals of which it is composed.

The Swedish life-tables show the progressive longevity in that country previous to the era of sanitary reform. The death-rate there for twenty-one years, 1755-75, was twenty-eight per thousand; for twenty years, 1776-95, it was twenty-six; and for twenty years, 1810-29, it was twenty-four; and in 1840 it had fallen to twenty-three per thousand annually. France shows the same progressive diminution of death-rate. In the sixteenth century the mean duration of life in that country was seventeen years; in the eighteenth century, as late as 1789, it was twenty-eight years; in 1817 it was thirty-one; and in 1860 it was about thirty-eight years. The mortality in France previous to the Revolution was more than forty per thousand annually. Between 1817 and 1836 it had fallen to twenty-five per thousand. We might reason from this that the French Revolution was a great sanitary measure, and that when Dr. Guillotine invented his deadly instrument he provided something which was favorable to human longevity.

The death-rate in Denmark in 1800 was 29.8 per thousand. It fell continuously, except in epidemic years, until in 1842 it was 20.5, and in 1885 it was 17.9 per thousand. In 1891 we are told * that less than two-thirds of the communes, outside of Copenhagen and sixty-six large towns, have no sanitary laws, and the sanitary measures which have been undertaken in the towns seem to be accompanied with no diminution of contagious disease. Diphtheria, which in 1870 caused one hundred and ten deaths in the cities, destroyed nine hundred and seventy-five people in 1889. One physician says, † " It is remarkable, considering the advance of public and private hygiene in Denmark during the last thirty years, that the mortality of those periods of age most liable to epidemic diseases, against which hygiene mostly directs its efforts, is not influenced by sanitary improvements."

Turning to Great Britain, we have ample and complete evidence of the steady gain in human life for two hundred and fifty years before the sanitary excitement. In London the average mortality in the middle of the seventeenth century, not counting still-born, was fifty per thousand. This represents the mortality of years free from pestilence, but not the absolute mortality, which for twenty-four years, 1620-43, was seventy per thousand. From 1660 to 1679 the death-rate was eighty; from 1728 to 1757 it was fifty per thousand,

---

* Denmark: Its Medical Organization, Hygiene, and Demography.
† Ibid., p. 426.

and continued so until 1780. Between 1801 and 1810 London's death-rate had fallen to twenty-nine, rising again to thirty-two per thousand in the five years, 1831–35, which period included the cholera epidemic of 1832. The percentage of infantile to that of total mortality in England was seventy-four between 1730 and 1749; twenty years later it had fallen to sixty-three, and in 1810 it had declined to forty-one per cent. Dr. Farr says, "If the Carlisle observations even approximately represented the mortality of England, the waste of life in the five years of infancy has almost diminished one-half during the last hundred years; other observations support this probability." According to Dr. Price's tables, the average death-rate at all ages in Northampton for the forty-five years, 1735–80, was thirty-four per thousand, and this probably very nearly represented the mortality of England and Wales at the time. In 1841 the death-rate in England had fallen to 21.6 per thousand. The mean duration of life in England in 1735 was about thirty years; in 1845 it had increased to about forty years. Dr. Farr estimates that the death-rate for England for eighteen years, 1813–30, was about 21.2 per thousand. So far as we have material to form an opinion, it is probable that the conditions of life and health in Great Britain had been for a long time more favorable than those of any continental state except Norway and Sweden.

We see that for nearly three hundred years the death-rate in all civilized countries was steadily declining, and that the falling off in mortality was principally in infant deaths, and in those from contagious disease. Yet during those three hundred years the sanitarians tell us that the precise causes of such mortality were accumulating, and it was not until about 1840, when the aurora of Sanitary Science appeared, that any effort was made to remove them.

Since 1838 there has been a complete system of registration of vital statistics in England. The sanitary movement began about the same time, and it is extremely interesting to note its progress and its influence on the death-rate in that country for the next forty years. For the three years, 1838–40, the death-rate for England was 22.4 (it was 21.8 in 1839) per thousand.

From 1841–45 the death-rate was 21.4	From 1861–65 the death-rate was 22.6
1846–50 " " " 23.3	1866–70 " " " 22.4
1851–55 " " " 22.7	1871–75 " " " 22.0
1856–60 " " " 21.8	1876–80 " " " 20.9

An average of 22.1 per thousand.

In 1876–77–78–79–80 the death-rate was 20.9, 20.3, 21.6, 20.7, 20.5, being in 1878 almost exactly what it was forty years before, in 1839, when it was 21.8. In 1845–50–56 it was 20.9, 20.8, 20.5. It was never so low again for more than twenty years. Registration shows the amazing fact that for nearly forty years there was practically no diminution of death-rate in England, notwithstanding the fierce sanitary excitement, the sanitary laws and persecutions. This is all the more astonishing when we observe that, except the Crimean war, England for those years enjoyed profound peace; that all of those influences which affect the well-being of a people, and which are believed, indirectly at least, to promote longevity, were in operation there to a greater extent than in other countries; that there was a steady decrease of illegitimate births; a steady increase in the ages of those marrying,—a sure indication that their offspring would be reared with more judgment; a diminution of the hours of labor; a more varied food, and a more plentiful amount of the same; better wages, clothing, and shelter; a steady decrease in pauperism; a steady increase of education among the common people; yet the death-rate for these forty years showed no tendency to decline.

As early as 1846 the registrar-general says, "The returns prove that nothing effectual has been done to diminish suffering and death. The sanitary improvements have been going on. They are showy," he says, "but they have not reached the homes and habits of the people." And more than ten years later,* he comes near openly expressing his doubts of "sanitary improvements," for he says sixty-four districts are found in England where reside a million of people, and where the mortality ranges from fifteen to seventeen per thousand; these people follow agricultural pursuits, do not drink poison in gin-palaces, and their minds are not overwrought by dissipation, passion, and intellectual effort. But visit their dwellings, and you find "the bedrooms are often small, close, and crowded; personal cleanliness is not much studied; the dirty pig and filth of various kinds lie here in close proximity to the house; the land there is imperfectly drained; ignorance yields its baneful fruits; medical advice is ill supplied and unskilful; yet the annual mortality of this million of men, women, and children year after year does not exceed seventeen per thousand." In fact, here was complete proof that in sixty-four districts in England a million of inhabitants were exposed to what the sanitarians call impure air, water, and soil, and yet they enjoyed a condition of health vastly superior to that of the country at large, and to that of those districts which were not reviled for their uncleanness.

That these sanitary measures had no influence on disease did not escape the eye of the English physicians, but for the most part they seemed to have been so cowed by the violence of the reformers that they seldom alluded to it in a public manner. Dr. Farr, in his letters to the registrar-general, which covered a period of forty years, wherein he analyzes with a masterly mind the vital statistics of England and Wales, hardly alludes to sanitary measures as affecting the general death-rate, or that from zymotic disease. In

---

* 1857.

1879 Dr. Fergus * ventures to review the registrar-general's reports. He shows that since the sanitary measures have been in progress there has been a marked increase in certain diseases which are declared to be due to defective drainage and filth; that diarrhœal deaths, which were only 203 per million living in 1838, and which averaged for the next five years only 298, had increased for the five years 1867–71 to 1161, and for the five years 1872–76 they were 998 per million living, annually, and in none of these years was there the disturbing element of cholera. Dr. Fergus called attention to the fact that deaths from the two fevers—typhus and continued—which are not caused by filth and by want of drainage, have fallen off in eight years, 1869–76, the first from 193 to 49, and the second from 245 to 83 per million living; while the only fever, typhoid, which is said to be more than any other due to bad drainage and filth had been least affected by the sanitary improvements; that is, its decline in the eight years being only from 390 deaths to 311 per million living. Dr. Fergus dares to quote Mr. A. H. Bailey,† who asks if there is any evidence that sanitary improvements affect the duration of life; large sums of money have been spent, and what has been the result? "It happens, curiously enough, that in each of the three decennial periods the rate of mortality has been identical,—namely, 22.35 per thousand of the population."

In 1874, Dr. Letheby, himself the president of the society of medical officers of health,‡ said, in despair, that there were influences outside of the sanitary condition which affected the death-rate; that "for thirty years the public health-officers had been working towards improving the sanitary condition of the country and the chief towns, par-

---

* *Medical Press and Circular,* 1879.
† Journal Institute of Actuaries.
‡ Journal Statistical Society, London.

ticularly the metropolis, and yet they could not touch the death-rate. If anything, it had increased. Speaking for himself and his brother-officers of health, he could say that they had worked for the sanitary improvement of the metropolis with an earnestness that was hardly to be found in any other part of the kingdom, and yet how had that affected the death-rate? Between 1841 and 1850 it was 25.29 per thousand; in the second decade it was 24.94, and in the third it was 25.11. There could be no remedy until medical officers of health had given to them *unlimited powers to enable them to place the poor in exactly the same sanitary conditions as the rich.*" (Italics ours.)

In 1875 the death-rate for England and Wales was 22.8; the mean rate for thirty-six years, 1838–74, had been 22.3 per thousand. In so far as we can ascertain, for the previous three hundred years, if we except plague-years, there was never a period of forty years when there was not a marked decrease in the death-rate in England. This halt for forty years in the decline of mortality, right in the face of the sanitary delirium and persecution, is unaccountable; there is nothing to do but to record the fact.

The registrar-general now * takes comfort that there has been no great increase of death-rate, and he assures us that " it is less than it would have been if no sanitary measures had been administered." He thinks, however, that it would be well to consider "how far the mortality is due to inevitable causes, and how far to movable causes;" we must know how far epidemic diseases are affected by "meteorological conditions," and also inquire "how the germs of some zymotic diseases are communicated." Yet for more than thirty years we had been told, until our ears were deaf from its repetition, that these epidemic diseases were caused by filth, and it was only a year or two before that Mr. Simon

---

* 1875.

had demonstrated to the satisfaction of all Eminent Sanitarians that the "septic ferment," which was generated in filth, was the cause of these diseases. This steady maintenance of mortality-rates in England for forty years is all the more astonishing because elsewhere, in the isles of the British seas, in Ireland, and on the Continent, the decline of disease, and especially of infectious disease, which had been going on for three hundred years, was still proceeding, in the absence of so-called sanitary measures, sanitary laws, and sanitary persecutions. In the island of Jersey the death-rate was equally low with that of England; while that of Guernsey was much lower, being an average of 19.7 per thousand annually for the decade 1871-80, and 17.3 for the decade 1881-90.

According to a special sanitary survey of the Isle of Man,[*] "a medical officer of health is a being unknown in that island;" there is "a very general ignorance of the duties and object of sanitary authorities." Douglas, the chief town, which has nearly doubled its population since 1851, is in the worst possible sanitary condition; the inspector had seldom seen a dirtier town. "The state of the narrow streets and back yards is horrible in the extreme; the sanitary condition of the houses is as bad as possible, and must render them hot-beds of disease." The drinking-water was horribly polluted; "human excrement was scattered all around the public water-supply;" "I saw such dirt and misery as would make the average Englishman shudder." In the more rural parts of the island sanitary matters are utterly unknown; "the well-water of the cottages cannot fail to be polluted by surface and excremental soakage." The inspector says, "I have never seen cottages so generally and so hopelessly unsanitary;" it is "a foul blot on the island." But when he inquired for disease, and especially infectious disease, he was

---

[*] *Sanitary Record*, 1879.

unable to find any. One of the chief medical men told him that "diphtheria was a very rare disease, and that there had not been an epidemic of typhoid fever for twenty years." Mr. Cadman* writes that great numbers come to the Isle of Man for "health's sake;" long experience had proved its salubrity; that "no more healthy country can be found," and he shows by the records that the inhabitants there are remarkable for their longevity. Turning to the vital statistics of the filthy island, we find that its population had been nearly stationary for thirty years, but that its birth-rate was six or eight per thousand greater than its death-rate, showing that emigration was considerable. It is fair to presume that those who emigrated were young and middle-aged people, just the ones to help furnish a low death-rate if they had remained. But with the departure of this important quantity, we find the death-rate below that of England and Wales, and that it was decreasing in the Isle of Man, as in the other islands of the British seas, in those years when it was stationary in England, and it is notable that the former were specially exempt from that class of diseases which their "unsanitary condition" is said to produce. Ireland, whose sanitary condition is represented to be fully as bad, if not worse, than that of the Isle of Man, had a death-rate for seventeen years, 1864–80, of seventeen per thousand inhabitants. Ireland was destitute of all health laws until 1878.

The registration of vital statistics in Ireland was begun in 1866. The returns from that island make a most interesting study, not only to the physician, but to the political economist. Here was an average population for twenty-five years of about five millions. During those years there was an average annual emigration of about seventy thousand. Contrary to popular belief, these statistics of the Irish people

---

* *Sanitary Record*, 1879.

show that, illiterate as they are, compared with their neighbors, fewer among them are married under age; that they have a lower marriage-rate, a lower birth-rate, a lower rate of illegitimacy (this being not one-half that of England, and not much more than one-fourth that of Scotland), and a lower death-rate than any other people except, perhaps, the Jews. None of these low rates can be explained by the emigration of only seventy thousand annually of that portion of the people who are likely to marry and bear children. The average annual death-rate for twenty-five years was about 17.5 per thousand. The condition and surroundings of large portions of the Irish people were so base that the average Englishman would have been ashamed to stable his horses or kennel his dogs among them. Dr. Tucker * describes the sanitary condition of the home of a nobleman's tenant: "The domestic circle consisted of a sick man, his wife, four daughters, one son, three cows, one horse, two calves, two pigs, and poultry, all in one common individual house; generally the pigs dwelt beneath the bed, the people in them, and the poultry overhead."

The drainage and the mode of living throughout the island were wretched in comparison with England, Wales, and Scotland, yet the general death-rate in Ireland for the twenty-five years under consideration was far below that of Great Britain. The proportion of infant to total mortality was equally low, and zymotic disease was not much more than one-half in proportion to what it was in England, Wales, and Scotland. Yet the so-called causes of such disease prevailed in Ireland far more than in those countries. These low death-rates led to the belief that registration in Ireland was imperfect, though there was no reason why the law should be violated here any more than in England and Wales. The percentage of infant and zymotic mortality,

---

* *Sanitary Journal*, Glasgow, 1887-88.

however, could not be affected by imperfect registration.

The registrars of the different districts frequently report to the registrar-general the bad sanitary condition of the towns. In 1869 nothing could be worse than the sanitary condition of Goleen. But Goleen that year had a death-rate of 12.2 per thousand. In 1874 the registrar of Emly wrote that "it is not an uncommon thing to see four or five human beings, two pigs, a goat, a cow, and a score of fowl, with perhaps a donkey, inhabiting in common one small room. That year Emly's death-rate was fourteen per thousand. Again, in Banbridge there was a shocking sanitary condition: of seventeen pumps not one yielded pure water, and four were actually dangerous; all refuse was poured into the streets, and when it rained it oozed into the wells, so that, the registrar says, "in plain language, people are drinking a solution of human excrement." The death-rate of Banbridge that year was fifteen per thousand. The registrar-general says (1874) that the reports show that in the majority of districts there is "utter neglect of sewerage, cleansing, or any other method which might tend to the healthfulness of the localities." A long list of places is given where the sanitary condition is abominable. Portglenore, Hillsborough, Fintona, Ballymate, Kilmedan, and Bangor are in a loathsome state. The death-rate of these places is recorded as fifteen, seventeen, fourteen, nineteen, seventeen, and fifteen per thousand. In Mulltown Malbray the wells are defiled with sewage. The death-rate here is 11.3 per thousand. In Dromore there were "abominable cesspools and collections of filth about the doors and under the windows of the houses;" here were "pestilential vats, the stench from which was truly unbearable." The death-rate for Dromore that year was 14.8 per thousand. In 1872, Castleneagh was reported in horrible sanitary condition. Its death-rate was 11.5 per thousand. Indeed, it appears that about in proportion to the filth of these localities, which the registrars seem to gloat over in their descriptions, do we find the good health of the people not only in the general death-rate, but in those sensitive and pretended unerring tests of sanitary condition,—namely, proportion of zymotic and infant mortality to that of total mortality. If we consider typhoid fever, which the sanitarians tell us is the disease of all others amenable to preventive measures, we find in the reports of the Irish registrar-general that for twenty-five years it was two or three times more prevalent in Dublin, which had the benefit of sanitary law and measures, than it was in other parts of the island where the sanitary condition received little or no attention. Dr. Grimshaw * wrote, "When it is remembered that enteric fever is generally considered by sanitarians to be dependent upon bad drainage or impure water-supply, it is difficult to account for the sudden increase of that form of fever in Dublin, as we know that the water-supply is nearly perfect, and that the drainage has been steadily, though slowly, improving." A writer in the *British Medical Journal* of November, 1891, says, "The steady increase of enteric fever for about a quarter of a century has been well known and often commented on by Dublin sanitarians;" yet in Ireland, just where the drainage was most imperfect and the water-supply most impure, do we find typhoid fever at its minimum, and its maximum is shown to be where the water-supply was purest and the drainage the best.

Registration of vital statistics was begun in Scotland in 1855. The registrar-general says, in his first report, "No one can have visited the rural districts of Scotland without being struck with the filthy state in which most of the houses of the peasantry are kept: damp earthen floors, low roofs covered with thatch, a fixed window of small size which

---

* *British Medical Journal,* November, 1871.

gives little light and no ventilation, no chimney for the smoke from the fire to escape, a pool before the door into which every fluid refuse is thrown, and which exhales a choking putrid exhalation during all the warmer months of the year, a pigsty or dunghill close to the wall of the house behind." Yet these rural communities gave a death-rate, year after year for thirty years, of about seventeen per thousand for the mainland, and about sixteen per thousand for the insular districts, against about twenty-five per thousand for the cities where sanitary improvements were being carried out. Infant mortality was about twice as great in the cities as in the country, and zymotic diseases were much more fatal in towns than in the country districts, which had no sanitary supervision. What was still more remarkable, after sanitary laws were well in operation and sanitary measures well established, zymotic diseases suddenly rose. The registrar-general, in the tenth report, says that the mortality from this class of diseases, which had been low and uniform for eight years, took a sudden rise in 1863–64.

The average annual death-rate for Scotland for the

Five years ending 1859 was 20.4.
" " " 1864 " 22.1.
" " " 1869 " 21.9.
" " " 1874 " 22.4.
" " " 1879 " 21.1.
" " " 1884 " 19.6.
" " " 1889 " 18.6.

The sharp rise in death-rate in Scotland after 1859 was maintained for nearly twenty-five years, yet during those years the sanitary excitement was at its height, sanitary laws were passed, and sanitary measures were carried out. The death-rate fell from 23.2 per thousand in 1875 to 20.8 in 1876. This year there was a sudden decline of zymotic deaths, but it could not be accounted for any more than the sudden increase thirteen years before in 1863. Meantime, the birth-rate declined from 35.1 per thousand in 1875 to 30.8 per thousand in 1889.

The death-rate in Sweden was steadily declining, as it had been for the previous century, during the forty years that it was stationary in Great Britain. In 1840 it was twenty-three per thousand; for the five years 1851–54 it was 21.3; for the twenty years 1852–72 it averaged 20.4; and for the five years 1873–77 it was nineteen per thousand. No sanitary laws were passed in Sweden previous to 1876.* Turning to France, we also find a reduction in the death-rate during the very years it was stationary in England. It was twenty-five per thousand between 1817–36. France had two disastrous years of war, 1870–71, when the death-rate ran up to twenty-eight and thirty-four per thousand; but, including these, her death-rate decreased to 23.6 for an average of twenty years, 1861–80. France was entirely destitute of sanitary codes. The London *Lancet* in 1891 informs us that there is a health law pending in France, and it congratulates the French that its adoption will remove the reproach that, with the exception of Turkey and Spain, France is behind all other European countries in sanitary legislation. The inflexible law of Sanitary Science, that two and two make five, is now so firmly established in our minds, that a survey of France's vital statistics occasions no surprise. Bearing in mind this law, we find, just what we ought to expect, that France, including the two disastrous years of war, had for twenty years a death-rate of from two to seven per thousand below that of any other Continental European country except Norway, Sweden, and Denmark (three almost entirely rural countries), and of Belgium, and excluding the two years of war, France's death-rate was lower than that of Belgium. In 1863 and 1864 the death-rate was lower in France than it was in England. For the twenty years 1861–80 the death-

---

* Palmberg, "Hygiene."

rate of Norway, Sweden, Denmark, Austria, Prussia, Holland, Belgium, Italy, and France was 16.9, 19.2, 19.7, 31, 26.8, 24.7, 22.8, 30, 23.6. Still having no sanitary law, the death-rate of France continued to fall, so that the average for nine years, 1881-89, was 21.9, that for 1889 being 20.5 per thousand.

The English sanitarians checked all criticism by offering as an explanation for the steady maintenance of the death-rate of forty years before, that the greater aggregation of people in English towns—density of population being an important law of mortality—defeated the good effect of sanitary measures. Nobody seems to have questioned this explanation, or to have publicly noticed that the same urban aggregation was going on in those countries where the death-rate was declining. Neither did any one twit the sanitarians with the answer that they had all along promised that sanitary law would overcome the ill results arising from density of population. They had said over and over again that a closely-populated region, with proper sanitary supervision, was much more healthy than a sparsely-settled locality without sanitary care. Besides, especially in England, the very cities which had increased had taken on a suburban character. Men and women who were employed in the heart of towns were making homes for themselves and their families in the suburbs. The day census of the city of London in 1866 showed that one hundred and seventy thousand one hundred and thirty-three people were employed in its limits; in 1881 this number had increased to two hundred and sixty-one thousand and sixty-one; while the night population, which in 1871 was seventy-four thousand eight hundred and forty-seven, in 1881 was only fifty thousand five hundred and twenty. The rate of increase in suburban districts for twenty years, 1861-80, was one hundred and twenty-seven; while that of the city was only thirty-six per cent.

There was one reply which the sanitarians made that could not be disputed, and which, as a last resort, they never failed to use,—namely, that the death-rate was less than it would have been if they had not been elevated to power.

In 1876 the death-rate in England and Wales was twenty-one per thousand, having been 22.8 in 1875, and having averaged 22.3 for thirty-six years; it had not been so low before for twenty years. It sank to 20.4 per thousand in 1877. The registrar-general says, "The meteorological conditions were more favorable to health."

This was a period of general health throughout Europe. The death-rate that year fell abruptly in Denmark, Sweden, Austria, Prussia, France, and Italy, to 18.7, 18.5, 31, 25.5, 21.7, 28.1, against an average for twenty years, 1853–72, of 20.3, 20.4, 31.9, 27.2, 23.6, 30.2. The death-rate suddenly rose in England and Wales in 1878 to 21.7 per thousand. From this time forward it continued to decrease, so that for the ten years ending 1890 it was 19.2 per thousand for those countries. It was natural to inquire if there were any known causes for this decline in death-rate. The sanitarians were now just as ready with their reasons for its decline as they were before with reasons for its continuance at its steady figure for the previous forty years. The same aggregation of population was going on, but the sanitarians said it was plain to every one why the death-rate decreased,—it was owing to sanitary law. The health act of 1875 was passed, and sanitary effrontery was now equal to the task of asking us to believe that taking the yeas and nays on a health bill in the British House of Commons not only caused an immediate fall in the death-rate in Great Britain of two per thousand, but had such a far-reaching effect as to cause almost precisely the same in France, Italy, Austria, Prussia, Denmark, and Sweden. There was an almost steady decrease of deaths in the Continental countries from 1875 to 1890.

	Death-rates for five years, 1871-75.	Death-rates for five years, 1886-90.
England and Wales	21.9	18.8
Scotland	20.7	18.8
Ireland	17.7	17.9
Denmark	19.5	18.7
Norway	17.5	16.8
Sweden	18.2	16.2
Austria	32.7	28.8
Switzerland	23.5	20.4
Prussia	27.7	23.9
Netherlands	25.6	20.7
Belgium	23.3	20.0
France	24.9	21.9
Italy	30.3	27.2

Though internal improvements were going on in these countries, hardly any of them had the benefit of so-called sanitary measures. France was entirely destitute of sanitary law; none of them had suffered sanitary persecution; and yet all, except Ireland, experienced about the same reduction in death-rate as England and Wales. Italy's death-rate had fallen off, proportionately, about the same as that of England; her "sanitary condition" is described in the London *Lancet*, December, 1891. Of the eight thousand two hundred and fifty-eight communes in that country, six thousand four hundred and four had no drains in the main streets, even for rain-water; refuse of all kinds is flung out to await the convenience of the scavenger. Only ninety-seven communes had a regular sewer system. In three thousand six hundred and thirty-six communes, containing eleven millions of people, the vast preponderance of the houses had no latrines. English water-closets are confined, even in the chief towns, to the hotels or abodes of English and American residents. In twelve hundred and eighty-six communes, latrines are unknown. Communes representing nine million inhabitants had drinking-water that was distinctly tainted and unwholesome. Yet, with these detestable conditions, Italy participated in the general decline in death-rate.

Since 1881, when the causes of death were first registered in Italy, the vital statistics show a steady decrease in all infectious diseases in that country except diarrhœa and dysentery. Instead of increasing, as in England, diphtheria, which in two hundred and eighty-four principal towns of seven and a half million people caused 11.2 deaths per ten thousand living in 1881, diminished almost steadily, so that only 4.7 deaths are recorded for the same disease in 1890, and only 4.1 for the whole country. Typhoid fever rose slightly in the same places until it reached 10.5 deaths per ten thousand in 1885, when it began to fall steadily, so that it only caused 6.7 deaths for the same number of people in 1890. In 1863 Dr. Hughes Bennet could hardly find a trace of typhoid fever in the hospitals at Naples. In 1881 it caused 9.4 deaths per ten thousand living in that city. This number dropped suddenly the next year to six, and continued steadily to fall to four in 1887, to three in 1888, and to 2.5 per ten thousand in 1890. Yet Naples had no benefit from "sanitary measures" until after 1886. Typhoid fever fell in Catania from eighteen per ten thousand in 1881 to seventeen in 1884, and to six in 1890. It fell in about the same proportion at Palermo, and all over the country; in those places where improvements were going on as where none were undertaken; showing clearly the spontaneous rise and fall of this disease in Italy without regard to "sanitary measures." The vile condition of Rome is familiar to all travellers; open sewers and untrapped drains are here the rule; yet typhoid fever in that city is no more fatal than it is in carefully-sewered and rigidly-policed Washington. Malarial fevers, which caused in two hundred and eighty-four principal towns, containing seven and a half million people, 6.2 deaths per ten thousand in 1881, fell to 3.2 for the same number in 1890. The mortality from puerperal fever was about the same in 1881 in Italy as it is now in England. It had fallen, however, more than thirty per cent. in 1890. There is also a marked decline in the two hundred and eighty-four towns in tuberculosis from 31.9 in 1881 to 27.5 per ten thousand in 1890. Measles, which shows no decline in England, has diminished one-half in Italy during the last ten years, and scarlet fever has fallen off in about the same proportion, so that in spite of the disgraceful condition which pervades all parts of Italy, and which is not at all overdrawn in the *Lancet's* account, as all travellers will attest, infectious diseases seem to be steadily falling off to a most striking degree, and that, too, over the entire country.

The fact is, that what is called the sanitary condition of England and Wales did not differ in 1876 and 1877, when the death-rate was 21 and 20.4, from what it was in 1875 and 1878, when it was 22.8 and 21.7. There was no repeal of the health act in 1878 which could cause the sudden rise from 20.4 to 21.7 per thousand, which was almost exactly what it was forty years before; neither was any addition made to this act which caused it to fall two years later to 20.5 per thousand.

What was the health law of 1875? It was the consolidation of previous acts with the addition of some new provisions; but it is simply ridiculous to allege that any of its clauses, old or new, had any bearing on public or private health. It prescribed the cleaning of roadways and pavements; the removal of snow; cleaning earth-closets and cesspools; it fixed the number of lodgers, and provided for the separation of the sexes in lodging-houses; it arranged for the laying out of streets and erection of buildings; the regulation of markets; the inspection of slaughter-houses; the care of cemeteries, forbidding the pasturing in them of cattle; it prescribed how the cabmen should feed their horses in the street; that cab-lamps should be lighted at night; it fixed the rate of fares; supervised the public baths, and set forth the limits for the approach of the two sexes in them; forbade the use of indecent language or the indecent exposure of the person; it ordered how people should conduct themselves in pleasure-grounds, and regulated the hiring of asses and mules in the parks; it controlled the carrying on of soap-boiling, tanning, glue-making, etc.; in fact, it took cognizance of pretty nearly everything that had been proved for centuries to have no relation to health or longevity.

The health act of Ireland was passed in 1878. It was almost precisely like that for England and Wales. The death-rate of Ireland, which previous to that year had no sanitary law, for the preceding fifteen years was 17.1 per thousand. And now, to sharpen the irony, her death-rate for the next twelve years rose to 18.2 per thousand. The average death-rate of England and Wales for twenty years, 1861–80, while they were enjoying the benefit of most despotic health laws, was 21.9, and for the ten years ending 1890 it was 19.2 per thousand.

The percentage of deaths from zymotic diseases in Ireland, 1870–79, was 11.7; that for the previous decade was eighteen, showing clearly that decline of contagious disease was not dependent on sanitary law or sanitary measures. For the same period the percentage of zymotic to total mortality in England averaged from 18 to 21 per thousand.

The futility of sanitary law or sanitary measures in controlling those diseases which are registered as zymotic is clearly shown in the vital statistics of England and Wales. These diseases caused 4.52, 4.25, 4.21, and 4.06 deaths per thousand for the four years, 1838–39 and 1841–42, which preceded the sanitary eruption. Twenty years later, 1859, these diseases caused 5.46 per thousand; in 1862, 5.88; and in 1863, twenty-five years after the outbreak of sanitary reform and the beginning of sanitary measures, zymotic deaths, which these measures were to prevent altogether, had risen from 4.25 to 5.88 per thousand living. As late as 1871 zymotic disease caused 5.44 deaths per thousand living. In 1874 zymotic deaths were 4.76 per thousand, and 4.47 in 1875. In 1876 this mortality fell to four per thousand, to rise again in 1878 and 1880 to 4.24 and 4.13, almost exactly what it was forty-two years before. It continued to fall from this on. Small-pox was nearly blotted out, scarlet fever showed a marked falling off, and diarrhœal diseases, though widely fluctuating, as will be seen, had declined. For the decades—

1841–50 zymotic disease caused 5.20 deaths per thousand living.
1851–60   "   "   5.05   "   "
1861–70   "   "   5.20   "   "
1871–80   "   "   4.23   "   "

For the next ten years zymotic deaths fell off to an average of about 2.60 per thousand living. But for forty years, when the sanitary excitement was the fiercest and sanitary laws the most severe in England and Wales, there was no decline in zymotic disease, while in Ireland, which had no sanitary law, the percentage of zymotic to total mortality, which was twenty-three for the decade 1851–60, fell to eighteen for the decade 1861–70, and to less than twelve for the decade 1871–80. Very early in this century zymotic diseases had begun to decline in Sweden. The deaths from the class "putrid fevers" fell off in that country from 7284, the average for five years, 1806–10, to an average of 205 for the five years 1826–30. Between 1811 and 1815 there was a sudden fall from the first number to 3668. For the five years 1806–10 there was an average of 1728 deaths from small-pox in Sweden; this number fell to an average of 327 deaths for the five years 1826–30.

Turning to individual diseases which are classed zymotic, we find that their presence or absence has no relation to sanitary measures. In 1838 small-pox in England and Wales destroyed 1064 people per 1,000,000 living; the number of deaths from that disease fell to 168 per 1,000,000 in 1842. In 1852 the number of small-pox deaths rose to 401, falling in 1861 to 64 and rising again to 1012 and 821 per 1,000,000 living in 1871 and 1872, falling to 21 in 1879, and rising to 119 and 103 in 1881 and 1885, falling to 10 and 1 per 1,000,000 living in 1886 and 1888. The most thoughtless sanitarian will not claim that these fluctuations of mortality bore any relation to the "sanitary condition" of England and Wales; that these countries were very dirty in 1838, when 1064 per 1,000,000 died, or that they were very clean in 1842, when only 168 per 1,000,000 were destroyed by small-pox, and that the deaths varied for the next fifty years in proportion as the filth was greater or less in amount. Neither did these fluctuations in mortality represent the vaccination of the people; the disease always came in waves; the unknown influence which controlled all epidemics guided this disease. In 1839 the registrar-general says five die weekly in London of small-pox when it is not epidemic; but "the question is not why these five die weekly, but why they become 10, 15, 31, 58, 88, weekly, and then fall progressively back the same measured steps." For the twenty years 1760–79 the average number of deaths from small-pox in London was 2333; for the next twenty years the number fell to an average of 1740. Dr. Farr says that this disease was declining before vaccination was discovered, " indicating, together with the diminution of fever, the general improvement of health then taking place." Dr. Farr says that the reduction in mortality from this disease is "the result, at least in part, of the vaccination act." But he notices with surprise that the deaths in France from small-pox were in 1842 only 91 per 1,000,000 living, and that for the years 1841, 1842, and 1843 the deaths in Austria from that disease were 4619, 5189, and 4411 out of a population of 22,000,000, while the deaths in England, where vaccination was discovered and which had a vaccination act, in 1838 and 1840 were 16,268 and 10,434 out of a population of less than 16,000,000. During the last great sweep which small-pox made over Europe,* the faith in vaccination to prevent its epidemical visits was terribly shaken. In Bavaria, of 30,742 cases, 29,429 had been vaccinated. Prussia was the best vaccinated country in Europe, and yet the mortality from small-pox was higher in that country in 1871 than in any other northern European State, and the death-rate in the German army, where revaccination is rigidly practised, was sixty per cent. greater than in the civil population of the same age, and the Bavarian Contingent in 1870–71, every one of which had been revaccinated, had five times more deaths from small-pox than the civil population of the same age. In 1870, at Cologne, the first unvaccinated person taken was the one hundred and seventy-fourth case in the town, and in Bonn and Liegnitz the unvaccinated ones seized were the fifty-second and the two hundred and twenty-fifth cases. Of 6533 admissions into Eastern Metropolitan Hospital, between 1871 and 1878, 5076 had been vaccinated. In 1871 and 1872, 14,808 patients were treated in hospitals in England for small-pox; of these, 11,174 had been vaccinated. Dr. Farr said, "Vaccination, therefore, does not afford entire immunity against attack or even death by small-pox," and about the only comfort that was derived during this furious epidemic from vaccination was, that it rendered the disease much less fatal.

Measles shows nearly the same fluctuations as small-pox; it caused 426, 705, 266, and 594 deaths per 1,000,000 in 1838, 1839, 1853, and 1887. For the ten years previous to 1875 measles caused an average of 401 deaths per 1,000,000

---

* 1870–72.

living, while for the ten years ending 1890 the average was 440 per 1,000,000. Scarlet fever caused 380, 1256, 747, 1030, 652, 1212, 451, 1478, 546, 1446, 515, and 1050 deaths per 1,000,000 living in 1838, 1840, 1849, 1852, 1857, 1858, 1861, 1863, 1866, 1870, 1872, and 1874, gradually falling to 691, 694, 675, 472, 215, and 231 per 1,000,000 living in 1876, 1879, 1880, 1883, 1886, and 1889.

It cannot be possible that these widely-vacillating figures represent the "sanitary condition" of England; that the country was in fairly good "sanitary condition" in 1838, when scarlet-fever deaths were 380, but that in 1840 it had undergone a frightful change for the worse, so that 1256 per 1,000,000 living died of this disease, or that it still continued to grow worse, until in 1870 the sanitary condition was so bad as to cause 1446 deaths per 1,000,000 from scarlet fever, and that the sanitary condition suddenly grew better in 1872, when only 515 died per 1,000,000, again growing so much worse in 1874 that it destroyed 1050 people per 1,000,000 living, or that the country was only half as dirty in 1886, when it caused 215 deaths, as it was in 1883, when it caused 472 deaths per 1,000,000.

The folly of ascribing the rise and fall of mortality from these diseases to the absence and presence of "sanitary measures" was, after a few years, apparent to the most unreflecting sanitarian, and the theory that these were caused by filth was abandoned. The reformers, however, were all the more obstinate in maintaining that diarrhœal diseases, typhoid fever, and diphtheria were the offspring of filth. Diarrhœal diseases caused 203 deaths per 1,000,000 living in 1838; this number was not higher than 260 for the next four years, when it rose to 372, leaping in 1848 to 780, and in 1852 to 1117 per 1,000,000 living. It did not fall again to much below 600 per 1,000,000 until 1885,—reaching in eleven of the years between 1852 and 1885 to 1000 and over,—when the number dropped suddenly to 487 deaths per 1,000,000. In 1884 diarrhœal diseases caused 971 deaths; in 1886 and 1887 they caused 888 and 717 deaths per 1,000,000. The deaths from diarrhœal diseases in 1889 were 635 per 1,000,000, more than three times as many as they were fifty-one years before, in 1838. Yet no disease has been more persistently held up to us as a filth-disease by the sanitarians. They could not fail to see, however, that it had doubled, quadrupled, sextupled even, in spite of, and about in proportion as, sanitary measures were adopted. Some explanation was required. The ingenious sanitarian was equal to the occasion. He said diarrhœa depended on a certain temperature of the filthy soil, which quickened the germ that lay therein into activity; he did not say how this ground did not happen to get heated in England until about 1848, neither did he claim that the germ had ever been found in the soil; in fact, it had not even been sought for. Nobody disputed the theory. Although the proof of its absurdity lay right before the eyes of the American sanitarian, in the comparative statistics of Philadelphia with New York, and New Orleans with Boston, no parrot ever responded with greater fidelity to his instructor than did our native sanitarian to his Anglican tutor in this instance; without any examination he accepted the theory, and not only actually published it in this country as one of the "settled principles of Sanitary Science," but demanded that the law-making power should shape its sanitary legislation to correspond with this theory. Sanitary measures had been established for nearly twenty years before diphtheria was recorded at all in the vital statistics of England. In 1855 it caused 20 deaths per 1,000,000 living. Four years later it caused 517 deaths per 1,000,000. The number fell to 93 in 1872, rising again steadily, as sanitary law became more stringent, to 189 per 1,000,000 in 1889. The death-rate in London from diphtheria for ten years, 1881–90, was 259 per 1,000,000 living against 122 deaths in the preceding ten years.* Typhoid fever was hardly known in England before the adoption of sanitary measures; it was not separated in the registrar-general's report from typhus and continued fevers until 1869, when it caused 390 deaths per 1,000,000.

This increase of typhoid fever in England during the progress of so-called sanitary measures is exactly what is taking place now in China and in India. Dr. Jamieson, in his studies of enteric fever in Shanghai,† says its existence was doubted in China twenty-five years ago, but that looking back for a quarter of a century it has steadily grown in importance as sanitary improvements have been going on. Malarial fevers have declined, and he says it may be asked whether subsoil drainage which removes malarial fevers " may not contribute to the increase of enteric disorders." The only cause of typhoid fever in China that he mentions is the predisposing one of age.

The number of typhoid-fever deaths in England remained nearly stationary from 1869 until 1876–78, when it fell to 309 and 306 per 1,000,000 living. It dropped sharply off to 231 in 1879, and continued to fall with slight intermission until 1889, when it caused 173 deaths per 1,000,000. In 1890 typhoid fever caused 179 deaths per 1,000,000 living in England and Wales. The number of deaths in three districts was as low as 31, 45, 64, and as high as 250, 367, 516 per 1,000,000 living in three other districts. Do these figures represent the filth of these respective districts, or do they represent the " unknown quantity" which the Massachusetts Board of Health say was necessary to produce the fever in that State?

This disease, which was unknown before the era of sanitary reform, and which the sanitarians say is wholly under

---

* Louis Parkes, Hygiene, April, 1892.
† China Imperial Maritime Customs, 1891.

our control, continues to cause more than five thousand deaths annually in England and Wales.

Mr. David Rennet * made an important study of typhoid fever in Scotland for twenty-five years, 1865-89. His inquiries extended over the whole mainland country, embracing a population of more than three million. In eight principal towns, except one, there was a marked fall in the death-rate from typhoid fever. In Paisley there was an increase of three per cent. But this decline in towns in typhoid fever is far from uniform, although all of them have improved water-supplies and efficient sewerage. Edinburgh and Leith have water and drainage system in common, yet their typhoid death-curves have no resemblance, as one shows a constant decline, while the other shows a steady decline and great variation. This decline in towns varies from sixty-nine per cent. for Aberdeen to twenty-two per cent. for Perth; in none of the cities does the fall bear any relation to water-supplies and drainage. In some cities there was, for a few years after the introduction of improved water and sewerage, a constant rise in the typhoid fever death-rate. Besides, he says the last few years, which is "precisely the period which has been marked by the greatest zeal for sanitary matters in the cities," the typhoid death-rate has remained stationary. The decline of typhoid fever in the rural districts of Scotland corresponds for five-year periods almost exactly with the decline in cities.

	1865-69.	1870-74.	1875-79.	1880-84.	1885-89.
Rural districts	45	36	38	27	19
Cities	37	36	34	26	19

per one hundred thousand living. No reason can be given for this decline in the rural districts, which is greater in proportion than in the cities. The sanitary work has been confined entirely to the latter. With rare exceptions, no attempt

---

* *Sanitary Journal*, Glasgow, January 19, 1893.

has been made to get purer water-supplies in the country, and a large majority of the people live in undrained houses. "The old wells in use thirty years ago are in use now, and we know they are far from good. The methods of excreta removal are the same as thirty years ago." Mr. Rennet says, "We cannot, then, well account for the fall in the typhoid death-rate by the sanitary improvements of the last twenty-five years." He thinks it may be accounted for, in that the people everywhere are better fed and clothed; they have less work and more wages; there is a higher grade of intelligence and education among them, and they have a better knowledge of the benefits of cleanliness. He illustrates the difference in death-rate between the poor and the rich by Aberdeen, which has two parishes,—one inhabited by the poor, the other by the rich. The water-supply and drainage are alike; but the poorer part shows an average for ten years of one hundred and five cases of typhoid fever per hundred thousand, while the well-to-do parish has sixty-three cases per hundred thousand from typhoid fever.

Typhus-fever deaths declined in England and Wales from 193 in 1869 to 36 in 1878, and from that to 5 per 1,000,000 living in 1889. The deaths from continued fever were 239 in 1869, 71 in 1878, and only 14 per 1,000,000 in 1889. It was never claimed that these fevers depended on filth or want of drainage; their disappearance and their appearance were equally mysterious. The returns of the registrar-general of England show that the mortality from three of the most important zymotic diseases steadily mounted higher for from twenty to forty years as sanitary laws were made more stringent and sanitary measures were made more thorough, and nothing seems plainer from the study of these reports than that the rise, progress, and decline of these, along with others of the zymotic class, bore no relation to such laws and measures. That there has been in the main a falling off of infectious diseases in Great Britain is not denied; but this decline has been no less pronounced everywhere, so far as we can gain information, alike in places which have had sanitary laws and measures as those which have not. Ireland is a most notable instance. Indeed, the decline in contagious diseases had been going on in Great Britain for a century until about 1840, when they appear to have received a new stimulus and to have increased for more than thirty years; and during that time diseases reappeared which had been absent so long that they had been almost forgotten. For the thirty years 1728-57, fever caused 148 of every thousand deaths; for the ten years 1771-80, it caused 124; for the ten years 1801-10, 89; and for the five years 1831-35, it caused 34 of every thousand deaths.* Fever, plague, and dysentery caused four-tenths of the fatal diseases in the sixteenth century. But while the sanitary excitement was the fiercest, the fever death-rate was rising; it fell from 10.41 deaths per ten thousand living in 1852, to 6.63 in the following year, when it again rose to 9.30, 9.77, 11.9, and 10.05 deaths per ten thousand in 1860, 1864, 1865, and 1866.

We have seen on a previous page how contagious diseases had declined in Sweden during the first forty years of this century.

Although the "sanitary condition" of London must have been worse than the country at large, the zymotic death-rate in that city in 1841-42 was less, being only 3.59 and 3.39 per thousand. In spite of fierce sanitary laws and thorough sanitary measures, this death-rate was not so low again for thirty-one years, when in 1873 it was precisely what it was in 1842. The percentage of zymotic to total mortality for the decades 1841-50, 1851-60, 1861-70, 1871-80 was 20.9, 21.3, 21.3, and 17. The general death-rate for the five years 1865-69 was exactly what it was for the five years 1841-45,—viz., 24.4.

---

* London *Lancet.*

London's death-rate since 1880 has steadily declined to an average of 20.5 for ten years, 1881–89, and 21.1 for 1890. But her sanitary condition was no better in 1885 than it was in 1875, before the passage of the health act. Her water-supply was even worse; as we have seen on a previous page, it was more and more polluted. In 1887 * the Thames was in " a disgraceful state;" the river was only a " huge sewer;" the "water-supply is fraught with incessant danger," and disinfection of the sewage in the river was pronounced a "costly sham." The whole "tidal portion is in a horrible and dangerous state." The water of one company † " on the 5th of February, 1891, surpassed in respect of organic impurity any sample of Thames water examined during the past twenty-five years;" for " five months of the year 1891 the water supplied to London from the Thames was for the most part of an inferior quality, and on several occasions for days together not fit for dietetic use; that is to say, not fit to drink," and Surgeon Hogg ‡ writes that " Thames water is most shamefully, nay, dangerously, contaminated." Professor Ewing, under the heading, "Actual State of London Houses," says, " By examination of the best houses in London it is no uncommon case for a house to be so completely without sewer-connection that all its own sewage sinks into the soil under the basement, and seventy-five per cent. of the houses inspected have failed to pass the smoke-test." A distinguished sanitary writer in this country very pertinently asks, " What must be the general condition of the houses in London if the best ones are so unsanitary?" Yet her death-rate was decreasing, and notably from that class of diseases which these conditions of air, water, and soil are said to cause. But a variety of circumstances have been present in that city the last twenty years which were

---

* London *Lancet*. † Louis Parkes, Hygiene, 1892.
‡ Hygiene, April, 1891.

calculated to materially diminish its death-rate. Since 1876 the birth-rate there has decreased from thirty-six to thirty per thousand. The marriage-age has increased; all of those influences which have been before mentioned as adding to the comfort of mankind have been augmented in London; her working-population has steadily sought the suburbs to live and sleep, and is becoming more and more semi-rural in its surroundings.

During the last ten years there has been in the City, Westminster, the Strand, St. Giles, St. George's-Hanover Square, a decrease in population of 25.5, 19.9, 18, 12, 10 per cent., while some of the outlying districts have increased in the same time from 95 to 133 per cent. London, for the last twenty years, has been more and more sought for as a residence by people in easy circumstances. When these are attacked by illness, if a change in climate promises relief or recovery, their departure from town is hastened. London, too, is recruited more and more by ambitious and vigorous men and women from other portions of the country.

The death-rate of London in 1891 was 21.1 per thousand. The districts of Hampstead, Lewisham, and Wordsworth showed a death-rate of 12.4, 15, 15.05, while that of St. Giles, the Strand, St. Luke's, and Holborn was 27.6, 29.6, 30.3, and 30.8 per thousand. The infant mortality,—the number of deaths under one year to each one thousand births,—for Hampstead, Wordsworth, and Plumstead was 104, 119, 120, while it was 201, 213, 228 in St. Saviour Southwick, Strand, and Holborn, that for the city at large being 154. Do these figures indicate that the health act of 1875 is a dead-letter in the districts with a high death-rate, and is rigidly enforced in the others; or that the health-officers in the former are ignorant and careless, while those in the latter are educated and vigilant? Do they not rather show that the people in the districts with the low death-rate enjoy ease of body and tranquillity of mind; that they are well fed, clothed, and sheltered; that they are sober and industrious; that they are slow to enter the married state; that, having assumed this condition, they are prudent in begetting offspring, and that they do not consider that their duty to society ends in the act of copulation, but that they rear their children with care and discipline, enlighten their minds, and educate their consciences? On the other hand, does not the high death-rate of the other districts represent poverty, ignorance, laziness, overwork, and maybe vice and intemperance; that the people there are largely unreflecting in taking on the responsibilities of wedlock; that procreation, rather than preservation and education, is the rule? As Dr. Letheby despairingly said in 1875, when medical officers have the power to place the latter class of human beings in the same mental, moral, and physical condition as the former, they can correspondingly reduce the death-rate. When this change will have been brought about it is safe to affirm that health acts which regulate the deodorization of ignoble privies, or which forbid cattle to graze in cemeteries, or which remove snow from sidewalks, or fix the hiring-prices of asses and mules in public grounds, or the location of a tanner or a soap-boiler, may, so far as regards the death-rate, be dispensed with altogether.

Liverpool presents an interesting study by which to judge of the influence of " sanitary condition" in causing disease and of sanitary measures to prevent it. A health act was passed for this city in 1846; before 1859 eighty miles of sewers had been built, the drainage of dwellings had proceeded slowly but steadily; many thousands had been forcibly ejected from the cellars; urinals had been established; public baths had been opened; there had been a gradual abandonment of cesspools; an increased water-supply; all privies in the interior of houses had been got rid of, and the midden entirely prohibited; yet, with all of these sanitary improvements which had been going on for twenty years, the death-rate was increasing, and the committee which reports on the sanitary condition of Liverpool squarely recants the filth pathology of disease, for it says, "The town then being in respect of drainage and the other matters above enumerated in a much better condition than heretofore, and still advancing in this kind of sanitary improvement, and yet the death-rate ascending higher and higher, it was necessary to look for other than the causes belonging to the first group to account for this increasing mortality." The committee now say, "The proximate causes of the increased death-rate are intemperance, indigence, and overcrowding." One great cause of the high mortality in Liverpool, they said, was the "stupefying of infants with opium by drunken mothers." For the five years 1857–61 the death-rate of Liverpool was 29.6; for the five years 1862–66 it was 35 per thousand. For the ten years ending 1863 there were one hundred and thirty thousand deaths in Liverpool; sixty-five thousand of these were under five years. The birth-rate in Liverpool in some years had been as high as 42 per thousand for an average of five years. Looking at the deaths from fever in Liverpool, we find that they had no relation to "sanitary condition." Typhus and infantile fevers in 1865, after twenty years of energetic sanitary measures and thirty years of sanitary excitement, caused 2338 deaths, against an average of 835 for the previous ten years. There were 1523 deaths from the same diseases in 1866; this number dropped off suddenly the next year to 656, only to rise to 841 in 1868, falling to 335 in 1873, and to 248 deaths in 1879, to rise again in 1882 to nearly 600 deaths,—about as high a mortality as fifteen years before, in 1867. Can anything be more silly, or can anything be more uncomplimentary to our intelligence and common-sense, than for the sanitarians to tell us that these figures represent the "sanitary condition" of Liverpool? Or that the diarrhœal deaths represented her "sanitary condition" any more than fever did, when they were 1145 in

1866, 796 in 1867, 1151 in 1870, 605 in 1877, 1028 in 1880, 841 in 1884, 422 in 1885, and 781 in 1886? Liverpool shared with the rest of Great Britain and with the Continent in the period of health that followed 1875; the average death-rate for the ten years 1866–75 was 30.6; in 1876 it was 27.5; it rose again in 1878, as it rose elsewhere in England, when it reached 29.3; but fell from that time almost steadily to 1890, when it was 23.3,—averaging for the decade 24, but rising in 1891 to 27 per thousand. But in none of these changes in the death-rate of Liverpool is there any relation between it and " sanitary condition," except that for more than twenty years it rose steadily as the " sanitary measures" progressed.

It is interesting to note that the birth-rate of Liverpool has fallen from as high as 42 in some previous years (for five years, 1844–48) to 27 per thousand in 1890. If the intemperance, indigence, and overcrowding which the Health Committee of Liverpool said were the causes of disease in that city have not been removed, there are at least more than thirty per cent. fewer infants born there to be " stupefied with opium by drunken mothers." Hardly any city had known such a rapid increase in population as Liverpool; in 1831 it had two hundred and five thousand, and—not estimating the suburbs—in 1881 it had five hundred and fifty-two thousand. The poverty, destitution, and intemperance of large masses of this population were extreme. During the last fifteen years, Liverpool, like all English cities, owes its increase almost entirely to its suburban population; its municipal population showing in the ten years 1881–90 a decline of six per cent., while the suburban population increased sixty per cent.; and since 1881 there has been actual decrease even outside the civic boundaries, and this decrease was particularly in the slums. The last census shows that English cities more and more are taking on a suburban character, and the death-rate of the inhabitants is decreasing in proportion, in this respect comparing more and more favorably with rural populations.

The birth-rate throughout Great Britain has now * decreased from 36, in 1876, to 30 per thousand. For more than fifty years there has been a steady increase in the age of women and men marrying; a steady decrease since 1875 of the number who marry. For the five years 1871–75 the average number was 17; for the five years 1885–89 the average number was 14 per thousand; a steady decrease in illegitimacy, being 4.4 per hundred births in 1890, as against 6.7 fifty years before; a steady increase in intelligence, as shown by the number who sign the marriage register with their names instead of with their marks. In 1846, 326 males and 482 females per thousand, who married, signed the marriage register with their marks; in 1890, only 72 men and 83 women per thousand married failed to write their names in the marriage register. During these fifty years there had been a steady decline in pauperism. As late as 1849, 62 persons per thousand received poor-law relief in England; 24 per thousand were receiving it in 1889. In short, the people of Great Britain were steadily advancing in the enjoyment of material comforts, in intelligence, decency, decorum, independence, and self-respect; they were approaching that condition of ease which Dr. Casper said was synonymous with vitality, and they were preparing themselves to make the most of those favorable but inscrutable changes which were going on in the types of disease; congratulating themselves that perhaps the period of devastating epidemics had passed away never to return, when they were reminded of the helplessness of the human race by the appearance of the blasting scourge of influenza, which swept over the globe with incredible swiftness, but which seemed to tarry longer in Great Britain than elsewhere, and to ravage that country

---

* 1890.

with special virulence, carrying off its people by tens of thousands, and increasing her death-rate in 1891 to 20.2, against an average of 19.1 per thousand for the previous ten years.

It is probable that these moral and economic influences had been in operation long before the period of the beginning of registration. As early as 1840, Dr. Farr remarks that moral restraint is in practical operation in England to an extent which will hardly be credited when stated in numbers. One-fifth, he says, of the people who attain the age of marrying never marry. "The reproductive force is repressed by prudence;" and in considering the early marriages which prevail in certain parts of England, where intelligence is the lowest, and where the competition for work and wages is so severe, he asks, "What would be the course pointed out by nature? Would it not be to defer the present early marriages until they have gained some experience in life, and accumulated some of the means of living?" Physiologists and economists would probably agree in saying that this increase of the prudence which is now in operation would not, except in special cases, be calculated to deteriorate the health and intelligence of their families, and he asks, "Can any one fear for the conduct and fate of this people if they should feel called upon to rear fewer children—to marry less early—than during the last forty years?"

It is a notable fact that during the last ten or fifteen years the birth-rate has declined markedly in those Continental states in which the death-rate has diminished.

The more one studies these tables of mortality the more one is convinced of the utter ignorance of mankind of any laws which prevent the appearance of any disease, or which affect its progress or removal.

Between 1851 and 1860 cancer caused in England 195 deaths per 1,000,000 living. This number has steadily increased until 1890, when it caused 676 deaths per 1,000,000; these were very unequally divided, however, even over this small country; the number of deaths from cancer in Lancashire in 1890 was 555; in Rutlandshire 1036, and in London 786 per 1,000,000. Diseases of the circulatory system, which caused, between 1858 and 1860, 900 per 1,000,000, had steadily risen to 1641 for four years, 1886-89, and to 1757 for 1890, varying widely, however, in different localities; for these diseases caused 2234 deaths in Wiltshire, and 1445 per 1,000,0000 living in Middlesex.

Phthisis deaths had steadily fallen off from an average of 2447 for the five years 1866-70 to an average of 1635 per 1,000,000 living for the five years 1886-90. The health act of 1875 was not passed until more than ten years after the sharp decline in phthisis began. This decline in deaths from phthisis at first took the sanitarians by surprise, for no measures had been directed to subdue the disease; they very quickly told us, however, that it was the sewers which had caused this lessened mortality by relieving the ground of its moisture; but we find that phthisis caused 2100 deaths per 1,000,000 in London in 1890, while only 1133 per 1,000,000 died of that disease in Herefordshire. Are we to understand by these figures that Herefordshire is nearly twice as well sewered as London?

Notwithstanding the sewerage had relieved the ground-moisture, diseases of the respiratory system increased from 3394, the average for the five years 1866-70, to 3639 per 1,000,000 for the five years 1886-90. The mortality from rheumatism had largely increased, and diseases of the urinary organs had nearly doubled in thirty years. The average number of deaths from puerperal fever was 55 per 1,000,000 for three years, 1858-60; this had increased to 76 per 1,000,000, the average for five years, 1886-90. The steady rise in mortality from this disease is startling, in view of what we have been led to believe of the efficacy of antiseptics in midwifery. Besides, many times we have been told, and most of us really believed, that this disease of all others was most amenable to hygienic measures.

Erysipelas, a disease which, we are told, is specially controlled by sanitary measures, maintained an average of about 90 per 1,000,000 from 1858 to 1885, when it suddenly dropped, without any apparent cause, to an average of 54 per 1,000,000 living. For three years, 1858–60, the number of deaths from old age was 1422 per 1,000,000; the average for five years, 1886–90, was 960 per 1,000,000.

In looking over the vital statistics of our own country, we find no relation between disease, and especially zymotic disease, and "sanitary condition." Dr. Billings, in his elaborate work on the vital statistics of the United States, 1880, says, pointedly, of diphtheria, which has probably given rise to more sanitary persecution in our country than any other disease, that "the disease cannot be due to any peculiarity of climate, of geological formation, of topography, or of methods of filth-disposal." Typhoid fever in 1880 was most prevalent in the Appalachian regions of North Carolina, Georgia, Alabama, and Tennessee, and in Central Ohio, all Indiana, the southern part of Missouri, eastern part of Texas, the western half of Kansas, and in Oregon. It was comparatively low in New England, Central Pennsylvania, the Mississippi Valley, and Gulf coast. Does any intelligent person believe that the privies were more leaky, or that there was greater carelessness in rinsing milk-cans, or that turtles tumbled oftener into the wells, or that decaying twigs and stumps were more plentiful in the water-courses in the first than in the second group of districts? Do we know why this disease chose those regions for its ravages any more than we know why cancer, which the same year caused 18.34 out of every 1000 deaths for the whole country, should cause 33.23 deaths in the Northeastern hills, while in the South Atlantic coast, in Charleston, it should cause only 5.70 of every 1000 deaths? It is nothing to say that cancer is twice as prevalent in the white as in the colored race; this gives no explanation why the white race in our country is two or three times more likely to have cancer in one locality than in another five hundred miles away.

In 1880 there died of typhoid fever per 10,000 living in Brooklyn, Philadelphia, Baltimore, Pittsburg, New Orleans, and Charleston 1, 4.13, 5.14, 10.99, 1.98, and 9.40. Do these figures represent the cleanliness of these places? Albany, said the Eminent Sanitarian, as he mounted the sanitary platform and waved his hand with august serenity, Albany has 4.51 deaths from typhoid fever per 10,000 living. Sanitary Science explains this problem: the city takes its water from the Hudson, which is contaminated with the sewage of Troy and the towns above. But we look at Troy, which takes its water from the pure Piscauin Creek, and find that this city yields 8.45 victims to typhoid fever per 10,000; and after the 3,000,000 of people in New York and Brooklyn have eaten, the year round, the ice which Albany, Troy, and other towns have contaminated with typhoid-fever germs, Brooklyn and New York have 1 and 1.71 typhoid-fever deaths per 10,000 living. Continuing his allocution, the Eminent Sanitarian said it was plain why Newark should have 3.51 deaths per 10,000 of its population from typhoid fever: its water-supply was from the Passaic, which was polluted by the sewage of Paterson and other towns; but when the same water, still taking up sewage in its course, reached Jersey City, it only caused 1.73 typhoid deaths per 10,000. It was just as clear why the cesspool city of Philadelphia should have 4.13 typhoid deaths per 10,000. It drank the dirty water of the Schuylkill. But when we reach Washington, which is thoroughly sewered and supplied by the pure water of the Potomac, we find 4.88 typhoid-fever deaths per 10,000; and when we get down to well-drained Charleston, with water eternally flowing from artesian wells 2000 feet deep, we have 9.40 typhoid-fever deaths per 10,000; and when we arrive at filthy New Orleans, with its scanty, polluted water, its pestilential air, and loathsome soil, we have only 1.98 typhoid deaths per 10,000 living. Pittsburg, which draws its water from a comparatively pure source of the Allegheny River, has 10.99 deaths per 10,000 from typhoid fever, while Allegheny City, whose water-supply, as we have seen on page 124, is horribly polluted by sewers, slaughter-houses, soap-factories, tanneries, etc., yields 2.79 typhoid deaths per 10,000. In whichever direction we turn we find, both in the vital statistics of our own and of other countries, the beautiful workings of that inexorable law of Sanitary Science that two and two make five.

BIBLIOGRAPHY.

Short, Bills of Mortality; McCulloch's Statistical Account of British Empire; Malthus, Essay on Population; Mallet, Recherches Historiques sur la Population de Genève; Daubrien, Études Démographiques; Casper, Die Wahrscheinliche Lebensdauer des Menschen; Quételet, Recherches sur la Réproduction et de la Mortalité des Hommes; Bertillon, Dictionnaire Encyclopédique des Sciences Médicales, Article " Mariage et Mortalité;" Tenth U. S. Census, Vital Statistics; Journal of the Statistical Society of London; Journal Institute of Actuaries, London; Bertillon, La Démographie figurée de la France; United States Census Bulletin No. 19, December, 1892; Rivista d'Igiene et Sanitá, Roma, 1892; Denmark: Its Medical Organization, Hygiene, and Demography; Statistique de la France; Palmberg, Hygiene in Sweden; Glatter, Ueber die Lebens Chancen der Israeliten; Encyclopædia Britannica, Articles "Vaccination," "Ireland," "Sweden," "Sewerage;" Report of Sub-Committee on Sanitary Condition of Liverpool; Registrar-General's Reports England and Wales, 1st to 53d; Registrar-General's Reports Scotland, 1st to 34th; Registrar-General's Reports Ireland, 1st to 26th; Revue Scientifique, 1881; Sonstaff, Studies in Statistics.

CHAPTER XIX.

Conclusion.

It is the duty, as well as the privilege, of the medical man to co-operate with his fellow-citizens, to the end that the air we breathe may be chemically of the normal quality; that, besides, it shall not be tainted with obnoxious odors; that our streets, parks, dwelling-houses, and places of public assembly shall be so constructed that there shall be a frequent renewal of the atmosphere; that the abhorrent slaughter-house shall be banished to a point where it shall not agonize tender sensibilities, or harden those which are less delicate; that all offensive trades shall be carried on where the community cannot be disturbed by their sight, or smell, or sound; that our sources of water-supply shall be cared for by practical men or women, who will be guided by common sense and experience in their selection, in order that the water we use and drink shall be as free as possible from extraneous matters, overflowing in quantity, sparkling to the eye, agreeable to the taste, and acceptable to the stomach; that the soil on which we build and tread shall not be encumbered with refuse which is offensive to our senses; that, while waiting for the time to come when, stimulated by the hope of gain, or harassed by the fear of starvation, the present lavish system of the disposal of sewage shall be replaced by one more frugal and more consonant with the processes of nature, our sewers shall be so constructed that they will perform the service for which they were intended; that our cemeteries shall be so located that they shall not wound persons of acute and nervous organization, and that our dead shall be buried with tenderness and respect; that the pomp, and folly, and extravagance, and idle curiosity of public funerals shall be discountenanced; that if the self-interest and natural decency of those who are striving for the honor of supplying our tables are not sufficient to move them to keep their establishments in becoming condition, but if they persist in exposing their wares in a manner disgusting to eye and nose, then it is the privilege, if not the duty, of the people to place these dealers under restraint and supervision.

If a certain portion of the community, who are pricked by a perverted or over-refined taste, or who are compelled by necessity, or incited by penuriousness, desire to eat diseased or putrid game, mutton, beef, fish, or vegetables, which are repugnant to the great mass of citizens, the latter have the right to enjoin that buyer and seller of such goods shall meet where no annoyance can reach the rest of the inhabitants.

For the furtherance of all these measures, which promote comfort, decorum, and self-respect, and which tend to dignify the human race, through ministering purely to its material wants, the physician has a duty and a privilege, in common with his fellow-citizens. It is to the honor of the medical profession that its members have never shunned this duty, nor failed to assume this privilege, but have let their light so shine that it has been a guide to the pathway of others.

But, if a body of men, for speculative purposes; or if a certain number, destitute of industry and talent to investigate the phenomena of nature, and titillated by irritable mediocrity to aspire to be nominated scientists; or if amateur philanthropists, whose chief object in life is to break the monotony of an elegant leisure,—if any or all of these, to appease their thirst for the public money, or to satisfy their lust for power, or to delight their vanity, combine to impose on the popular mind, to excite fear and panic, by pretending that the air, water, soil, food, and public improvements are so beset by perils to health and life that they cannot be enjoyed by the public unless they are controlled, through legislative enactments, by the speculators, pseudo-scientists, and pseudo-philanthropists, it is the privilege of the medical practitioner to declare the humbug. If its promoters take the ground that, unless the terrors of disease and death are portrayed as the penalty, the danger is so great that the common herd will relapse into barbarism and prefer filth and misery to elegance and comfort,—then it is for each physician to judge how far he will lend himself to extend and perpetuate the delusion.

Since the beginning, king, and noble, and pseudo-scientist have all been leagued and pledged to perpetrate pious frauds, which they said were for the people's spiritual and temporal good. It may be thought by some that it is better that the collective nervous system of civilized mankind shall not be prematurely shocked by the knowledge that the sanitarians, organized or unorganized, notwithstanding their pretensions, are ignorant of the causes of epidemic diseases, and powerless to prevent them; and that, in upholding the fiction that some board is looking out for their health, the people are amply compensated for the frequent drafts on the public purse which are made by the sanitarians. Indeed, it may be true that mankind is happier in contentedly acquiescing in that which is false than in being distressed by the suspense of uncertainty. The medical man, however, need not partake of any of these illusions. It is nothing to him that the reputation of professors of Sanitary Science is in danger if their wrangling is not quieted, as we saw on page 39; or that some speculator will "not have so good a case" in establishing his mortuary chapel if the truth should be known respecting the emanations of a dead body, as we saw on page 211; or, as we saw on page 289, that the removal of certain errors regarding the cause of yellow fever "will paralyze sanitary legislation," and thereby prejudice the interests of sanitary experts.

But it is of immeasurable importance to mankind that the

physician should know the true cause of disease, and that, while laboring and waiting for the light to dawn, he should distrust and resist the theory of every man—no matter how famous he may be—respecting such cause which is unsustained by facts; that he should elect to be forever tossed and agitated on a sea of doubt rather than to repose for a moment in the haven of error. In reality, this is the course which is pursued by the medical man who is engaged in the daily practice of his profession; who is watching the inception and progress of disease and studying its various symptoms. While he listens patiently to many fine-spun hypotheses respecting the origin and treatment of disease, and perchance at times quotes them approvingly, yet, when he reaches the bedside of his patient, he lays them aside and trusts to his learning, common sense, and experience. It is not to his discredit that he prefers to confess his ignorance of the Unknown rather than to lead others or be himself led astray. Is the fame of Louis and Chomel, and Bretonneau and Trousseau, and Bartlett and Jackson, and many others like them, any less bright or less enduring than if they had indulged in airy speculations regarding the causes of those diseases whose symptoms and morbid anatomy they so matchlessly displayed, but whose origin they frankly avowed was to them unknown? Their candor saved them from the gentle scorn, if not from the lively ridicule, of posterity.

Above all, it is essential to the honor and dignity of a profession whose members rely for their support on their education, talents, industry, character, and fidelity to their clients, that it should decline, even at the risk of being charged with indifference or with enmity to the public health, to embellish the train of a body of men who have diagnosed themselves as Sanitary Experts, and whose sole claims to public recognition consist in noisy demonstrations on the platform of a Sanitary Convention.

INDEX.

A.

Air, The, 50.
    of cemeteries, 186, 195, 198.
    of sewers, 160, 170.
    of soil, 143, 149.
Awakening, The Great Sanitary, 29.

B.

Bacillus, cholera, 105, 135, 325.
    typhoid, 132, 135, 151.
Bacteria in the soil, 148, 151.
    in water, 134.
Board of Health, Metropolitan, 365.
    of Connecticut, 395.
    of Massachusetts, 387.
    of Michigan, 401.
    of Minnesota, 40.
    of New Hampshire, 398.
    of New Jersey, 400.
    of New York City, 369.
    of New York State, 399.
    of Pennsylvania, 400.
    of Rhode Island, 398.

C.

Cemeteries, 37, 181.
    dangers of, 189, 199.
    gases of, 186, 188, 195.
    of England, 185.
    of Long Island, 203.
    of New Orleans, 202.
    of Paris, 196.
Cholera, 292.
    epidemics of, 293, 296, 301, 303, 308, 316.
Cholera, filth-theory of, 292, 315, 318, 325.
    in India, 323.
    in U. S. army, 307.
    reports of epidemics of, 294, 297, 299, 303.
Contagion from dead body, 206.
Cremation, 190, 199.

D.

Death-rate of Denmark, 415.
    of England and Wales, 429.
    of European States, 429.
    of Ireland, 433.
    of Liverpool, 446.
    of London, 442.
    of New Haven, 92.
    of New Orleans, 99.
    of Newport, 89.
    of New York City, 99.
    of Sacramento, 101.
    of Washington, 100.
Diarrhœal diseases, 437.
Diphtheria, 329, 438.
    causes of, 330.
    filth-origin of, 329.
    from use of milk, 327.
    in Boston, 334.
    in California, 341.
    in Connecticut, 338.
    in England, 342.
    in France, 342.
    in Massachusetts, 331.
    in Minnesota, 340.
    in New Hampshire, 337.

460   INDEX.

Diphtheria in New York, 335, 373.
    in Ohio, 340.
    in the Northwest, 341.
Duration of life, mean, 413.

E.

Emanations, putrid, 50, 58, 61, 62, 79, 184, 256, 277, 316.
Epidemics, 345.
Exhumations, 191, 193, 200.

F.

Filth-diseases, 21, 52, 248.
Food, animal, 216.
    of various tribes and nations, 215, 218.
Funerals, Public, 205.
    contagion at, 206, 213.

G.

Germ of disease in putrefaction, 105.
Gibbons, Dr., on filth-causation of disease, 105.
Graves, Dr., on filth-causation of disease, 105.

H.

Heredity, 404.
Hippocrates, 17.

I.

Ice, contaminated, 129.
Influenza, 361.

J.

Jews, sanitary code of, 16.

L.

Lewis, Waller, on cemeteries, 192.
Liebig on agriculture, 250.

M.

Markets of New York, 246.
Meat, 215.

Meat, diseased, 219, 226, 229, 244.
    in Berlin, 225.
    in China, 224.
    in England, 219.
    in Massachusetts, 221.
    in New York, 228.
    in Paris, 222, 229.
    poisoning from, 227.
    putrid, 222.
Middle Ages, hygiene of, 19.
Milk, 231.
    cause of diphtheria, scarlatina, typhoid fever, and consumption, 235, 238, 241, 243.
    tuberculous, 232.
Mortality in town and country, 35, 36.
    law of celibacy, 407.
        of education, 408.
        of heredity, 404.
        of legitimacy, 404.
        of race, 405.
        of sex, 404.
    of plumbers, 163.

P.

Plague, 345.
    in Florence, 347.
    in France, 348.
    in Geneva, 350.
    in Marseilles, 355.
    in Milan, 352.

S.

Salles de dissection, 209.
Sanitarians, Ancient, Mediæval, and Modern, 15.
Sanitary condition of Belgium, 78.
    of Brighton, 79.
    of Canton, 76.
    of China, 71.
    of Chinatown, 102.
    of Constantinople, 63.
    of Great Britain, 29, 35, 432.

Sanitary condition of Hamburg, 318.
    of Hunter's Point, 94.
    of Ireland, 423.
    of Isle of Man, 421.
    of Italy, 430.
    of Japan, 76.
    of Liverpool, 445.
    of London, 443.
    of Mount Vernon, 97.
    of New Haven, 91.
    of Newport, 85.
    of Philadelphia, 379.
    of Sacramento, 100.
    of Scotland, 425.
    of Sea Cliff, 95.
    of summer resorts, 83.
    of Tivoli, 97.
    of Tunis, 70.
    inspections of public institutions, 82, 94, 101.
Sewage farms, 67, 165.
Sewer-gas, 153.
    composition of, 156.
    diseases caused by, 158.
    germs in, 172, 174, 178.
    pressure of, 175.
    properties of, 155.
    in public institutions, 165.
Sewers of Bristol, 172.
    of London, 162.
    of Paris, 160, 164.
Small-pox, 435.
Soil, The, 142.
    pollution of, 145.
Spartans, hygiene of, 18.
Statistics, Vital, 403.
    of Denmark, 415.
    of England and Wales, 433, 441.
    of France, 412, 427.
    of Geneva, 412.
    of Great Britain, 415.
    of Ireland, 422.

Statistics, Vital, of Isle of Man, 422.
    of Italy, 431.
    of Jews, 405.
    of Scotland, 425.
    of Sweden, 414, 427.
    of United States, 451.

T.

Typhoid fever, causes of, 81, 251, 252, 253, 279, 439.
    in Australia, 258.
    in Brighton, 80.
    in Colorado, 274.
    in Connecticut, 279.
    in Dakota, 274.
    in Dublin, 263.
    in India, 258, 261, 262.
    in Massachusetts, 267.
    in Michigan, 272.
    in New Hampshire, 271.
    in New Orleans, 277.
    in North Boston, 137.
    in Plymouth, 279.
    in Scotland, 440.
    in Vienna, 264.
    in Zurich, 244.

W.

Water, The, 106.
    analyses of, 110.
    diseases caused by, 108.
    impure, in Alleghany City, 124.
        in Baltimore, 121.
        in Boston, 120.
        in Brooklyn, 125.
        in Connecticut, 121.
        in Gloucester, 118.
        in Jamestown, 122.
        in New York City, 124, 126.
    polluted, of the Elbe, 320.
        of the Seine, 322.
        of the Thames, 37, 44, 46, 47, 114.

Y.

Yellow Fever, 280.
    causes of, 280, 288, 289.
    filth-origin of, 286, 290.
    in Charleston, 281.
    in Florida, 291.
    in Mississippi Valley, 288.
    in Mobile, 289.

Yellow fever in New Orleans, 284, 288, 291.
    in New York, 286.
    in Philadelphia, 282.

Z.

Zymotic diseases, 9, 21, 433, 441.

THE END.

www.ingramcontent.com/pod-product-compliance
Lightning Source LLC
Chambersburg PA
CBHW031958300426
44117CB00008B/816